DISCARDED
JENKS LRC
GORDON COLLEGE

Fundamental Astronomy

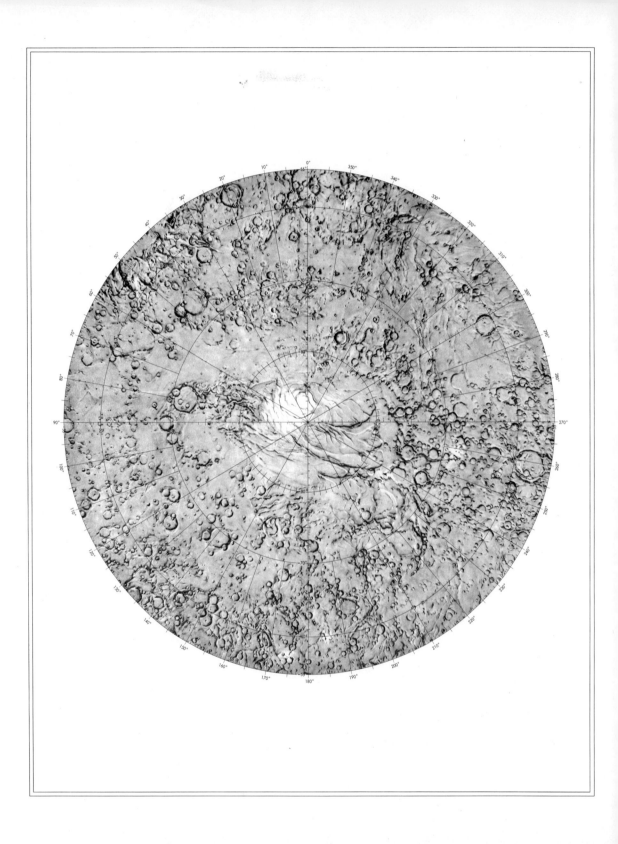

Fundamental Astronomy
Solar System and Beyond

FRANKLYN W. COLE

Professor of Astronomy and Meteorology
Foothill College, Los Altos Hills, California

J   S L.R.C.
GORDON COLLEGE
255 GRAPEVINE RD.
WENHAM, MA 01984-1895

JOHN WILEY & SONS
New York   London   Sydney   Toronto

QB
45
.C65
1974

Cover photograph by permission of Royal University Observatory, Lund, Sweden.

Part I frontispiece: Reconstructed from the original Aztec Calendar Stone, National Museum of Archaeology and History, Mexico City, D.F.

Part II frontispiece: Courtesy NASA.

Part III frontispiece: Courtesy the Hale Observatories.

Copyright © 1974, by John Wiley & Sons, Inc.

All rights reserved. Published simultaneously in Canada.

No part of this book may be reproduced by any means, nor transmitted, nor translated into a machine language without the written permission of the publisher.

Library of Congress Cataloging in Publication Data:

Cole, Franklyn W.
  Fundamental astronomy: solar system and beyond.

  Bibliography: p.
  1. Astronomy. I. Title.

QB45.C65   520   73-12146

ISBN 0-471-16472-0
ISBN 0-471-16473-9 (pbk)

Printed in the United States of America

10 9 8 7 6 5 4 3 2 1

Preface

This book is for the beginning college level course in astronomy typically offered during the first two years of an arts and letters curriculum, or for partial satisfaction of the physical science graduation requirement. Thus it will provide an adequate grounding in astronomy for elementary and secondary school teachers and should be of value to anyone who desires or needs a knowledge of fundamental astronomy.

Now, more than at any other time in our history, a knowledge of science is fundamentally important in the education of the literate man. Society has moved a long distance from an agrarian-based economy toward its present technological orientation. Perhaps no one, especially in the culturally advanced societies, has escaped the impact of scientific advancement.

Astronomy enjoys a unique role in contemporary society. The earliest records kept by man—the cave writings and hieroglyphics, scrolls and tablets—in folklore, mythology, and religion, all reflect his powerful and consuming interest in the celestial universe in which he found and finds himself. Hence the modern study of astronomy has cultural relevance; man's interest in the universe, no less today than in past ages, excites his wonder and speculation.

Fortunately, astronomical knowledge has progressed to a point where the average nonscientist can profit from its study in numerous ways. Our broad interest in the sun, the moon, the planets, and the stars is transported to new heights as astronomical knowledge converges upon a solution to the central problem: What *is* the structure and probable future of the universe?

Astronomy, of all branches of science, provides a central core for scientific interrelations. As such, it deserves a central position in man's intellectual pursuits. Indeed, as so cogently put by W. S. Krogdahl, ". . . astronomy contributes uniquely to intellectual perspective in that it can assist one to estimate both the significance and insignificance of man."

The organization of this textbook proceeds naturally from the above considerations and from my own first premise in writing, that the reader's interest will be captured, stimulated, and challenged to new heights by pursuing a step-by-step course from the astronomically familiar to the philosophically inscrutable. As new tools and methods become necessary to the investigation they are developed, appropriately introduced into the discussions, and brought firmly to bear on the central problem.

From the earth and sky with their familiar phenomena, outward through the solar system, the local stellar populations of our galaxy, and to the populations of galaxies and beyond, the subject matter of this book's ever widening horizons eventually ends on a speculative note: From whence? And whither? And what is it all about?

I have kept the mathematical treatment to a minimum, although I have not contrived to avoid its use. I have emphasized the elegance and simplicity of mathematical symbolism and its advantage over verbal exposition wherever such treatment is appropriate. In every case where a mathematical statement or formula is employed for the first time, I have provided a transition from the verbal to the symbolic language, giving each term and operations symbol a thorough explanation. The student and teacher will quickly come to consider the mathematical description in the category of support for the verbal text analogous to the use of figures, graphs, and diagrams. An ample, yet simple, mathematical review is assembled in Appendix 1 for those who feel this need.

I have deliberately kept the descriptions of astronomical instruments and their use as brief as is consistent with a general understanding of their role in astronomy. Appendix 3 provides additional detail for those who are interested in the instrumentation.

While the constellations and other asterisms are incidental to the significance of astronomy as a science, they do assist in finding one's way around the sky. I have provided a list of the principal constellations with the coordinates of a convenient reference star for the 1970–80 epoch in Appendix 7. This appendix also contains star charts showing the relative positions of prominent astronomical features as a support for observational practices.

I believe Appendix 6 is an innovative feature for college astronomy textbooks. I have provided a short, simplified guide for the use of a telescope, so that the newcomer to astronomy can begin his adventures in the night skies on the very first clear evening following acquisition of this book.

A very complete Glossary of terms employed in astronomy will be found following the appendixes. I have also provided a graded bibliography as a guide for further reading. These features, together with selected questions and problems at the end of most chapters will, I am hopeful, make this a complete presentation of a most fascinating subject.

I am greatly indebted to John W. Kern, of the University of Houston, and Paul E. Trejo, of De Anza College, who studiously and critically read the completed manuscript. Their suggestions for improvement were valuable and timely; most were incorporated into the finished product. I also thank my colleagues who blew on the embers when necessary, and thus kept the fires of determination burning.

I would have been at a loss without the mechanical assistance with the manuscript of Barbara Godwin, Foothill College, and Bette Wooster, Martin-Marietta Corporation of Denver. To the Wiley production staff headed by Bernard Scheier, and including Linda Riffle, Elodie Sabankaya and the Wiley Illustration Department, I convey my gratitude for a superior production achievement.

To my many hundreds of students, who provided the inspiration for this effort, and to my friends and colleagues who suffered my agonizings in the face of deep personal tragedy, I express my heartfelt thanks. Often I was ready to abandon the effort, but their encouragement to continue prevailed. It is to them that I dedicate this book.

Los Altos Hills, California   *Franklyn W. Cole*

Contents

Appendix

PART ONE
Beginnings
and Methods

Prologue

Time: 13 billion B.C. Our universe (may have) had its inception with the explosion of a cosmic primeval "atom," a fireball of energy out of which all matter ultimately was formed. Time: 10 billion B.C. The first stars of our Milky Way Galaxy began to shine by energy released from nuclear reactions deep within their cores. Time: 5 billion B.C. Our sun, newly born, joined the galaxian assemblage of stars. Time: 4.5 billion B.C. The consolidation of the earth out of cold, solid debris within the gas and dust clouds of the Milky Way was completed.

Time: 3 billion B.C. The oceans, covering 71% of the earth's surface, were formed, and the first breaks occurred in the cloud cover that had obscured the surface of the earth for hundreds of thousands of years. Time: 2 billion B.C. Oxygen released by living plants made its first appearance in the atmosphere. The present earth's atmosphere was not completely formed until 1000 million years after the first oxygen appeared. Time: 600 million B.C. The abundant fossil records in the seas and sedimentary rocks were first laid down. Time: 250 million B.C. The sun commenced the round trip around the center of the galaxy that it just completed. Time: 150 million B.C. Dinosaurs and related reptiles dominated the animal world. Time: 80 million B.C. The Rocky Mountains were uplifted, forming the backbone of the North American continent.

Time: 35,000 B.C. Modern (Cro-magnon) man made his appearance; he was not destined to leave historical records of his presence for another 30,000 years. But now the time intervals in the chronology of the universe shorten dramatically. During January 1974 the planet Pluto completed the revolution around the sun that it started in 1726 A.D. Time: 8½ minutes ago. The heat and light that warms us and enables us to see the far horizons at this moment had just left the sun. Time: 1.3 seconds ago. The light by which we see the moon at this instant had just left there. And during that 1.3 seconds, the central star in the Crab nebula spun on its axis through 39 complete revolutions!

This book is about the astronomical events that occurred during the whole of this time span. It is both the history of man's intellectual assault on the mysteries of the universe, as witnessed by his scientific discoveries, and an account of *how* he solved many of the perplexities of nature. It is also a testimony of how much more remains to be learned.

Introduction

1

The Scale of the Universe

We are inhabitants of one member of a "double" planet (the earth and moon are now considered by most astronomers to be a binary system, not just a planet and its satellite) that has been around for several billion years. Even though the changes we effect in our social structures and our environment during one lifetime seem large, the whole span of evolution of terrestrial life-forms is as a moment in the life history of the earth.

Our planetary system—earth and moon—is one of six known to astronomers of antiquity.

We presently know of three other planets, although one of these appears to have been a late comer when measured against the history of the others. Indeed, one of the first six (Mercury) may have had a similar inception. Most of the planets are attended by one or more satellites, and all—planets and satellites together—follow essentially predictable paths around a very different kind of object, the sun. There are countless numbers of other minor bodies attendant upon the sun. In turn the sun interacts with countless numbers of other suns in again an essentially similar and predictable fashion.

An enormous problem confronted ancient as well as modern astronomers in securing data upon which to base presently accepted notions of the nature of the universe. We can partially appreciate the magnitude of the problem by

Mercury Venus Earth Mars

Jupiter

Saturn

Uranus Neptune

Pluto

FIGURE 1-1 The relative sizes of the planets with respect to each other and with respect to the sun. The sun's diameter is more than 109 times that of the earth.

contemplation of several scale models. One of these reduces the solar system and its immediate environment to reasonably familiar dimensions.

We select a scale of 1 inch representing 250,000 miles as our reference. Even with this tremendous reduction we find it impossible to fit a scale model of the solar system into an average classroom (Figures 1-1 and 1-2). We can suspend a scale sun, 3½ inches in diameter, at the front of the room, and a scale earth, 1/32 inch in diameter at the back of the room, 31 feet away. We experience some difficulty in manipulating the scale moon to a position 1 inch from the scale earth, mainly because of its minute size — 1/128 inch in diameter. Sixteen objects of this size, if placed side by side just touching each other, would extend over a distance of only ⅛ inch. The scale moon is totally invisible from a distance of only 1 foot.

On the other hand Jupiter, the planetary giant of the solar system will be comparatively prominent since it will be about 5/16 inch in diameter, about the size of a pea. Unfortunately Jupiter's scale distance from the sun requires that we devise a suspension for it some 160 feet away. Pluto, the outermost planet will be about one-half the size of the scale earth, smaller than a period on this page. Its scale position is hardly neighborly; it is over 1225 feet or nearly ¼ mile distant.

Now, if we wish to introduce a scale model of the star nearest our sun we reach a logistical impasse. This scale star must be suspended 6700 times as far from the front of the room as Pluto, or over 1300 miles distant. This is one-half the airline distance between New York and San Francisco. Indeed, such vast separation between its members is one of the chief characteristics of the universe.

Even within the solar system, such huge distances take on added significance when we relate them to familiar speeds. At 60 miles per hour (mph) it takes over 167 days to traverse a path equivalent to the average earth-moon distance. Even at 2000 mph, a speed attained at present by only experimental jet aircraft (if we exclude missiles or satellite launch vehicles), the journey would span 120 hours, or 5 days.

But now if we wish to travel a distance equivalent to the average Earth-Mars distance at their closest approach to each other, the 2000 mph speed would entail a travel time of 25,000 hours, or just under 3 years. The time of travel to Jupiter under the same conditions is 27 years; to the nearest star the trip time exceeds 1,430,000 years. Even at speeds representative of the orbital speed of low-level satellites, nearly 18,000 mph, a trip to the nearest star would encompass 160,000 years. At this great speed, Jupiter is still 3 years distant.

One of the fundamental constants of nature which we will examine in more detail in later chapters, the speed of light, can be used to advantage in our assessment of the magnitude of the universe. Let us accept as the approximate value of this constant 186,000 miles per second (mi/sec), or nearly 670 million miles per hour. At this speed the earth is 8 minutes from the sun; Pluto is 5¾ hours away. The star nearest the sun will receive a light beam from the latter 4⅓ years after its departure. The size of the local collection of stars, the Milky Way Galaxy, becomes faintly understandable from

FIGURE 1-2 Relative distances of the planets from the sun, drawn to scale. Uranus (U), Neptune (N), and Pluto (P) are too distant from the earth to be seen with the unaided eye.

the realization that light takes over 100,000 years to travel across it. Nearly 2 million years are required for light to reach us from our nearby sister galaxy, Andromeda.

Admittedly the magnitude of the universe is awe inspiring and only vaguely comprehensible. Perhaps this is the frame of reference with which we should approach the study of astronomy, yet never losing sight of the fact that man's intellect and ingenuity have provided us with the means for even this limited comprehension of our universe.

Historical Beginnings

Except for his ability to modify it by his mode of dress, design of habitations, and agricultural practices, man is subject to the vicissitudes of his environment. Many of these changes are derived ultimately from the astronomical characteristics of the earth. Hence it is only natural that astronomical knowledge was for millenia most widely diffused among uncivilized peoples whose very existence depended upon unswerving allegiance to the dictates of the natural environment. The face of the sky was their clock; the relative position of the stars and sun was their almanac. The variations in their daily pursuits were determined by the orderly succession of astronomical events. To primitive man the celestial cycles and the responses evoked by them in the living environment had a mystical aspect. Barbaric familiarity with the heavens became associated with or gave rise to many superstitions. In time religious beliefs became deeply rooted in these superstitions and their associated mysticism.

In order to understand his relationship to his gods, man turned to a fixed system of observations, many of which were extraordinarily ingenious and accurate. It was only natural that in time control of such activity should become concentrated in the hands of the priesthoods. But until the Greek civilization made the transition from experience to theory, ancient astronomy remained largely a superstitious cult, not a true science.

The earliest systematic astronomical records are found in ancient Chinese, Egyptian, and Babylonian writings. In China calculations of the time of the equinoxes (days and nights of equal length) and of the solstices (those times when the sun is at its extreme distance north and south of the equator) were routinely accomplished as early as 2300 B.C. In 1100 B.C. Chou Kung, an accomplished mathematician, made a surprisingly accurate determination of the inclination of the earth's axis of rotation relative to its orbital path around the sun. In Peking as late as 1881, two astronomical instruments constructed around 1280 were still in existence. Both instruments anticipated Tycho Brahe's (q.v.) most important inventions by over 3 centuries.

In Egypt the stars were observed that they might be properly worshipped. Other than systematic observations of the stars which, in connection with the annual flooding of the Nile river, resulted in an accurate determination of the length of the year, Egyptian astrolatry (star worship) contributed little or nothing to the progress or development of astronomy.

The situation in Babylon was far different. Records dating from as early as 3800 B.C. imply that even then the varying aspects of the sky had long been under expert investigation by the Babylonian priest-astrologers. The names and configurations of many of the constellations appear to have been transmitted to the West around 2800 B.C.

There was intense interest in the pseudo science, astrology, from which astronomy purified itself in time. Babylonian astronomy became adapted to the needs of civil life. Astronomers computed cycles for the periodic return to the same approximate position among the stars for the planets Mercury, Venus, Mars, Jupiter, and Saturn. In their ephemerides (tables giving the coordinates of one or more celestial bodies at a number of specific times during a given period) published year by year the Babylonians listed the times of the new moon and the times of the first appearance of the crescent, from

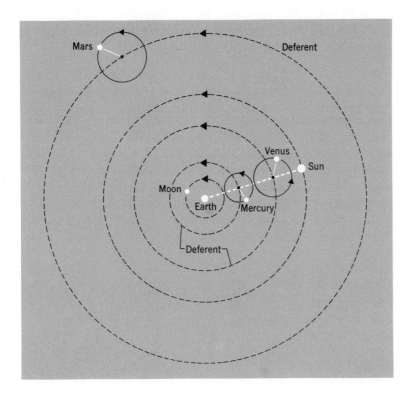

FIGURE 1-3 Partial, schematic diagram of the Ptolemaic geocentric (earth-centered) system. Only one epicycle is shown for each planet; the actual system has many. Observe that epicycles are not required for the moon and the sun since they do not have retrograde motion.

which the beginning of each month was calculated. The astronomers were able to predict the circumstances of both lunar and solar eclipses with consistent accuracy.

A steady flow of knowledge from East to West began in the 7th century, B.C. It seems undeniable that the great advances in astronomical knowledge, which were Greece's contribution during the succeeding 1000 years, were founded on Babylonian astronomy. Among the more important Greek contributions was the first genuine heliocentric (sun-centered) system developed by Aristarchus of Samos sometime between 280 and 264 B.C. Even though the scheme received recognition by Archimedes in the *Arenious* it met with considerable disfavor, and it was soon discarded in favor of the Eudoxian system.

Eudoxes was the father of the homocentric (man-centered, hence earth-centered) solar system which attempted to reconcile the ob-

served planetary, lunar, and apparent solar motions to circular movements. The Eudoxian system was successively modified by a number of investigators into a system made famous by Ptolemy in the *Almagest*. This complicated system of motions (Figure 1-3) consisting of cycles and epicycles and deferent motions, indeed, a system of "wheels within wheels," was so successfully used in the prediction of observed planetary, lunar, and solar motions that it dominated astronomical thought for the next 1200 years.

The Greek astronomy culminated in Alexandria and flourished there from 320 B.C. until 400 A.D. It was at the Alexandria Museum that Euclid wrote his *Elements of Geometry*, so essential in the development of the mathematical sciences. It was here that Apollonius of Perga wrote his treatise on *Conic Sections* (see Appendix 2) which was to become one of the catalysts for the new astronomy ushered in by

Copernicus and Kepler some 1800 years later. It was here that Eratosthenes demonstrated the first effective means of measuring the curvature of the earth's surface. By computing the difference in direction of the arrival of the sun's rays at a given instant at two widely separated points of known distance apart, Eratosthenes was able to calculate the circumference, and therefore the diameter, of the earth (see Chapter 5). If our modern estimate of the length of his unit of measure is correct, Eratosthenes obtained a very accurate approximation of the earth's size.

The Eastern astronomy was eventually transported by the Moors to Spain after Alexandria was razed by the Arabs in 641 A.D. Yet it was not until nearly 400 years later that astronomy began to take root in western Europe.

FIGURE 1-4 Nicolaus Copernicus (Yerkes Observatory Photograph).

Astronomy in the Middle Ages

The rational picture of the universe provided by Greek astronomy and exemplified in the Ptolemaic system experienced little improvement throughout the period of the Dark Ages. Astrology, even though banned by the Christian church, continued strong in western Europe. Its form, however, would have been scarcely recognizable in Babylon. The Babylonian priest-astrologers devoted great attention to the heavens and made many astronomical discoveries, whereas the post-Christian type of astrology was more concerned with divination of the future. The scholasticism of the times neglected the legacy of Greek learning, and in the West people again lived on a flat earth under an inverted bowl of the sky.

Toward the close of the medieval period interest in astronomy was revived among certain European rulers. In Spain Alphonso X commissioned a translation of the *Almagest* from the Arabic, the only version then available. At the same time scholars, whom he had working on tables for predicting the positions of planetary bodies, had found it necessary to represent their motions with as many as 60 epicycles. The Ptolemaic system was collapsing under the weight of its own complexity. Alphonso is supposed to have remarked that had he been present at the Creation, he could have provided the Almighty with some excellent advice.

By the early part of the 16th century science was enjoying a resurgence from its millenium of slumber. One eager scholar of this period became convinced that the complex Ptolemaic system could be essentially discarded if only the observer was considered to be on a planet revolving with the others about the sun. That man was Nicolaus Copernicus.

The heliocentric system which bears his name accounted for nearly all of the observed motions of the universe. Yet Copernicus was apparently unaware of the similar system invented by Aristarchus over 1000 years earlier.

FIGURE 1-5 Tycho Brahe (Yerkes Observatory Photograph).

FIGURE 1-6 Tycho Brahe's observatory (Yerkes Observatory Photograph).

According to the Copernican system, the complexity of the observed motions was merely a question of the effects of perspective as the orbiting planets changed positions relative to an orbiting earth. The diurnal (daily) rotation of the celestial sphere was simply an effect produced by the earth rotating about an axis passing through its center which pointed toward the celestial pole.

The Copernican theory was soon placed on the *Index* of forbidden reading by church authorities who could not tolerate the demotion of man from his hitherto central position to one of lesser importance. It remained on the list for over 200 years.

In Denmark, and later in Prague, Tycho Brahe (1546–1601) laid the groundwork for modern astronomy. His work, consisting of voluminous records of precise observations, preceded the invention of the telescope, and

the accuracy he achieved was thus all the more remarkable.

Tycho never accepted the Copernican concept. Indeed, he endeavored to refute it by showing that if the earth revolved about the sun, some stars should display periodic displacements resulting from parallax (the apparent displacement of an object as seen from two different points). The meaning of parallax becomes clear if the observer holds a pencil vertically at, say, arm's length and views it against a convenient background alternately with one eye and then with the other. According to Tycho the brighter stars, which he reasoned were nearer than the fainter ones, should appear to move against the background in a manner analogous to the pencil's apparent displacement.

Such motion was not detected. What Tycho failed to realize was that all stars are so distant that parallactic oscillations could not be observed with the existing instrumentation. Thus he erroneously concluded that the earth was stationary, and he carried this belief to the end of his life.

Tycho Brahe's most significant contribution to astronomical progress was his consummate skill as an observer. The fruit of his labors was realized in the subsequent work of Johannes Kepler, Tycho's mathematical assistant in Prague, who fell heir to the old astronomer's records.

Kepler, at first antagonistic to it, ultimately accepted the Copernican theory although he was careful to conceal this fact from his employer. Now the Copernican system required a few epicycles in order to fit some of the observed variations in the planetary motions to strictly circular paths. Even Copernicus felt it necessary to preserve the notion of a perfect path, which was circular according to the scholastics, as the only allowable one in a perfect Creation.

Kepler was unable to reconcile Mars' observed motion, as recorded by Tycho Brahe, with a circular orbit around the sun, even with the aid of epicycles. He eventually reached the conclusion that the planetary path was an ellipse (Chapter 4) and in 1609 he announced the first two of three laws of planetary motion bearing his name. The third law was promulgated in 1619. Thus, Kepler's great break with tradition can be considered as the beginning of the transition to modern astronomy, as well as exemplifying medieval astronomy at the height of its perfection.

While Kepler was thus engaged in his research on planetary motions, a contemporary in Italy, Galileo Galilei in 1609 applied the telescope, an invention of Hans Lippershay, an obscure spectacle maker in Holland, to his study of the heavens. What Galileo observed seemed in harmony with the Copernican theory, and he ably defended the proposition in his *Dialogue of the Two Chief World Systems*.

FIGURE 1-7 Johannes Kepler (Yerkes Observatory Photograph).

In spite of the prominence Galileo achieved in defense of the Copernican theory (mostly through his difficulty with the ecclesiastics over his theological assertions), Galileo's chief contribution to astronomy was somewhat indirect. He was an experimenter rather than a philosopher. Many of his investigations refuted the traditional Aristotelian views that things were as they were because they fit, that the natural state of matter was that of rest, and that all motion was enforced. Else why, argued the Aristotelians, did the observed motion of an object require an explanation of how it was kept in motion? That the observed motions of the heavenly bodies lacked such an explanation did not seem to deter the scholastics from their notions.

Galileo's experiments led to a radically different viewpoint, that only *change* in motion,

FIGURE 1-8 Galileo Galilei (Yerkes Observatory Photograph).

FIGURE 1-9 Sir Isaac Newton (Yerkes Observatory Photograph).

either in speed, or in direction, or in both, required an explanation. Here was the foundation upon which Sir Isaac Newton was to formulate his laws of motion of bodies on earth and in the heavens.

Newton was born in 1642, the year of Galileo's death. Perhaps no other man in history succeeded to the degree that Newton did in matching a persistent curiosity in a wide variety of perplexing matters with the ability to supply valid reasons for them. Two remarkable achievements stand out from Newton's many accomplishments in respect to the impact they had on astronomical progress. One was the crystallization of the concept of *force*, a concept we shall examine in Chapter 4. The other was the formulation of the universal law of gravitational attraction, which solved the problem of orbital planetary motion. These dis-

coveries marked the end of the transition from medieval to modern astronomy that was initiated by Johannes Kepler.

Our focus on the present horizons of astronomy concerns the very nature of the universe: its extent, its configuration, its content, its past history, and its future prospects. Is the universe limited, or infinite? Is space curved and closed, or curved and open, or neither? Besides stars and planets and galaxies and clouds of gas and dust, there exist those mystery members of the universe—the quasars, the pulsars, and perhaps black holes in space. Perhaps the latter are not really black holes; maybe they are simply "tunnels" leading from this to other universes in other, unknown dimensions.

The quasars: What are these quasi-stellar objects that appear to be the remotest objects

Astronomy in the Middle Ages **11**

known, and yet individually seem to be brighter than entire galaxies of ordinary stars? How can pulsars, estimated to be no more than 10 miles across, emit prodigious bursts of energy at microwave and visible light frequencies as often as 39 times per second? Finally, where did the universe come from, or has it always been here? Will it have an end, a final curtain, or will the unfolding of the plot keep the show going forever?

These are some of the significant questions that astronomy is striving to answer. This book traces the successes and failures of the quest, and as might be anticipated, it arrives at no single conclusion.

The Value of Astronomy

If we were to employ the same criteria in assessing the value of astronomy that we apply to other branches of learning, or even just to other branches of science, its study would rate a very low priority. This is because we tend to place the greatest values on those lines of inquiry which contribute to our well being or our pleasure. Lest, however, this may be too harsh a judgment, we can include those pursuits which are esthetically satisfying. Astronomy rates better here.

As a species, man throughout his history has been a predator against his environment and against other men. So we have a world which seems threatened with over population, with environmental pollution or outright destruction, with politico-military ambition, with inflation, and, some claim, with declining moral values. Yet there are areas of man's knowledge which can enable him to eliminate, or at least mitigate, the negative influence of these forces: knowledge which represents man's efforts toward satisfying his innate curiosity about the universe he lives in; knowledge leading toward a better understanding of other men and man's relationship to his Creation; knowledge leading

toward, ultimately, a true assessment of his real worth. Astronomy certainly has value as a speculative science in most of these areas. The sheer grandeur of the cosmos as disclosed by astronomical discovery will influence many to search for a Grand Purpose, and for man's place in that Plan. And man is bound to become more unselfish in the process.

Astronomy has its practical side as well. Until recent years astronomical observations were the ultimate authority for timekeeping. They still provide the most widely used, practical method. Astronomical observations are still indispensable in navigation, and in some cases also in land surveying. Astronomy is a fascinating and esthetically rewarding hobby engaged in by many thousands of devotees.

The space age has opened new horizons in practical astronomy. Travel to the moon is already an accomplished fact. Unmanned deep space probes have brought Mars and Venus within range of electronic telemetry, and the future holds great promise for even more exotic exploration. Yet without a knowledge of celestial mechanics and of the outer space environment as disclosed through astronomy, these successes could not have been brought about.

Finally, astronomy is one of those few sciences in which research is not primarily motivated by the opportunity for profit. Astronomical knowledge is an end in itself, although it is true that the knowledge thus gained is frequently put to unanticipated uses. For example, some atmospheric processes are now known to be related to the astrophysical behavior of the sun. This knowledge may in time contribute to our successes in environmental control. Indeed, man's ability to probe the secrets of the universe and to secure confirmation of his discoveries through alternative astronomical methods certainly assists us in appreciating man's remarkable intellectual capacity. From all of this we must conclude that astronomy has relevance in man's cultural evolution.

Questions for Review

1. Where and when did astrology originate, and how does it differ from astronomy?

2. Name some of the early Greek astronomers and indicate examples of their contributions to astronomy.

3. Discuss the role of the Alexandria Museum in Western scientific advances.

4. Who were the principal medieval astronomers who are identified with the transition from ancient to modern astronomy? Summarize the contributions of each.

5. Does astronomy have relevance in modern cultural growth? Explain.

Prologue

Motion, seemingly endless motion: Random motion and motion that is complex yet organized characterize man's earliest perceptions of his universe. Yet without the sense of sight, man would be unaware of the apparent, let alone the real, motions of the celestial population.

   With the solid earth taken as an unmoving, immovable frame of reference, we devise coordinate systems that enable us to relocate any point previously found. Not only on earth, but to the heavens as well do we extend our measurement techniques. And in so doing we have the crude beginnings of time reckoning, essential to man's agrarian pursuits and, seemingly, just as essential to his religious inclinations.

Earth and Sky

2

Most astronomy texts assume that the reader is familiar with at least two of the earth's motions: rotation about its own axis and revolution about the sun. Are these assumptions *essential* in establishing a frame work of observational astronomy, in defining useful astronomy-based timekeeping systems, and in the prediction of many astronomical phenomena? Certainly not. This was the position of the astronomers of antiquity, and they were very successful in spite of their ignorance of the earth's motions.

We shall find it both interesting and instructive, if not thoroughly intriguing, to imagine ourselves in the classical frame of reference in which the earth is a stationary platform at the center of the universe. There appears to be no better way to quickly and fully appreciate the drama of man's intellectual assault on the mysteries of the universe. And since we can compress millenia of intellectual growth into a few hours of study we can afford to spend some time in contemplation of the larger cosmic scheme. We might even choose to engage in the search for its meaning. In any event, until we are prepared to *prove* otherwise, let us take our fictitious position at the center of the universe. *But let us also be uninhibited about accepting new truths as they are revealed to us.* This attitude alone will give us a great advantage over our predecessors.

Apparent Motions from a Stationary Earth

A catalog of the readily observed motions of the principal astronomical bodies is a good starting point for our investigations. We shall exclude for the time being the motions of such transient objects as meteors and comets. Nor shall we for the present worry about fixing precise positions or charting the observed paths of astronomical objects. After we compile the list of observed motions we shall then select appropriate timekeeping and reference-coordinate systems. After integrating these with the observations, we shall attempt to distill from our impressions whatever systematic behavior there is to be found.

<u>Apparent Solar Motion.</u>   First, and most prominent, is the motion of the sun. It rises

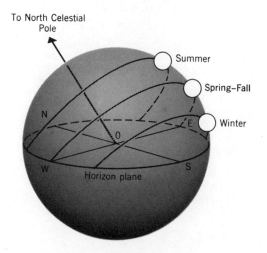

FIGURE 2-1 Diurnal paths of the sun across the sky in each of the seasons, as seen by a northern midlatitude observer.

(a)

(b)

FIGURE 2-2 The Big Dipper as it might be seen in the early evening (a) in January and (b) at the same hour in February, 1 month later. Observe that not only does the Dipper appear higher in the sky in February, it also appears to be rotating about the upper central star.

daily in a generally easterly direction and sets in a westerly direction. It moves along a path which appears to be the arc of a circle. For midlatitude observers in the Northern Hemisphere this arc appears lower in the southern sky in midwinter than it does in midsummer. Indeed, the sun rises and sets further to the south in winter than it does in summer. Figure 2-1 illustrates the relative paths. In all cases the sun appears to move with uniform speed along its daily path from horizon to horizon. Hence we relate the length of the day to the length of the path that is visible above the horizon.

All stars and star groups follow circular paths which appear to be parallel to the sun's daily path. We also note that relative to the sun, certain recognizable collections of stars rise and set approximately 2 hours earlier each suc-

ceeding month (Figure 2-2). Either the stars are moving in a direction so as to meet the sun, or the sun is moving eastward through the stars. We choose *eastward* to mean *the direction of the rising bodies.*

Apparent Lunar Motion. Next, we observe that the moon moves along a path similar to the sun, but with some significant differences. A careful check of the moon's position against the background stars discloses that it moves toward the east through the star field at a rate of approximately its own diameter per hour. Thus, on the average, the moon rises about 50 minutes later each day (Figure 2-3).

Other differences are apparent. The path followed by the moon suffers more variance north and south than that of the sun. Also, the moon

FIGURE 2-3 Relative positions of the rising moon at the same hour on two successive nights. The first rising is shown at (*a*); the second is shown at (*b*). Can you tell whether the time is a few hours after sunset, or a few hours before dawn?

18   Earth and Sky

FIGURE 2-4 "The old moon in the new moon's arms." The brilliant crescent, the portion of the moon's disk illuminated by the sun, is overexposed in the photograph in order to bring out the detail of the rest of the lunar disk that is illuminated only faintly by the earth (Yerkes Observatory Photograph).

is sometimes a daytime object, and sometimes it is seen only at night. Periodically, at intervals of approximately 29 days, we see the moon as a slender crescent with the "horns" pointing away from the sun. At one time it is an evening object, setting very soon after the sun. About 3 weeks later it is an early morning object, rising just before the sun. In between these times the crescent expands to a fully illuminated disk, at which time it rises very close to the time of sunset. Thereafter the full disk shrinks again to a crescent. Following this, and until the crescent is again observed in the west at sunset, the moon disappears from view. We are forced to conclude that it shines principally from light reflected from the sun, and one of our tasks

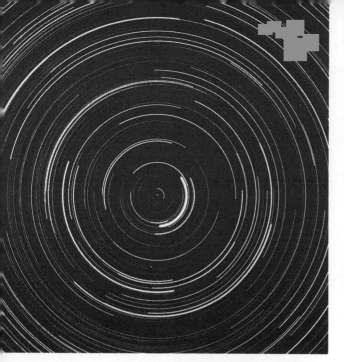

FIGURE 2-5 Circumpolar star trails, obtained with a time exposure taken by a motionless camera pointed at the celestial pole. Length of the exposure is about 8 hours, as revealed by the length of the star trails (Courtesy Lick Observatory).

FIGURE 2-6 Equatorial star trails looking due west. The photograph was taken in the same manner as that in Figure 2-5 (Courtesy Lick Observatory).

will be to determine the space geometry which causes the observed effects. One fact presents something of a puzzle to the uninitiated. The full disk of the moon can occasionally be very dimly seen as though cradled in the arms of the crescent (Figure 2-4). What is the source of this illumination? We cannot, at this stage of our investigation, rule out the possibility that the moon gives off a feeble light of its own.

Apparent Motion of the Celestial Sphere.
Next, what is the apparent motion of the celestial sphere? (We must agree with the ancients that it does resemble an inverted bowl with the stars appended like shimmering jewels.) Most of the stars rise and set each night, following paths which carry them from east to west. On the other hand, relative to the sun they appear to rise and set about 4 minutes earlier on the

average, each night. But whereas the sun and moon vary in their arcs across the heavens, the stars seem always to follow their same respective paths, each of which is parallel to all of the others. It is as though the whole celestial sphere is turning toward the west, with the sun and moon not rigidly attached to it.

We are soon able to locate one star (if, of course, we are in the Northern Hemisphere) which has very nearly a fixed position (Figure 2-5). All other stars make circular paths around this point, although the horizon blocks a portion of most of the paths from our view. Another set of motions of the celestial sphere is apparent to observers traveling north and south. If we travel north for several days or more, the "fixed" star is seen ever higher in the sky. All of the circular paths, including those of the sun and moon, become more nearly horizontal. Similarly, if we travel toward the south, the fixed star sinks toward the northern horizon, and the circular paths become more nearly vertical (Figure 2-6).

Apparent Planetary Motions. Finally, let us catalog the observed motions of several of the planets. Two are always seen in the general vicinity of the sun. The rest can be observed anywhere along, and not far above and below, the sun's path through the stars. The two planets that remain close to the sun appear first on the evening side of the sun, then again on the morning side. Their respective periods of oscillation from one side of the sun to the other are short, less than 1 year, but they are not of the same length.

As for the rest of the planets, their travel times (periods) around the celestial sphere vary from approximately 2 years for Mars to 30 years for Saturn. Jupiter, the second most prominent planet, has an apparent period of nearly 12 years.

To all except experienced observers, all of the planets display a puzzling phenomenon. Every now and then they seem to pause in their eastward motion among the stars and retrace their paths for several weeks or even months, after which they resume their relentless forward march. We call this *retrograde motion;* the path is a *retrograde loop.* The length and duration of the retrograde loop and the time of occurrence differ for each planet.

These, then, are the principal apparent motions that we observe. Our task now is to unravel the mysteries of these motions, and through the application of logic, assisted by a little geometry, determine whether or not the apparent motions are, or are not, real. In order to use geometry we must make measurements. And to make measurements we need suitable reference systems. And since motions are involved, the passage of time becomes another quantity which we must accurately determine. Hence we need suitable timekeeping systems. Let us establish the coordinate systems first.

The Reference Systems

Several reference, or coordinate, systems of utility on earth have analogs in celestial observations. We seek one or more in which a point can be located uniquely by a single set of two numbers. An even greater advantage will result if, in following the apparent motion of an object in the system, only one of the numbers need change.

Polar Coordinates. Suppose we are standing in a plane at a point marked O and we wish to locate uniquely another point in that plane. We measure the distance between the two points. Then we realize there could be an infinite number of points in an infinite number of directions all lying in the plane at the same distance from O. Further, a line connecting these points will lie on a circle whose radius is the selected distance. Hence the direction to the desired point must also be specified. This suggests another coordinate from which to measure an angle. According to convention an arbitrary straight line wholly in the plane is drawn through the point O. We next draw a line from O to the desired point. The two lines form an angle at their intersection at O. We start from the arbitrary reference line and measure the angle in a *counterclockwise* direction, the universally agreed upon direction for angular measure. We have reproduced the *polar coordinate system* shown in Figure 2-7. The coordinates are a line of length, say r, and an angle we might call θ; the set of two numbers which identify the magnitudes of these two quantities uniquely locates the point with reference to O.

Cartesian Coordinates. Another useful reference system was devised by René Descartes (1596–1650), a French mathematician and philosopher. It is named a *rectangular Cartesian coordinate system* after him. In the two-dimensional version two perpendicular lines intersect at an initial reference point marked O. Sometimes this point is termed the *origin*. For conve-

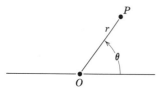

FIGURE 2-7 The polar coordinate system. The location of point P is uniquely specified by the distance r from the reference point O, and the angle θ measured counterclockwise from any specified reference line passing through O.

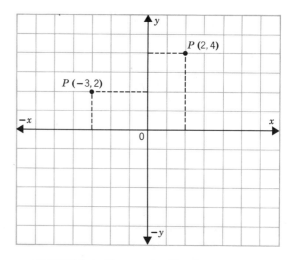

FIGURE 2-8 The rectangular Cartesian coordinate system. The location of a point P is designated by the x, y distances in that order. Thus, P (2,4) states that the point P is +2 units from the y-axis and +4 units from the x-axis.

nience one reference line is ordinarily oriented horizontally and is labeled the x-axis. Then the vertical reference line is termed the y-axis. Now we let directions to the right of the y-axis be considered positive in x, and directions above the x-axis be considered positive in y. If the remaining directions are considered correspondingly negative, then a pair of *signed* (plus or minus) numbers representing distances from the respective axes uniquely locate any selected point (Figure 2-8). This coordinate system can easily be extended to include three-dimensional space by the addition of another reference axis which is mutually perpendicular to the other two.

Altitude-Azimuth Coordinates. If we are interested in the direction of a point in three-dimensional space but distance is of no interest, we can devise a coordinate system based upon a pair of angular values which has considerable utility, provided the observed point remains stationary with respect to the reference system. Let us select a reference direction from an initial point O (usually north). The complete range of angular measure in the horizontal plane is 0°–360° with values increasing clockwise.† North is taken as the 0° reference. We let angular distances above and below the local horizon range from 0° to +90° and −90°, respectively. To locate the point we measure

† See Appendix 2 for a discussion of angular measure.

the angle from north around toward east to an imaginary vertical line passing through the point (Figure 2-9). The angle thus generated we term the *azimuth* of the point. We next measure the vertical angle from the horizon to the point, this angle we term the *altitude* of the point. The system is appropriately called the *horizon system*. We should note that if the object whose position we are measuring is moving in any direction other than horizontally or vertically, the use of this system has serious drawbacks. As we shall see, apparent motion other than horizontal or vertical generally prevails in observational astronomy.

Geodesic Coordinates. We might analyze one final coordinate system which has a very useful celestial analog. Let us mark two diametrically opposite points on a ball and draw a line around its circumference which is equidistant from the two points everywhere along its length. We assign the terms *poles* to the two points and *equator* to the line. We observe that

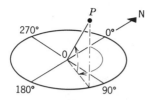

FIGURE 2-9 The altitude-azimuth system in relation to the location of a given point *P*. The vertical angle ranges between 0° and 90°.

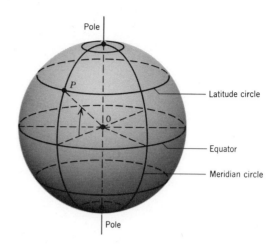

FIGURE 2-10 The geodesic coordinate system on a solid sphere. Only a few latitude and longitude lines are shown. This system is often referred to as the latitude-longitude system.

the equator defines a plane which divides the ball into two hemispheres. A line imagined to pass through the two poles is perpendicular to the equatorial plane.†

Now with reference to an imaginary observer at the center, let us define the angular distance of a point on the surface of the ball above (or below) the equatorial plane as the *latitude* of the point. Latitudes above the equatorial plane are designated *north latitude;* those below it are termed *south latitude*. We note that there are an infinite number of points which satisfy the condition for any given latitude, and a line connecting all points having the same latitude generates a plane which is parallel to the equator (Figure 2-10). This plane cuts the surface of the ball in a circle. In order to single out any one specific point on such a circle, which is termed a *latitude circle,* we need one other dimension.

We define a *great circle* as the line generated on the surface of the ball by any plane cutting it that *also* contains the center of the ball (Figure 2-10). Obviously an infinite number of great circles is possible. We are interested in one family of great circles: that which also contains the poles. We term each semicircle bounded by the poles a meridian of longitude. Obviously again, there are an infi-

† We could just as well have used the earth's poles and its equator. But since these imply rotation of the earth about an axis passing through the poles, let us preserve the fiction that we do not *know* the earth is rotating until we can satisfactorily prove it.

nite number of such possible meridians. But if we choose one of them as a reference meridian and appropriately select and number the rest, we have the means of locating a point on the surface of the ball with one and only one pair of numbers: the latitude and longitude of the point. Angular measure is again convenient, but the convention is 0°–180° east, and 0°–180° west of the reference meridian. This is the meridian which contains the poles and one other designated point (Figure 2-11). On earth the poles are simply designated points. (We still do not admit we know the earth is rotating.) These points uniquely specify the location of the equator and define the family of great circles called meridians of longitude. The reference meridian (0°) passes through the site of the original Greenwich (England) observatory. Meridians between 0° and 180° to the east of Greenwich are designated *east longitude;* the remainder are termed *west longitude*.

Celestial Coordinates. Let us briefly review our imaginary position with respect to the visi-

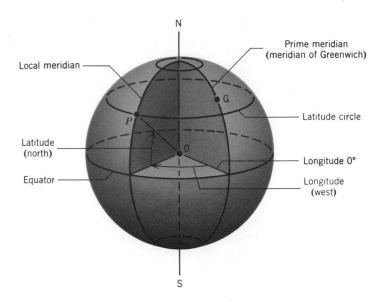

FIGURE 2-11 The latitude and longitude of a given point *P* with reference to the prime meridian (0° longitude). The prime meridian according to international agreement passes through Greenwich, England *G*.

ble universe. The sky appears to us, as it did to the ancients, as an infinitely large hollow sphere. The earth is at the center of this sphere and all of the celestial objects appear to be fixed to, or travel on, its inside surface. Furthermore, the celestial sphere appears to make one rotation about the earth in just under 1 day; during each rotation it turns through 360°. A new question now arises: Might we not find considerable utility in a celestial coordinate system which is analogous to, and concentric with, the geodesic system on earth? We shall see.

Let us select that point in the northern sky as the *north celestial pole* about which the celestial sphere seems to rotate. The *south celestial pole* will be diametrically opposite, although not visible from the Northern Hemisphere for reasons which will presently become apparent. We then define the *celestial equator* as an invisible but nonetheless real reference coordinate which divides the celestial sphere into two hemispheres.

Next, while we are on the subject of poles and equator, we shall orient a *geodesic coordinate system* on earth in this manner: we lo-

cate the point on earth that is vertically under the north celestial pole and mark this as the earth's north pole. The point directly under the south celestial pole we shall call the earth's south pole. Now by definition we shall find that every point on the earth's equator is directly under the celestial equator. Or, saying it another way, if we could now extend the *plane* of the earth's equator indefinitely, it would cut the celestial sphere in the precise location of the celestial equator.

In an exactly analogous manner we can establish latitude circles on the celestial sphere just as we established latitude circles on the ball; in fact we can now transfer our thinking from the ball to the earth. Figure 2-12 shows these relationships even though it is impossible to draw an infinitely large celestial sphere. It is interesting to note that there is a one-to-one angular correspondence between the terrestrial and the celestial latitude circles. Consistent with tradition, we shall name the latter *declination circles;* the angular distance north or south of the celestial equator on the celestial sphere is called *plus* or *minus declination,* written $+\delta$ or $-\delta$, respectively.

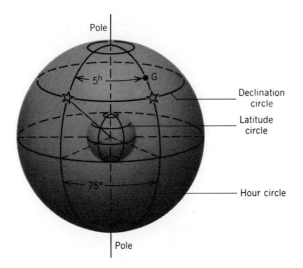

Pole

5ʰ

G

Declination circle

Latitude circle

75°

Hour circle

Pole

FIGURE 2-12 The celestial sphere with the earth at the center. The relationship between the geodesic (earth) and the equatorial (celestial sphere) coordinate systems is shown. The longitude of a place on earth can be determined from the time of transit of a given star across the local meridian relative to its time of transit across the prime meridian. See text for discussion.

We next consider the family of great circles on the celestial sphere which contains both celestial poles. These are the celestial analogs to the meridian circles on earth (Figure 2-12). A valid question now arises: How many such circles shall we employ, and how shall we name them? We can resolve these questions more intelligently after we establish another astronomical reference system which is directly related to the observer's position on earth. Let us imagine a straight line extending from the center of the earth through some point on the surface as in Figure 2-13. We further imagine a horizontal plane centered on the point; by definition of the word horizontal, this plane will be perpendicular to the line from the center of the earth. Now imagine that we are standing on the point. We then call the plane our *astronomical horizon plane;* its infinite extension is our astronomical horizon.

We call the point directly overhead the *zenith;* obviously the zenith is peculiar to every single individual; unless one person is standing on another's shoulders, no two individuals have the same zenith. Correspondingly but less important, the point directly beneath our feet, or more exactly the direction opposite the zenith, is called our *nadir.* Now we imag-

ine a circle passing through our zenith and through both of the celestial poles that is fixed with respect to the earth. We term this circle the *local meridian.* Interestingly enough, while no two individuals can have the same zenith, an almost infinite number can have the same local meridian. All that is required is that they be due north or south of each other.

We have now set the stage for the selection and naming of the celestial meridians. If we look upward along our local meridian, we can imagine the whole array of celestial meridians being carried across it by the rotation of the celestial sphere once approximately every 24 hours. (This rationalization becomes clear in Chapter 3.) Thus if we designate 24 such celestial meridians, one will cross the local meridian every hour. With this adoption we then name the meridians *hour circles.* Now just as we found it necessary to define a prime meridian on earth, so must we select a zero hour circle on the celestial sphere. We proceed to the task.

If we were to plot the annual apparent path of the sun as it journeys around the celestial sphere, we would find that it crosses the celestial equator twice: once as the sun ascends toward its maximum displacement of 23½°

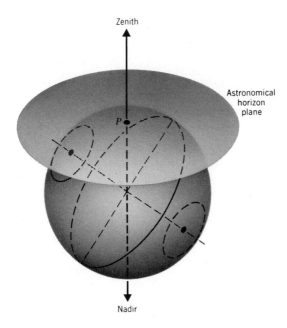

Zenith

Astronomical
horizon
plane

P

Nadir

FIGURE 2-13 The astronomical horizon plane for point P. An imaginary north-south line passing through the zenith point and the celestial poles but which is fixed with respect to the horizon plane becomes the local meridian for point P.

north of the celestial equator, and again as it descends toward its maximum displacement of $23\frac{1}{2}°$ south of the celestial equator. The sun's apparent path is called the *ecliptic* because it is only when the moon is also on or near this path that eclipses can occur.

Let us place the zero hour circle so that it passes exactly through the intersection where the sun crosses from the southern to the northern celestial hemisphere (Figure 2-14). The ancients called the projection of this point on the celestial sphere the *First Point in Aries*. The term is related to the constellation in which the celestial crossing occurred when the system was first defined. As we shall see later, this point has shifted westward on the celestial sphere nearly 30° in the last 2100 years. Finally, we number the remaining hour circles in ascending order from 1 to 23 as they successively cross our local meridian.

The construction we just derived is called the *equatorial system* because of the prominent role played by the celestial equator as a reference line. It enables us to uniquely locate any

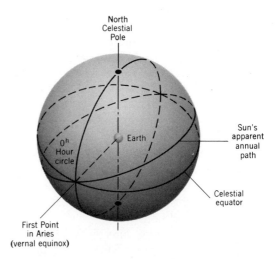

North
Celestial
Pole

Earth

Sun's
apparent
annual
path

0^h
Hour
circle

Celestial
equator

First Point
in Aries
(vernal equinox)

FIGURE 2-14 The zero hour circle passes through the point on the celestial sphere where the apparent path of the ascending sun annually crosses the celestial equator. The ancient term for this point was the First Point in Aries; the modern name is *vernal equinox*.

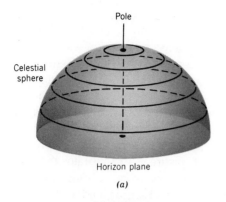

(a)

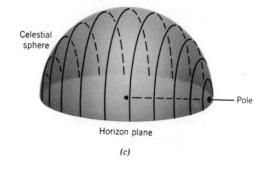

(c)

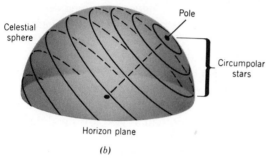

(b)

FIGURE 2-15 Diurnal circles traced out by stars for various relative positions of the astronomical horizon; (a) as seen from the pole; (b) as seen from a representative midlatitude position; (c) as seen from the equator.

celestial object which is fixed to the celestial sphere. With the addition of a celestial time-keeping system, we can extend the application of the equatorial system to the location of moving astronomical objects as well.

Since all 24 hour circles cross the local meridian during one 360° rotation of the celestial sphere, there must be 360° ÷ 24 = 15° between every hour circle. For reasons which become apparent when we review useful time-keeping systems, each degree is conveniently divided into 60 minutes (60') of arc, and each minute is further subdivided into 60 seconds (60'') of arc. Thus the celestial sphere turns through 1° every 4 minutes, through 15' of arc per minute, and through 15'' of arc every second. We must keep in mind one thing, however. The time we use here is in reference to the stars. We are not entitled to assume it is the same as the time we would use in reference to the sun. We have already noticed that the sun appears to move eastward relative to the star

field, and this may have an important consequence for time-keeping systems. We shall resolve the issue in Chapter 3.

Determining Longitude

We now have a means for determining longitude on the earth which in many cases is superior to direct measurement. Suppose a certain star on the celestial sphere is observed on the meridian of Greenwich at a certain time. Exactly 5 hours later the same star crosses the meridian of an observer in the United States. Since the celestial sphere turns through 15° every hour, 15° per hour × 5 hours = 75°. The observer must be 75° west of Greenwich. But Greenwich is located at 0° longitude. Therefore the observer is at 75° W longitude (Figure 2-12).

The celestial meridian which passes through a given star or reference point on the celestial sphere is traditionally called the *right ascen-*

sion of that star of point.† The term is also used as an alternate name for the hour circles. Thus, the hour circle which passes through the First Point in Aries is often termed the 0 hour of right ascension, written 0^h RA. The lower case Greek α is often used to signify right ascension. The right ascension and declination of a great many stars are listed in one or more star catalogs and ephemerides. This information enables us to compute longitude and latitude quickly, to find our way around the star field quite easily, and to navigate from place to place on the earth, even though no other references or landmarks can be seen. Some applications of right ascension and declination to observing practices are discussed in Appendix 6.

Another phenomenon of the celestial sphere surely becomes evident to those of us who are privileged to travel and who have the time to observe the heavens. We have concluded that the angular elevation of the north celestial pole above our astronomical horizon corresponds to our latitude. Hence all stars within the same angular distance of the pole as our latitude never set. Such stars are said to be *circumpolar;* the circumpolar collection of stars depends upon our point of observation. If we were at the equator, the north celestial pole would be on the northern horizon, and all stars would be below the horizon half of the time (Figure 2-15). If we were at the north pole of the earth, the north celestial pole would be directly overhead and no stars would set; all would describe horizontal, parallel circles around the celestial sphere.

Determining Latitude

The latitude of an observer is even easier to determine from the stars than the longitude, because the reference point from which it is de-

† *Right* in this context appears to have meant *immediately,* or *next.* Hence, right ascension means next rising, or next ascending.

termined is not moving. We will need the assistance of a theorem or two from elementary plane geometry, however, if we are to see how latitude is measured. Let us determine the angular elevation of the north celestial pole above our astronomical horizon. Then the latitude of the place of observation is just this angular number. To see why this is so, let us review the geometry needed, then apply it to the argument. Figure 2-16 illustrates the properties of a triangle as the apex is moved farther and farther from the base. Indeed, we can visualize that the two sides connecting the apex of a triangle to its base must become parallel to each other as the apex is moved infinitely far from the base.

We also need the theorem: when two parallel lines are cut by a third straight line (Figure 2-17a), the alternate (or corresponding) angles thus formed are equal. We also need the definition: two perpendicular lines, or the angle between a line perpendicular to a plane and

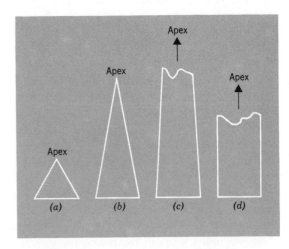

FIGURE 2-16 As the apex of a triangle becomes farther removed from its base, the opposite sides become more nearly parallel. In (d) the apex is imagined to be infinitely far removed; in this case the sides of the triangle would be parallel.

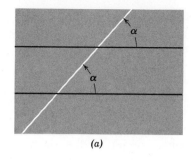
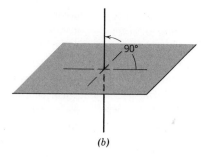

(a)

(b)

FIGURE 2-17 Some angle relationship from elementary geometry. In (a) two parallel lines are cut by a (straight) transverse line; the corresponding angles are equal. The relationship holds for other similar pairs of angles in the figure. In (b) a line perpendicular to a line in a plane is also perpendicular to the plane; the angle formed is a right (90°) angle.

the plane itself, form a right angle (Figure 2-17b). Finally, an angle of 90° is a right angle. A little study of Figure 2-18 will disclose all of these relationships. A line to the latitude point in question forms the latitude angle with the equator. This line when extended contains the observer's zenith and nadir (assuming, of course, that the observer is at the point). The observer's astronomical horizon is perpendicular to this line. Further, a line connecting both of the earth's poles will, if extended, pass through the celestial poles. Since the celestial pole is infinitely far away, the observer's line of sight is parallel to the line containing the poles. The reader should have very little trouble in verifying the angular relations between the various lines in terms of the geometry we employed.

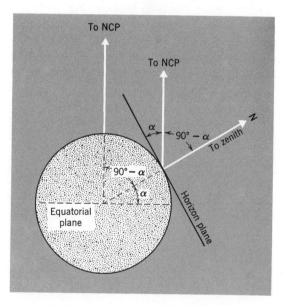

FIGURE 2-18 The angle formed between a line from the center of the earth through an observer's zenith and the equatorial plane is the observer's latitude. The line is also perpendicular to the observer's astronomical horizon plane. Since both lines, when extended to the north celestial pole (NCP), are parallel the angular relationships are easily seen.

The Seasons

Everything we have so far observed and recorded is perfectly independent of either frame of reference: a stationary earth (with all of the motion residing in the rest of the universe) or a rotating earth that is revolving about the sun. We can add two more major theses to our repertoire without violating our original premise, that of a stationary earth as the frame of reference. One of these will serve to conclude this section of our initial survey. The other forms the subject matter of Chapter 3.

We have implicitly accepted the notions of the day, the night, and the year. Let us list some additional details concerning the latter. Again the vantage point for our observations will be in the midlatitudes in the Northern Hemisphere. A logical item for first consideration is the motion of the sun as seen against the background of stars. Since we cannot see the stars through the blinding glare of the sun, let us measure, with any convenient device, the time interval between two successive crossings of the sun across our local meridian. Careful measurement discloses the time interval varies during the year, but let us for the moment ascribe this to errors in timekeeping. After determining the proper interval, we make our subsequent observations halfway between the sun's crossings. Thus we shall see particular portions of the star field cross the local meridian at local midnight.

Since any selected position of the sun will serve equally well for the starting point, we shall observe the portion of the star field which crosses the local meridian at the first local midnight after the sun reaches its *lowest* point in the southern sky. Night after night we observe stars which were previously to the east of the local meridian at the appointed observing time arrive on, and then pass to the west of, the meridian. All the while, the sun ascends ever higher at local noon, until finally it reaches its northernmost position and starts to swing south again.

At midnight of the day that the sun once again reaches its southernmost position in the sky, we note that 365 days have elapsed. But the original reference stars are not yet at the meridian where they were when we started the reckoning; they are still on the eastern horizon. We find that it takes approximately 6 additional hours, one-fourth of a day, for the celes-

tial sphere to regain its initial position with reference to the local meridian.

But *time* is reckoned with respect to the sun's crossing of the local meridian. We conclude, therefore, that the changes in the orientation of the star field as seen each midnight are a result of the sun's motion through the stars. It takes the sun very nearly 365¼ days to make one complete circuit of the celestial sphere. For the present this is an adequate definition of the year—the time interval between two successive appearances of the sun in the same place in the star field. By universal custom another reference point is used; the year is reckoned from two successive appearances of the sun at the vernal equinox. Any other point would do just as well.

We tentatively reject an alternate hypothesis that the celestial sphere spins westward to meet the sun, because of the latter's alternation north and south in declination as though it were a yo-yo on a string. Admittedly this is as much speculation on our part as anything else. At this stage of our investigations, however, it just seems a more reasonable choice.

Now let us examine nature's behavior during the yearly cycle, still from our Northern Hemisphere vantage. In general when the sun reaches its southernmost point the event marks the onset of the coldest part of the year. Indeed, custom defines this as the first day of winter, although the lowest average temperatures most generally occur from 4 to 6 weeks later. Similarly, when the sun reaches its northernmost point it marks the first day of summer. Warmest temperatures occur from 4 to 6 weeks after this event. The lag is associated with the time it takes to reverse the respective cooling and warming trends in the earth-atmosphere system.

When the sun reaches the midpoint on its apparent northward journey it crosses the celestial equator. This event not only signals the first day of spring, it also exactly locates the First Point in Aries (which is no longer in the constellation Aries; it is in Pisces). Considering a point on the exact center of the sun's disk, every place on earth on this date experiences day and night of equal length. Thus, the point in *time* and the point in *space* where the event occurs are *both* termed the *vernal equinox* (*vernal,* spring; *equinox,* equal nights). An analogous event occurs when the sun crosses the celestial equator on its apparent journey southward. This point in time and space is called the *autumnal equinox.* The time and place of the southernmost and northernmost points in the sun's travels are called the *winter solstice* and the *summer solstice,* respectively.

The Lag of the Seasons. Experience shows us that the temperature maximums and minimums do not coincide with the respective solstices. Thus, the summer days continue to get warmer for several weeks after the first day of summer even though the days begin to get shorter. Similarly the winter days get colder for more than a month after the first day of winter even though the days are getting longer. We may not be able to identify all of the phenomena which might affect the seasonal temperature

change, but let us pursue the problem as far as we can, being ever ready to modify or change our hypotheses if the need arises.

If we compare the altitude of the noonday sun on the first day of summer and winter respectively we observe a substantial difference: approximately 47°. In the midlatitudes the sun's rays arrive at the earth's surface much more nearly horizontal in winter than in summer (Figure 2-19). Hence a bundle of the sun's rays of a given imaginary cross section must heat a larger area of the earth than is the case in summer. If we now add to this the shorter daylight period in winter, we can account for the colder winter temperatures, at least with respect to these phenomena.

Now we know that the earth *on the average* is neither gradually getting warmer nor cooler. Therefore the heat energy absorbed at any given place must be lost in some way so as to preserve an *average* heat balance. We can rationalize that just as it takes time for a frying pan to heat up and cool off, so must it take time for the earth to do likewise. Thus this hypothesis is suggested: even though the days

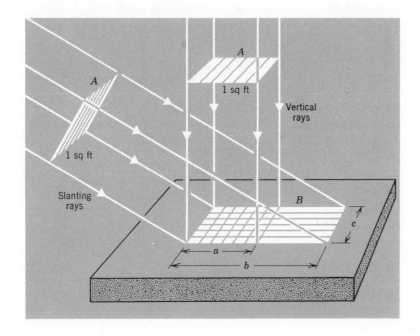

FIGURE 2-19 Energy distribution on an intercepting surface depends upon the angle of the incoming energy rays. Energy distribution is more concentrated on a perpendicular surface *A* than on a slanted surface *B*. A 1 ft² column of energy following a slant path must heat up the area *ac* which is clearly larger than an area *ab* which would be heated by a 1 ft² vertical column of energy (After Cole).

get shorter after the first day of summer, there is an excess of heat gain over heat loss for several weeks. Finally a balance is achieved, after which the cooling rate begins to exceed the warming rate. The situation is not reversed until several weeks after the first day of winter. We still must be aware, however, that other phenomena may make their contributions and we must be alert to the possibilities.

This brings us to a point in our investigations where improvements in our observational techniques are required. Two things must be done; we must employ improved observational instrumentation and we must devise and use the necessary timekeeping systems. We can assume the instrumentation problem is easily solved since it is principally one of engineering. We shall explore the nature of timekeeping systems in the next chapter.

SUMMARY

It is unnecessary for us to make any assumptions about the earth's motions in order to describe the readily observed motions in the universe. We are no worse off than were the ancient astronomers in imagining a stationary earth platform with all observed motion external to this frame of reference. Thus we ascribe the observed diurnal motions of the stars, planets, and the sun and moon to the westward rotation of the celestial sphere about us. We also ascribe the changing positions of the planets, sun, and moon on the celestial sphere to their independent revolutions about the earth.

We employ various reference coordinate systems in describing motions, or positions on the earth or celestial sphere. Two of these which are of great convenience are the geodesic (latitude-longitude) system on earth, and the equatorial (right ascension-declination) system on the celestial sphere. These can be mutually related through an appropriate timekeeping system. We established two primary reference coordinates: the meridian of Greenwich on earth, and the 0^h of right ascension passing through the vernal equinox on the celestial sphere.

Finally, we established a preliminary, though crude, basis for describing seasonal phenomena in terms of the sun's position on its annual journey along the ecliptic.

Questions for Review

1. List the celestial motions as observed from a stationary earth. Identify each.

2. Using polar coordinates, draw a figure showing the location of a point P (4,30°) with reference to a given point O. How does the position of this point compare with the position of a point P (4,390°)?

3. Construct a Cartesian coordinate system on a piece of graph paper ruled in quarter-inch squares. Locate on the graph points P_1 (1,1); P_2 (1,−3); P_3 (−3,2); P_4 (−2,−2).

4. Using a world globe (usually available in any library) find the names of the towns or cities having these approximate geodesic coordinates: (a) 123° W longitude, 45° N latitude; (b) 104° E longitude, 2° N latitude.

5. What determines the position of the equator? The prime meridian?

6. Given are the declinations of several stars. Determine which (if any) are, as seen from 40° N latitude, (a) always above the horizon, (b) above the horizon less than half of the time, (c) always below the horizon.

Star	Declination
Dubhe	+68°
Vega	+38°
Sirius	−17°
Canopus	−53°

7. With respect to the solar system define (a) ecliptic, (b) zenith, (c) nadir, (d) local meridian.

8. A star whose declination is +38° passes through an observer's zenith. What is the latitude of the observer? If the star had crossed the observer's local meridian 15° south of his zenith, what is his latitude?

9. Define (a) vernal equinox, (b) summer solstice, (c) autumnal equinox, (d) winter solstice. What is the right ascension of the vernal equinox?

10. Explain the lag of the seasons.

11. A star seen near the celestial equator and just rising at sunset on the first day of spring is near the point on the celestial sphere known as (a) the vernal equinox, (b) the summer solstice, (c) the autumnal equinox. (Select the correct answer.) Explain.

12. What are the average annual dates for the occurrence of (a) the vernal equinox, (b) the autumnal equinox?

Prologue

Whether we are conscious of it or not, timekeeping is one of our fundamental preoccupations. This has been the case ever since Cro-Magnon man first dimly perceived the motions of the celestial sphere. Indeed, timekeeping enjoyed an equal footing with the astrological and metaphysical inclinations of the first astronomers.

In this chapter we shall examine the intricacies of the apparent irregularity of the sun's motion along the ecliptic. We shall also discover by direct measurement the almost perfect regularity of the motion of the celestial sphere containing the "fixed" stars. We find, however, that we cannot regulate our lives according to star time. Because the sun moves through the star field, we would periodically be sleeping during midday and arising in the middle of the night. We would find the experience psychologically punishing, to say the least.

We therefore fictionalize the sun's apparent motion in such a way that the length of the days, measured from noon to noon, are all the same. In the process, we discover a need to fictionalize the number of days in a year in order to match the periodic repetition of celestial events. With our lives thus well-ordered (astronomically speaking), we are free to venture forth in search of more distant astronomical horizons.

Time in Astronomy

3

Thus far we have made only crude approximations of the time intervals between recurring astronomical events. For example we approximated the year as 365¼ days; we have taken the daylight portion of the day to be the length of time the sun is above the astronomical horizon. Differences in the duration of daylight during the year are strikingly noticeable for many observers. At 40° N and S latitude, for example, this difference can be nearly 6 hours—one-quarter of a day.

For practical reasons we need some means of accurately determining uniform intervals of time, derived perhaps from astronomical events. Having defined the unit of time and devised an instrument for measuring it, we can apply this technology toward increasing the precision of our astronomical observations.

The ancients, both civilized and barbaric, were aware of a similar need and they attacked the problem in a variety of ways. Slow-burning knotted cords, the sand (hour) glass, the hour candle, the water clock, the sundial, and the simple pendulum all are examples of the early efforts to determine uniform time intervals (Figure 3-1). While the details make fascinating leading, the historical development of both the clock and the fundamental timekeeping unit, the second, is beyond our scope. We must ignore centuries of effort, and turn only the outcome of that effort to our practical advantage.

Solar Time

Historically, trial-and-error methods eventually produced a clock which measures in seconds, the fundamental unit of time. Now if we use the clock to count the number of seconds in the interval between two successive appearances of the center of the sun on our local meridian, we find considerable variation, as much as 50 seconds per day, during the course of a year. Moreover, the effect is cumulative. If we analyze the data, we find the sun is sometimes as much as 16 minutes early, sometimes as much as 16 minutes late in arriving on the local meridian. Table 3.1 shows the differences for selected days during 1971. Time by sundial is correspondingly sometimes fast, sometimes slow, compared to a clock.

Local Apparent and Mean Solar Day. Civil uses, among others, require a uniform length of day. Thus, an averaging technique is suggested. If we *average* the number of seconds per solar day over a period of 1 year, taking every day into account, we find there are 86,400 seconds in an average, or *mean solar day* (MSD).† With respect to the MSD, the clock appears to keep time with respect to a fictitious sun that is never early or late in arrival at the local meridian. On the other hand, we define the *local apparent solar day* as the interval between the real sun's successive arrivals at the local meridian. Clearly, this is the time kept by sundials.

Now if we preserve our pretensions of ignorance of the motions of the earth, the question

† In reality we are "backing" into the problem with this approach. We should determine the length of a mean solar day first, then define the second to be 1/86,400 of this interval. However, the continuity of thought is better preserved through this bit of pedagogy.

(a)

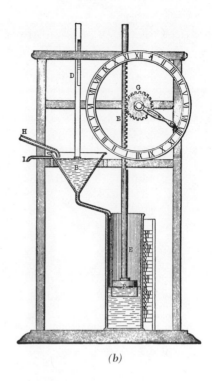

(b)

(c)

FIGURE 3-1 Some early devices for measuring time intervals: (*a*) the sand glass (still popularly used as eggtimers for soft-cooking eggs); (*b*) the water clock, or clepsydra; (*c*) the sundial. The sundial shown dates from 1550 A.D. (Parts *a* and *b* Courtesy of the American Museum of Natural History; Part *c* Courtesy of the Bettmann Archive).

TABLE 3.1   Reduction of Local Apparent to Local Mean Solar Time†

Date	Sun's local transit time††	Reduction to local mean time	Date	Sun's local transit time††	Reduction to local mean time
Jan. 1	$12^h03^m24^s$	-03^m24^s	July 1	$12^h03^m40^s$	-3^m40^s
Jan. 15	$12^h09^m19^s$	-09^m19^s	July 15	$12^h05^m49^s$	-05^m49^s
Feb. 1	$12^h13^m37^s$	-13^m37^s	Aug. 1	$12^h06^m18^s$	-06^m18^s
Feb. 15	$12^h14^m14^s$	-14^m14^s	Aug. 15	$12^h04^m32^s$	-04^m32^s
Mar. 1	$12^h12^m32^s$	-12^m32^s	Sept. 1	$12^h00^m09^s$	-00^m09^s
Mar. 15	$12^h09^m09^s$	-09^m09^s	Sept. 15	$11^h55^m53^s$	$+04^m37^s$
Apr. 1	$12^h04^m05^s$	-04^m05^s	Oct. 1	$11^h49^m51^s$	$+10^m09^s$
Apr. 15	$12^h00^m11^s$	-00^m11^s	Oct. 15	$11^h45^m55^s$	$+14^m05^s$
May 1	$11^h57^m09^s$	$+02^m51^s$	Nov. 1	$11^h43^m37^s$	$+16^m23^s$
May 15	$11^h56^m17^s$	$+03^m43^s$	Nov. 15	$11^h44^m31^s$	$+15^m29^s$
June 1	$11^h57^m40^s$	-02^m20^s	Dec. 1	$11^h48^m51^s$	$+10^m09^s$
June 15	$12^h00^m16^s$	-00^m16^s	Dec. 15	$11^h54^m54^s$	$+05^m06^s$

† Data are from *The American Ephemeris and Nautical Almanac* for 1971. The exact reduction for any given day varies slightly from year to year.
†† This is the clock time when the exact center of the solar disk crosses the local meridian; the minus and plus signs indicate how many minutes and seconds the real sun is behind or ahead of the mean sun in reaching the meridian respectively.

arises as to the cause of the variation in the length of the apparent solar day. Is the celestial sphere irregular in its rotational rate? Or is the sun alternately spurting ahead or lagging behind in its motion along the ecliptic? Suppose a time check is run on the celestial sphere's rotation by measuring the intervals between successive transits of a convenient star across the local meridian. We find the intervals to be remarkably regular—almost exactly 236 seconds *less* than the 86,400 seconds in a mean solar day. By elimination, it appears the sun is the culprit, but why? Let us pursue this perplexing question further.

Local Civil Day.   It is not convenient to begin a new apparent solar day with the sun's crossing of the upper meridian, the meridian which passes through our zenith. Not only is the event difficult to observe because of the relatively large image diameter and brightness of the sun's disk, but using it would require a change of the calendar date at apparent noon. This inconvenience is avoided by making the date change, and the start of a new apparent solar day begin, with the sun's transit across the lower meridian. This is the meridian passing through the nadir of our position.

Let us therefore define the *local civil day* as the mean solar day which starts at 0^h at or near midnight. Then the next succeeding noon coincides approximately with the passage of the sun across the local upper meridian. Now the *hour angle* of any celestial object is the angular distance measured westward along the celestial equator between the local meridian and the celestial meridian, or hour circle, which passes through the object. Thus if the celestial meridian containing the sun is 45° west of the local meridian, the sun's hour angle is 3^h00^m. At the meridian transit, the hour angle of the sun is 0^h, but clock time is 12^h. We therefore define *local civil noon* as the hour angle of the mean sun plus 12 hours. Local civil time is the same as local mean solar time except that the latter does not run through two cycles of 12 hours each day. Local mean solar time starts with 0^h00^m at midnight and runs through

23^h59^m at 1 minute before the next midnight. Thus, 2 P.M. civil time is the same as 14^h00^m mean solar time.

Greenwich Civil Time (GCT) and Universal Time (UT). In timekeeping as in any other measurement process, reference coordinates are mandatory. The hour angle of the mean sun referred to the meridian of Greenwich is the universally accepted reference. Thus, if the hour angle of the mean sun is 3^h, the local civil and local mean solar time is 3 P.M. and 15^h00^m respectively. Indeed, Universal Time (UT) derives its name from mean solar time on the Greenwich meridian. Except for the familiar 12-hour cycles in common use in civil time, Greenwich Civil Time (GCT) is identical to UT and the terms are frequently used interchangeably.

Standard Time. Both local apparent solar and local mean solar times give a different time for every independent location on earth except for those points due north or south of each other. In the absence of a universal time-keeping system, we can easily surmise the confusion which attended attempts to coordinate time schedules between several locations, to transact business where time is of the essence of contractual obligations, indeed, to conduct almost all civil pursuits. Yet it was not until 1884 that standard time was adopted by international agreement. In this system the earth is divided into 24 time zones. These are nominally 15° wide, centered on each standard (15°) meridian beginning with Greenwich. The standard time in each zone differs by 1 hour from the time in either adjacent zone, with any given time occurring later to the west.

Figure 3-2 shows the relative time zones for North America. Eastern Standard Time (EST) coincides with mean solar time for the 75° meridian, hence it occurs 5 hours later than Greenwich Civil Time. Central Standard Time is the mean solar time on the 90° meridian, Mountain Standard Time is mean solar time on the 105° meridian, and Pacific Standard Time

is mean solar time on the 120° meridian. The time in each of these time zones is successively 1 hour earlier than that in each zone immediately adjacent to the east.

Local modifications to the time zone boundaries are made in order to avoid different times in any one political subdivision or geographic marketing area. But in general, no point in a given time zone differs by more than 1 hour from the local mean solar time on the reference meridian for that zone.

International Date Line. Let us embark on a fictitious trip around the world traveling from east to west starting from a convenient place in the Greenwich time zone. As we enter each successive time zone we must set our clocks back 1 hour; when we arrive back at the time zone from which we started we will have set the clocks back 24 hours. At first glance it appears that we have lost 1 day. What really happens is that we have made the sun appear to overtake us by more than 1 revolution each day, or what is the same thing, by more than 1/24 revolution for each time zone we enter. Thus, the sun *appears* to lose one diurnal revolution over the duration of our trip. Obviously this cannot happen. Therefore, we must compensate for the apparent discrepancy by advancing our time by 1 whole calendar day. By international agreement this is always done at the *International Date Line* which, except for minor adjustments, coincides with the 180° terrestrial meridian (Figure 3-2). On a westward crossing of the date line calendars are advanced 1 whole day; on an eastward crossing they are turned back 1 day. The calendar is in itself a timekeeping device as we will see in a later section.

Sidereal Time

We have already noted that mean solar time and time by the stars is significantly different. Two successive transits by a given star across the local meridian occur in approximately 4

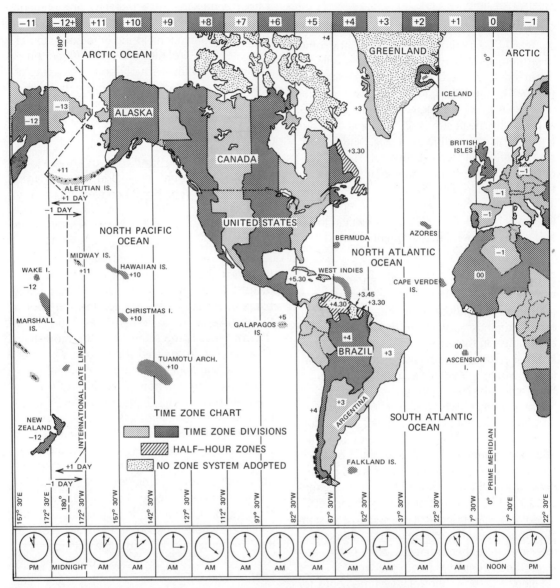

FIGURE 3-2 Standard time zones for the Western Hemisphere. The nominal boundaries for each zone are at the indicated longitudes. Geographic time zones are indicated by alternate shadings. These deviations are adopted for political, economic, and geographic convenience (Adapted from United States Hydrographic Office Chart 5192).

minutes less time than the corresponding transits of the mean sun. Moreover, converting the relatively irregular local apparent solar time to local mean solar time can be a complex and time consuming prelude to each series of astronomical observations. And since the sun appears to move through the star field during the course of time there exists the problem of knowing where any given object on the celestial sphere is in relation to the sun.

The great regularity with which given stars cross the local meridian suggests a solution to the timekeeping problem, that of devising a time system in which a special clock keeps pace with the rotation of the celestial sphere. Such a clock will measure *sidereal* time (L., *sidereus*, constellation). A sidereal day begins at 0^h with the transit of the vernal equinox across the local meridian. Hence sidereal time is a purely local time, and it coincides with the hour angle of the vernal equinox at the place in question. It runs through 24 sidereal hours of 60 sidereal minutes, each minute having 60 sidereal seconds. Since the sidereal day is shorter than the mean solar day, every time unit in this system is also proportionately shorter. For example, a sidereal second is approximately 0.997 solar second. Said another way, 1 sidereal hour is approximately 10 solar seconds shorter than 1 solar hour. Thus, the transit of successive hour circles on the celestial sphere is in tempo with sidereal, not solar, time.

Determining sidereal time by observing the transit of a star with known right ascension is relatively easy, since by definition the right ascension of any star is equal to the hour angle of the vernal equinox. Sidereal time can be calculated quite simply from prepared tables which convert mean solar time into sidereal time. Several methods are detailed in Appendix 6.

Since sidereal time is the hour angle of the vernal equinox at the place in question, when the mean sun is at the autumnal equinox, mean solar and sidereal times coincide. At the vernal equinox, mean solar and sidereal times differ by 12 hours. The sun moves through an angle of 24 hours each year, therefore it moves through an angle of approximately 2 hours every 30 days. Hence, sidereal time gains on solar time at the same rate. We can use this fact in a rough computation of sidereal time, at any time on any date. Suppose for example we wish to know the sidereal time at sunset, 5 P.M. on January 30. Between the time of the autumnal equinox (about September 23) and January 30 the sun moves eastward on the celestial sphere through an angle of approximately 8^h30^m. Thus at the time of the observation the 20^h30^m hour circle is just setting with the sun. But this hour circle is about 6 hours west of the local meridian, which is the reference circle for calculating sidereal time. Hence the 2^h30^m hour circle is on the meridian; sidereal time, then, is 2^h30^m into the next sidereal day. While this is just a crude approximation, it is accurate enough to enable us to determine whether or not a particular star or constellation is visible above the horizon, and about where to look for the object, providing of course, that we know the local time.†

The Calendar

The concept of the calendar (L., *calendarium*, an account book, therefore an accounting of the civil divisions of the year) has been deeply interwoven with all of man's cultural pursuits, certainly since the dawn of recorded history and probably long before that. Certainly the progression of the seasons has had great significance down through the ages, not only for agriculture, but also for nearly all secular activities. The seasons have been especially important in religious observances. Astronomical events related to the apparent motions of the sun and moon and some of the brighter stars in combination with each other had added metaphysical significance. It is little wonder that many attempts were made to devise fun-

† Some interesting practical applications of the use of sidereal time to amateur observing practices are detailed in Appendix 6.

damental bases for reckoning time over extended periods in one or the other, or in combinations, of these events.

Tropical Year. As we noted earlier, the ancients knew that a whole number of complete apparent revolutions of the sun about the earth (solar days) cannot be fitted into the interval between successive arrivals of the sun at, say, the vernal equinox. On the other hand primitive agricultural people were so dependent upon the seasonal aspects of their environment that the seasonal or tropical year became and survived as the basis for civil reckoning.

By means of a precision clock the interval between the sun's successive arrivals at the vernal equinox is measured as 365 days, 5 hours, 48 minutes, and 46 seconds of mean solar time. In decimal notation, which is more convenient for our immediate purpose, this interval, which by definition is the *tropical year,* is 365.24220 days. The reader should satisfy himself that there is no way to fit an integral (whole) number of days, of hours, or of months as we shall presently define them, into the interval of 1 apparent solar revolution around the celestial sphere.

Sidereal Year. Another, more mystifying occurrence is apparent when we compare the positions of appropriate prominent stars relative to the sun at two successive vernal equinoxes.† We find that stars are nearly 20 minutes late in arriving at their corresponding position relative to the sun, and this error is cumulative, year by year. Numerically, the interval between successive arrivals of the sun on the local meridian *coincident* with a given reference star is 365 days, 6 hours, 9 minutes, 9.5+ seconds, or 365.25636+ mean solar days. We make two conclusions here. First, the latter interval, called the *sidereal year,* suffers from the same defects as the tropical year with respect to con-

taining an integral number of time units based upon the apparent solar motion relative to the earth. Second, and more intriguing, the position of the vernal equinox appears to be moving westward on the celestial sphere relative to the sun. That is why the sun arrives at the vernal equinox some 20 minutes before it arrives back at the place in the star field it occupied at the previous vernal equinox. Let us keep this (apparent) anomaly in mind for future use. We shall seek the explanation in connection with determining the validity of our imagined reference system based upon a stationary earth platform.

It is interesting to note that Hipparchus in 125 B.C. was aware that the tropical and sidereal years were of different length. He was able, even without the aid of precise optical instrumentation, to compute the difference within 20% of the presently accepted value, a surprisingly good result!

Synodic and Sidereal Month. Let us next examine the moon's apparent motions, relative first to the sun and second to the stars. Observations disclose that the interval between two successive new moons, for example, is slightly more than 29.53 days.†† This interval is called the *synodic month* (Gr., *synodas,* a meeting). Similar observations disclose that the moon takes only 27.32+ days to complete one apparent revolution around the earth relative to the stars. The name *sidereal month* is given to this interval. We note that neither of these months is exactly divisible into either the sidereal or the tropical year. For example, there are 12 synodic months plus 11¼ days in the tropical year.

Modern Calendar. We have neither the time nor the need to trace the development of the calendar in detail. Interested readers wishing more information on this subject should consult the bibliography at the end of the text. Our

---

† Using the vernal equinox as the reference is purely arbitrary. Any other *identifiable* point on the celestial sphere would serve just as well.

†† The phases of the moon are governed by the geometry of the relative positions of the earth, sun, and moon. We shall examine the matter in detail in Chapter 7.

interest here is to gain some insight into the inadequacy of the existing calendar as a precision timekeeping device. The principal forerunner of the present calendar was mandated in 45 B.C. by Julius Caesar at the suggestion of his court astronomer, Sosigenes. The Julian calendar contained 365¼ days. In practice, 3 years of 365 days each were followed by a leap year of 366 days. This calendar accounting of the number of days in a year is very nearly the same as is used today.

The Julian calendar was a great improvement over earlier solar-lunar calendars. The principal defect in it was that the Julian calendar year was approximately 11 minutes and 14 seconds longer than the real tropical year. Hence the tropical year lagged behind the Julian year by about 3 days in every 400 years. In 1582 Pope Gregory XIII decreed the revision which gave us our current calendar. Among Gregory's adjustments, which included dropping 10 days of accumulated error, was the provision that the accumulating error of 11 minutes, 14 seconds be corrected by the expediency of omitting leap years every century year that was not divisible by 400. Thus, 1900 A.D. was not a leap year; 2000 A.D. will be.

Because none of the celestial timekeepers keep exact pace with the sun, such corrections are approximations only. A residual error still exists in the Gregorian calendar. One way of reducing it will be to eliminate leap year every millenium (thousandth) year that is not divisible by 4000. We will find it necessary to resort to this, or a similar adjustment, sometime in the next few thousand years. Even this correction will not be permanent.

Ephemeris Time

A discussion of time in astronomy cannot be considered complete without recognition of *ephemeris time.*† If we are to preserve the fic-

† Ephemeris: (Gr., *ephemeros*, ephemeral; L., *ephemeris*, a calendar or diary) any tabular statement of the assigned places of astronomical bodies for regular periods.

tion that we do not *know* the earth turns on its axis, then ephemeris time is not only meaningless, it is undiscoverable. Ephemeris time should be introduced and acknowledged at such time as we retreat from the classical stationary earth principle.

But in the interests of a complete discussion we must acknowledge that irregular fluctuations in the earth's rotation rate, or more in context, in the rotation rate of the celestial sphere, have been known for many years. The exact cause is unknown (Chapter 5). Ephemeris time is based on a presumed exact regularity in the length of each succeeding tropical year beginning with the epoch of 1900. The accumulated difference between ephemeris time and universal time since 1900 is slightly over 30 seconds of solar time. Modern astronomical ephemerides have the necessary correction incorporated in the data.

SUMMARY

Man's need for a uniform, accurate, and convenient timekeeping system eventually led to the invention of the clock which measures in seconds, each of which is 1/86,400 of a mean solar day. The latter was a contrivance made necessary from the discovery (with the aid of the clock) that apparent solar days are of different length. The mean solar day keeps pace with a fictitious sun which is assumed to cross the local meridian at regular intervals. Sundial time which keeps pace with the real sun thus varies as much as 16 minutes early or late during the year with reference to the clock which keeps mean solar time.

Greenwich Civil Time was selected as the reference coordinate in an internationally accepted timekeeping system. Time zones nominally 15° wide centered on the appropriate meridian starting with Greenwich, England were a necessary adoption. Within these zones civil time is successively 1 hour earlier to the west. Universal Time, used extensively in astronomical based practices, corresponds with GCT, but UT employs a 24-hour notation rather than two cycles of 12 hours. An International Date Line lies approximately along the 180° meridian. The calendar date is 1 day later to the west of this line than to the east of it.

We find that mean solar time does not keep pace with the celestial sphere because of the sun's apparent motion around the sphere toward the east. While the real sun (and the mean sun as well) arrives back at the vernal equinox after approximately 365¼ days, neither arrive back at the same place on the celestial sphere for an additional 20 minutes. This suggests the use of the more regular sidereal time, in which the day is 3^m56^s + solar time shorter than the mean solar day, as a basis for astronomical timekeeping systems.

The length of the sidereal second is chosen so that the same number of seconds, minutes, and hours appear in the sidereal day. The hour circles on the celestial sphere cross the local meridian in pace with sidereal, not mean solar, time.

Finally, we defined the tropical calendar which logs the passage of days cyclically with the seasons. These in turn are cyclic with the sun's appearances at the equinoxes. We similarly defined the sidereal calendar which has some specialized uses other than governing civil affairs. In neither calendar is there an integral number of days, or hours, or minutes during the interval of a year. This phenomenon requires adjustment in the calendar dates by the addition of "leap days" at regular intervals. Continuing finer adjustments will be a future necessity.

Questions for Review

1. Define: (a) local apparent solar day, (b) local civil day, (c) Greenwich Civil Time, (d) Universal Time, (e) standard time, (f) sidereal time.

2. It is 8 A.M. Thursday, March 2, at a place 5 miles east of the International Date Line. What is the time, the day, and the date at a place 5 miles west of the International Date Line?

3. List some of the disadvantages to everyday activities if *everyone* used local apparent solar time exclusively.

4. Suppose that a plane departs from Greenwich, England at 8 A.M. GCT and flies due west a rate of 15° longitude per hour. What is the local standard time when the plane crosses the meridian of New York City? Chicago? San Francisco? Explain.

5. A pilot flies his plane due west at just the right speed to keep the sun constantly on his local meridian. Does this make time "stand still" for the pilot? Explain.

6. A star who's right ascension is 5^h30^m is observed at the instant it crosses the local meridian. What is the sidereal time? At that instant, what is the hour angle of the vernal equinox?

7. What is the approximate sidereal time at sunset, March 21? Assume a sea-level horizon.

8. The sun is observed to be at RA 3^h0^m. What is the date (within 3 days)?

9. Define: (a) tropical year, (b) sidereal year.

10. Discuss the difference between sidereal period and synodic period. Can the sun have a sidereal period? A synodic period?

Prologue

Try as we might, we find we cannot reconcile experience, based on observations governed by meticulous timekeeping methods, with what well-behaved celestial bodies *ought* to do: not, at least, with respect to a reference platform consisting of a stationary, centrally located earth.

The early Greek philosophers and the scholastics headed by Thomas Aquinas had their problems in these affairs also. There is, however, an essential difference between us and them. Unlike many of our predecessors, we are unwilling to sweep the apparently inexplicable eccentricities of the universe under the rug.

We didn't just "get" this way in one grand and glorious quantum jump. Society grew into the change, and rather rapidly at that. Earth-centered models of the universe gave way grudgingly to sun-centered models. Not because the former were "wrong," but because the latter were so much simpler and more accurate.

We shall trace Kepler's development of his experimentally determined "laws" of celestial motions based on a sun-centered system for a number of reasons. Among them is the need to satisfy ourselves that such laws have validity as general, not singular, statements describing the behavior of revolving, gravitationally-bound systems of celestial bodies.

Moreover we shall scrutinize Newton's laws of motion with even more care because they constitute the mathematical verification of Kepler's laws, and also because all ordinary motion experience since Newton's laws were formulated reinforces the convictions that they are universally applicable. With them we can deal with the universe quantitatively, as well as philosophically.

These then, Kepler's and Newton's laws, are prominent in our armamant for the assault on the mysteries of the universe, including motion, seemingly endless motion. With these weapons we can confidently stride forth to "slay celestial dragons in their lairs, and beard the lions in their dens." Figuratively speaking, of course.

The Laws of Motion

4

In the preceding chapters we assumed the position of observers on a stationary earth platform at the center of the universe. There were two reasons for this. First, it placed us in the frame of reference from which man progressed from philosophical speculation about the universe to an experimental, scientific study of its nature. Second, it is a most familiar setting, for we experience no direct sense of motion. Most of the true facts about the nature of the universe and the mechanics of its behavior are not directly accessible to our sense perceptions.

In our progress we became aware of some puzzling aspects of the observed apparent motions which seemed to defy, or at least resist, ready, logical explanation. While remembering that these peculiarities may exist only because of the selected frame of reference, let us list them for convenience.

1. The stars, and other arbitrarily selected points on the celestial sphere, appear to follow repeatedly the same circular paths in a very regular fashion. The sun, moon, and planets depart substantially from this regularity and order.

2. In making its apparent annual path around the celestial sphere, the sun does so in an oblique, or slant, path relative to the celestial equator. This path carries the sun alternately 23½° north and 23½° south of the celestial equator during the course of 1 year.

3. The sun also appears to spurt, then falter with respect to the length of its diurnal trips around the earth, and in its speed along its path through the stars. This motion is periodic in that it is repeated annually, but the motion is not symmetrical. That is, the sun does not vary its apparent motion in a smooth, alternating fashion.

4. The length (in time) of days and years is disproportionate; the year, or any sum of years, cannot be divided evenly by a whole number of days. Furthermore, there is a difference in the length of the year as measured by the stars and as measured by the sun's rearrival at, say, a specific equinox. (Any other identifiable point will do.) It is as though the sun's oblique path through the star field is drifting slowly westward around the celestial sphere. This is another way of saying that the point of the vernal equinox, for example, occurs successively westward relative to the stars each year.

5. Most of what is said above about the sun also applies in principle to the moon; the angles and time scales are different. The moon's path around the celestial sphere is oblique not only with respect to the celestial equator, but also with respect to the sun's path.

6. The planets each have their own individual peculiarities of motion. Each has its own oblique path (relative to the ecliptic) through the star field, although in each case the obliquity is not very different from that of the sun's apparent path. This means that none of the planets is ever very far above or below the sun's path. The planets display other anomalous phenomena. They too spurt and lag in their respective journeys around the celestial sphere. To be sure, refinements in methods of observation and in timekeeping were needed in order to

make the phenomena quantitatively discernible. The retrograde loops periodically displayed by all of the planets vary substantially in size, in shape, and in duration from event to event for each planet. Each planet varies in brightness in a periodic but unsymmetrical way.

7. Finally, the sun, moon, and planets exhibit variations in apparent size. In the case of the planets, the larger apparent size coincides with the periods of greater brightness. While these latter phenomena are not discernible without the aid of optical instrumentation, and thus were not mentioned earlier in the discussions, they are included here for completeness. We will come back to these phenomena later in the chapter.

Unsymmetrical Periodic Motion

It is worthwhile to pause here for a moment and investigate the visible nature of the periodic but unsymmetric motion of the sun as previously mentioned in (3). This special case is a useful example of the meaning of periodic but unsymmetric events which we will encounter many times in our future investigations. A graph of the difference between the transit times across the local meridian of the real sun and the fictitious mean sun as defined in Chapter 3 is shown in Figure 4-1. We observe that the curve crosses the time axis in unequal time intervals during the year (Figure 4-1a). Moreover the heights of the curve above and below the zero line differ between each

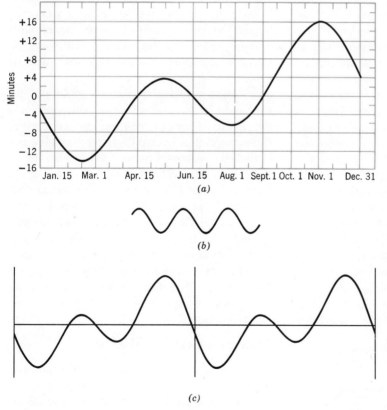

(a)

(b)

(c)

FIGURE 4-1 (*a*) Graph of the data of Table 3.1 for reducing local apparent to local mean solar time. Minus signs indicate that the real sun is later than the mean sun in reaching the local meridian; plus signs indicate just the opposite. (*b*) An example of a symmetric, periodic curve (called a *sine wave*). (*c*) Two periods for the curve in (*a*), indicating that the curve is *periodic* even though *unsymmetric*. The curve (*a*) is commonly called the *equation of time*.

crossing. Thus, the curve is unsymmetrical. A symmetrical curve is shown in Figure 4-1b. Figure 4-1c shows the periodicity of the phenomena since the curve is essentially repetitive each year.

In order to fully appreciate the significance of some of the other listed phenomena we need to retrace our steps in history and review the philosophical model of the universe of which Plato and Aristotle were probably the chief architects. What is now known as the Aristotelian scheme evolved through the continuing efforts of Aristotle's followers during the period from about 6 B.C. to 2 A.D. This was a period during which man's intellectual advances with respect to astronomy proceeded at a rate that was possibly the greatest of all time.†

Scholastic Mechanics

Thomas Aquinas and his followers, known as the schoolmen (or scholastics) redacted the Aristotelian schema both in astronomy and in physics in the 13th century. The term *scholasticism* is derived from this relationship. In the scholastic mechanics a sharp difference existed between earthly and celestial objects. Whereas the former consisted of four elements—fire, air, water, and earth—singly or in combination, the celestial objects consisted of only one, the *quintessence* (L., *quinta essentia,* the fifth element). The quintessence was not ordered as to its "natural" place as were the other four. Hence, the celestial objects were free to move in the simplest manner possible, endlessly in perfect circles about the center of the Aristotelian universe. This motion was essential according to the Platonic philosophy that " . . . the stars, regarded as divine, unchanging, eternal beings, move around the

† This heritage was lost to Western civilization, a victim of the Dark Ages. The vast store of Greek knowledge was reintroduced to the West from Arabic sources some 10 centuries later. Arabic influences left an indelible mark on the records.

earth once each day in that eminently perfect path, the circle." Somehow, reasoned Plato, the objects which appeared to stray from this order must move in combinations of ordered circles.

The Aristotelian earth-centered cosmological scheme consisted of concentric spheres the outermost of which carried the fixed stars around the earth from east to west once each day. The sun, moon, and planets were each carried on their own spheres, all concentric with the celestial sphere and with the earth at the center. The separate spheres rotated simultaneously about several sets of axes so oriented that the observed motions of the objects were accounted for.

This was a complicated, ponderous model indeed. Eudoxes, a pupil of Plato, thought that a total of 26 simultaneous, uniform rotations, properly distributed among the 8 spheres, would suffice. Aristotle added at least 29 more in order to correct gross errors which characterized the earlier model. Even so, the addition of more spheres, or the subtraction of some, or frequent adjustments in the rotations, failed to yield a perfect model. For one thing, the variable brightness of the planets could not be explained by variations in their distance from the earth. They were fixed to concentric spheres and thus were prohibited from changes in distance. An even more serious difficulty attended an alternative notion that the planets indeed changed in intrinsic brightness. This concept did unthinkable and unacceptable violence to the Greek philosophy of the immutability—the changelessness—of the celestial objects. Both of these flaws proved to be fatal defects. Yet Aristotle chose to ignore them on the grounds that the whole grand scheme could not be tainted by such trivial poisons.

Ptolemy's Celestial Mechanics

As we saw in Chapter 1, the Eudoxian-Aristotelian hypothesis was eventually replaced by the Ptolemaic model of the universe. Retained were the essentials: uniform circular motion and a stationary earth. Gone were the concentric spheres. Gone also was the requirement that every rotation be earth-centered. Added were epicycles, the centers of which moved in the circular orbits. Ptolemy was eventually forced to add another device, the *equant* (Figure 4-2), in order to account for the fine details of planetary motion. It is interesting to note that in essence the introduction of the equant replaced the heretofore purely circular motion with *eccentric motion*. It would be interesting to know the degree of equanimity with which Ptolemy thus broke with the Greek traditions.

The fact that large numbers of epicycles and an equant were needed to force a given planetary orbit into the circular paths required by Greek philosophy paved the way for dissident hypotheses. The notion of an equant was difficult to accept. Another defect was the inability of the system to predict planetary positions and motions for an extended period without considerable arbitrary manipulation of the data. On the other hand the long period of success of the Ptolemaic system can be partially attributed to the undisturbed slumber of intellectual inquiry in the Western world during the Dark Ages. During this period the Ptolemaic system was secure from attack. At least there is no certain record that the Moslem world made any significant changes in the model during the 1000 years it was entrusted to its keeping.

Copernicus' Celestial Mechanics

Copernicus, in his attack upon the Ptolemaic hypothesis in the 16th century, acknowledged, as we must, that a geocentric frame of reference was entirely possible. Such a model, however complex, should be tractable to mathematical analysis. To be sure, the observed motions must be possible within the limits of known physical laws. But Copernicus, in one of the quantum jumps which has so characterized man's assault on the mysteries of nature, reasoned that a heliocentric universe, in which

the earth has a circular path along with the other planets, was a far simpler model (Figure 4-3). In his reasoning Copernicus was undoubtedly influenced by Plato's classical problem of depicting a planetary system with the *fewest* possible circular motions. For Plato, all nature was essentially simple, and therefore it ought to be simply described.

Not only was Copernicus concerned by the considerable discrepancies between observations and predictions in the Ptolemaic system, he found the use of the equant abhorrent as it was then being taught in the Ptolemaic geocentric theory. To Copernicus, changing the point of view to a sun-centered system, while still preserving the concept of circular orbits with just a few epicycles, accomplished several things. (It now appears that Copernicus may have employed even more epicycles than did Ptolemy.) It avoided the complexity and the forcing techniques of the Ptolemaic system. It rendered prediction of astronomical events with a much higher degree of success than did the Ptolemaic system. More important to Copernicus, who was a highly-placed churchman, it was theologically more pleasing to have the sun, the "giver of life," centrally enthroned rather than enjoying the dubious distinction of being attendant upon the earth.

Opposition to the heliocentric theory was as powerful as it was lacking in rationale. Opponents asserted that the immobility of the earth was a self-evident truth since the senses failed to perceive any motion. Not only that, but according to the scholastic physics, a colossal force would be needed to move the sluggish earth. Not so for the heavenly objects, since the quintessence was supposed to have no weight. Yet another objection and a more scientific one was the lack of an observable parallax (p. 9) for the near stars, which ought to be associated with an orbiting earth.

In truth, but not a factor in the ultimate rejection of the Copernican system, both it and the Ptolemaic system were about equal in prediction capability. The true explanation for the

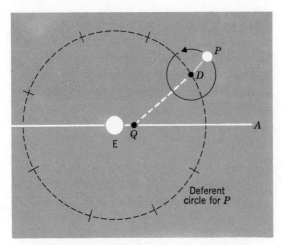

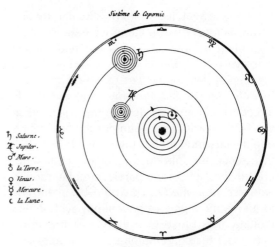

FIGURE 4-2 One planetary orbit in the Ptolemaic system which has an equant Q. As in the earlier model, the center of the planetary epicycle D moves around the earth E on a deferent circle. However the planet also moves *uniformly* with respect to the equant. This means the angle AQD changes uniformly with time. The effect causes the planet to appear to travel various portions of the path along the deferent circle with different speeds. This is equivalent to an eccentric motion when viewed from E. The tics on the deferent circle show where the center of the epicycle would be at the end of successive equal time intervals.

FIGURE 4-3 Copernicus' heliocentric system according to his treatise *On the Revolutions of the Celestial Spheres* (Yerkes Observatory Photograph).

demise of the heliocentric system stemmed from the resistance of the theologians, who would not surrender their beliefs that scripture proved the universe was earth-centered, or more importantly, man-centered. When Copernicus' treatise on *The Revolutions of the Heavenly Spheres* was placed on the *Index*, the heliocentric theory retreated into the limbo of unpopularity, awaiting its rescue by Johannes Kepler, some 50 years later.

Kepler's Laws

We must be careful not to acquire a distorted picture from the popular rejection of the helio-

centric theory as being a true representation of the universe. It was the *theory* that died, not its utility as a computational device. Many astronomers and mathematicians continued to use it because it was so much less cumbersome mathematically than the Ptolemaic theory. Most of the objections to the geocentric frame of reference that we listed earlier do not apply to the heliocentric system. Yet there remain in Copernicus' system problems which give us a considerable amount of intellectual discomfort. First, what mechanism or what agency can be hypothesized which can explain how the center of a planetary epicycle is held to a circular path on the deferent circle? Second, to account for the variations in the apparent brightness of the planets, principally Mars, Copernicus had to place the center of his system a little to one side of the sun. How, then, can the system be accepted as being truly heliocentric? Why must there be such a gross departure in the system from the perfect order envisioned in Greek philosophy and preserved in the scholasticism of the times? The correct answers to the puzzle were discovered by Kepler.

The System According to Tycho Brahe. Before we turn our detailed attention to Kepler's laws one more model of the universe deserves mention. This model is one hypothesized by Tycho Brahe in connection with his long research on planetary motions (p. 9ff). His failure to detect parallactic displacement of the stars enabled him to reject without qualms the heliocentric theory on scientific, not metaphysical grounds. But Tycho was not satisfied with the Ptolemaic theory either. He proposed his own *Tychonic theory* in which the earth was still fixed at the center of the universe. The sun and moon revolved about the stationary earth as before, but all the other planets revolved about the sun. Mercury and Venus were closer to the sun than to the earth; the other planets were farther away than the earth. In principle each outer planet revolved in an epicycle centered on the sun. The epicycle in turn moved on a deferent circle around the earth. Mercury and Venus behaved as in the Ptolemaic theory (Figure 4-4).

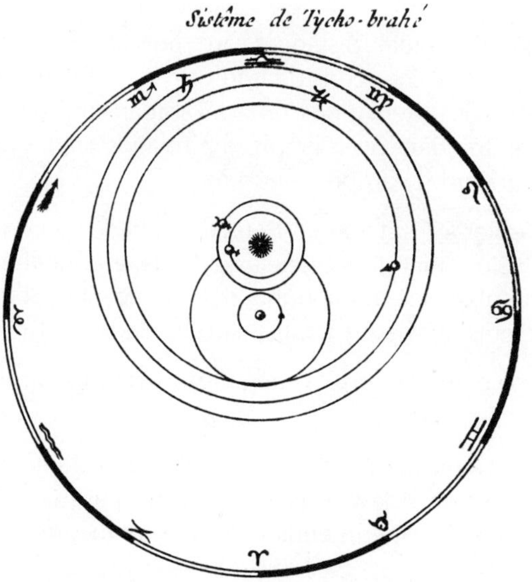

Sistème de Tycho-brahé

FIGURE 4-4 The universe according to Tycho Brahe (Yerkes Observatory Photograph).

The Tychonic theory provided every bit as good an explanation of the motions as did the Copernican theory. Tycho's had the advantage that the parallax of the stars was not required, and fewer epicycles were needed. The theory did not gain wide acceptance, probably because Kepler's proposal followed so soon on its heels. As we are about to see, Kepler's model, while still lacking proof of the earth's motions, nevertheless was eminently more satisfactory than the others because of its great simplicity. It did, however, complete the transition away from the geocentric concept and the requirement for circular motions.

Kepler fell heir to Tycho Brahe's voluminous data, and he kept working with it after the old astronomer's death in 1601. But 4 years of intensive effort trying to fit Tycho's superb data on Mars' motions to a circular, epicyclic, and heliocentric orbit convinced Kepler that it could not be done. The errors in it were more than 8 times as great as the precision in Tycho's observations. The method Kepler used was ingenious, and it deserves our detailed attention because in the end it led Kepler to the correct hypothesis.

Determination of a Relative Orbit. Kepler first *assumed* that the planets *including the earth* all revolved about the sun in nearly the same plane. Thus the apparent return of the sun to the same place among the stars marked one complete orbital revolution of the earth around the sun. Next Kepler determined that, on the average, a 780 day interval (approximately) occurred between the successive returns of Mars *to the same alignment with the earth and sun*. During this interval the earth completed approximately $2\frac{1}{7}$ revolutions around the sun. In net angular displacement the earth's position at the end of 780 days was about 50° east of its orbital position at the beginning of the cycle (Figure 4-5). That is, the net displacement was a full circle of $360° + 50° = 410°$.

Next, Kepler reasoned that Mars had gone

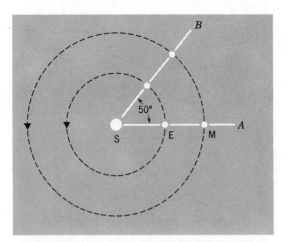

FIGURE 4-5 The synodic period for Mars M with respect to the earth E. At the beginning of the period the sun, the earth, and Mars are on the line S*A*; at the end of the period they are on the line S*B*. See text for discussion.

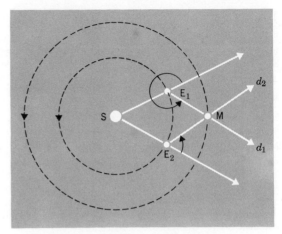

FIGURE 4-6 Kepler's method for determining the orbit of Mars. E_1 and E_2 are the respective positions of the earth relative to the sun, (determined with reference to the celestial sphere) 687 days apart. The direction of Mars M as seen from earth, relative to the earth-sun direction line, is plotted from the observed angles. Repeating this construction for a sufficient number of pairs of observations 687 days apart will permit the graphing of Mars' relative orbit.

through the same angular displacement in 780 days, since at the end of this period it and the earth and the sun were realigned as they were at the start of the cycle. Therefore, the number of days it takes Mars to complete revolution (360°) in its orbit could be calculated from the simple proportion: the number of days is to 360° as 780 days is to 410°. We can write this as an equation.

$$\frac{\text{no. days}}{360°} = \frac{780 \text{ days}}{410°}$$

We can solve for the unknown number of days by multiplying both sides of the equation by 360° and performing the indicated divisions.† The solution is 687 days. (Actual computation with the given data will yield 685 days, merely because the data used are rounded off and are thus only approximate.)

Figure 4-6 illustrates a geometric construction which yields the relative position of Mars with respect to the earth and the sun even

† A brief, simple, mathematical review is provided in Appendix 2 for those who feel the need for it.

though actual distances are not known. By repeating the construction for a number of pairs of observations 687 days apart and plotting the data on a graph, the relative *shape* of Mars' orbit can be determined.

Kepler's First Law. Kepler reached the conclusion that the planetary orbits are neither circular nor compounded of epicycles, deferents, and equants. Instead, he deduced that:

A planetary orbit is an ellipse with the sun at one focus.

This constitutes the empirical statement of Kepler's first law. It is empirical because it was based on experimental data, not on theoretical proof.

The ellipse is so important in astronomy that we should briefly examine some of its properties. Suppose two pins are driven into a flat surface a short distance apart. Then an un-stretchable string is knotted into a loose loop and pulled tight against the two pins by a pencil point as shown in Figure 4-7. Obviously the total length of string in the loop remains constant. So does the portion of the loop between the two pins. Now if the pencil is moved so as to trace a line on the surface all the while keeping the loop tight, the curve so traced is an ellipse. We note that one of the chords of the loop containing the pencil point gets shorter while the other becomes longer. They both do this in such a way that the *sum* of the two chords remains constant. We can use this information to define an ellipse: *an ellipse is a closed-plane curve such that the sum of the distances from two fixed points to any point on the curve remains constant.* Each of the two fixed points is called a *focus* of the ellipse. It lies on a line connecting the two points on the ellipse which are most distant from its center; these points are singly called a *vertex* of the ellipse. An important property of an ellipse is its *eccentricity*. This is defined as the ratio between two particular line segments: from the center to a focus, and from the center to the corresponding vertex. The latter line segment is

known as the *semimajor axis* of the ellipse. Several ellipses all having the same semimajor axis but with different distances from center to focus are shown in Figure 4-8. The eccentricity of the ellipse can range from 0.00, in which case the curve is known as a circle, to 1.00, when the curve ceases to be closed and is then called a *parabola*.†

The Second Law. In working with the various pairs of data which led him to the conclusion that Mar's orbit was an ellipse, Kepler made another significant discovery. He found that a planet travels at different speeds in different portions of its orbit. From the definition of speed—distance traveled per unit time—the differences in the speed of a planet can be interpreted as differences in distance along the orbit with respect to a fixed time interval. Kepler found that the speed of a planet is greatest when it is in that portion of its orbit nearest the focus containing the sun, and least when it is in that portion farthest removed from the sun. When Kepler plotted Mars' orbit and drew lines (called *radius vectors*) from the sun to two positions of the planet, he found that the *area* enclosed by the radius vectors and the in-

† The reader who is unfamiliar with the family of curves known as *conic sections*, of which the ellipse and the parabola are members, is urged to consult Appendix 2 for further details.

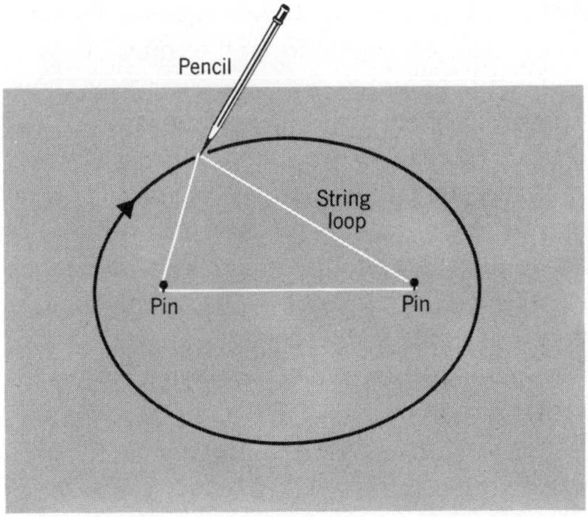

FIGURE 4-7 Constructing an ellipse as described in the text.

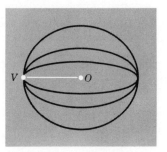

FIGURE 4-8 Three ellipses with the same semimajor axis (line OV) but with different eccentricities.

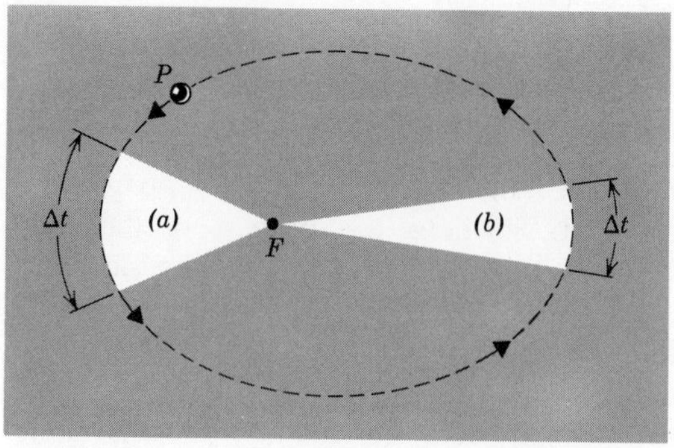

FIGURE 4-9 Kepler's equal area law. Area (*a*) is equal to area (*b*). See text for discussion.

tercepted portion of the orbital path was always constant provided the time intervals were always the same (Figure 4-9). A formal statement of this discovery becomes Kepler's second law:

The radius vector joining the sun and planet sweeps over equal areas in equal intervals of time.

Kepler published these two laws in 1609 after the deaths of Copernicus (d. 1543) and Tycho Brahe (d. 1601), but prior to the time that the telescope was employed in astronomical observations by one of Kepler's contemporaries, Galileo Galilei. Kepler knew from the geometry involved and from his observations that the more distant a planet is from the sun, the longer it takes that planet to complete one revolution about it. Not only is the path length longer; *the planet travels more slowly in its orbit.* The slowest planet is the one that is most distant from the sun.

The Third Law. Formulating the comparative rates of motion proved to be a formidable task; it was not until 1618 that Kepler was able to express the relationships in what is known as his third law:

The square of the period of revolution of a planet about the sun is proportional to the cube of its mean distance from it.

We can write this statement in the form of a proportion

$$p^2 \propto d^3$$

provided the proper choice of units is made. The symbol means "proportional to." For the present let us take the period to be years, and the distance to be units of the mean earth-sun distance. Then the proportion holds for the earth: $1^2 \propto 1^3$, since any power of 1 is still only 1. The proportion holds for Jupiter whose period is 11.86 years and whose distance from the sun is 5.2 units, thus, $(11.86)^2 = 140 = (5.2)^3$.

Kepler's laws do not constitute proofs of the earth's orbital and rotational motions. Kepler's model was just one more in the series of astronomical hypotheses. No one of them is more "right" than any other simply because of the chosen frame of reference. The governing restriction in a choice between them should be that *all* of the observed phenomena must be accounted for. Kepler's hypothesis alone has stood this test; his laws are as valid today as they were over 360 years ago. Kepler's model has the added advantage of being by far the simplest of them all.

Galileo's Support for Kepler's System. Galileo's discoveries with the telescope succeeded in crumbling the last vestiges of scientific opposition to the heliocentric theory.

Coming as they did after the formulation of the first two laws, his announcements lent great weight to the Keplerian model. Newton's success in reformulating the physics of motion sounded the death knell of the scholastic physics; his triumph in putting Kepler's laws on a sound mathematical basis clinched the case. Only the church persisted stubbornly in its doctrinal opposition. It, too, finally relented officially in 1835.

Galileo helped to demolish the geocentric theory of Ptolemy in several ways. With the aid of the telescope he discovered details on the moon's surface which he interpreted as being mountains and plains or seas.† He reported dark spots or blemishes on the sun which traveled slowly in parallel paths centrally across the sun's disk. Galileo discovered four of Jupiter's satellites whose orbital motion was centered on Jupiter, *not* on the earth (Figure 4-11). And he discovered that Venus exhibits phases similar to those of the moon (an impossibility in the Ptolemaic system), and that both Jupiter and Saturn were flattened — not perfectly spherical. All of these discoveries were in direct contradiction to the premise inherent in the geocentric hypothesis that the celestial objects were " . . . perfect (meaning spherical), unblemished, unchanging, and perfectly ordered." With respect to this last requirement, Galileo pointed out that if the geocentric system were indeed valid, the planets had to rotate very slowly about the earth in *opposition* to the daily motion of the celestial sphere. Furthermore the periods of all of the bodies increased outward, from approximately 28 days for the moon to 30 years for Saturn. Yet the celestial sphere which was even still further out made a complete revolution in a single day! Both of these phenomena are departures from the perfect order. Lastly, the discovery of the four Galilean satellites of Jupiter invalidated the no-

tion that there were only seven celestial bodies in the heavens, because "there are only seven windows in the human head, two eyes, two ears, two nostrils, and a mouth . . . "; a one-to-one correspondence was required!††

Newton's Laws

The gathering evidence all pointed to the fact that Kepler had finally deduced the correct form of the solar system and had expanded the bounds of the universe beyond all expectation. In spite of this, the Keplerian hypothesis did not become firmly entrenched until Newton mathematically demonstrated that an elliptical orbit was a logical consequence of a central force system.

Isaac Newton was born in 1642, 12 years after Kepler's death. He was a somewhat frail, not very precocious child. But after his admission to Trinity College at Cambridge at the age of 19 his talents came to full flower. By the time he was 24 he had made profound discoveries in mathematics, optics, and mechanics. Among his mathematical accomplishments was the invention of the calculus, made necessary by the inability of algebra to express the relationship between changing quantities in terms of the rates of change. Newton needed the new mathematical concepts in his attack upon the Aristotelian mechanics. His three laws of motion and his universal law of gravitation are still valid today in elementary mechanics and dynamics. The Newton story is as fascinating as it is diverse. All we can do here is highlight those of his discoveries that are pertinent to our work in astronomy.*

By the year 1400 A.D. most European scientists were satisfied that the Aristotelian physics

---

† Contrary to popular belief, Galileo *did not* invent the telescope. He constructed one from the published notes of an obscure Dutch spectacle maker, Hans Lippershey, describing the optical properties of combinations of lenses.

†† From a quotation attributed to the Florentine astronomer Francisco Sizi, quoted in Holton, G., and Roller, D. H. D., *Foundations of Modern Physical Science*, Addison-Wesley, Reading, Mass., 1958, p. 160.

* Several references which cite Newton's life and work are listed in the bibliography.

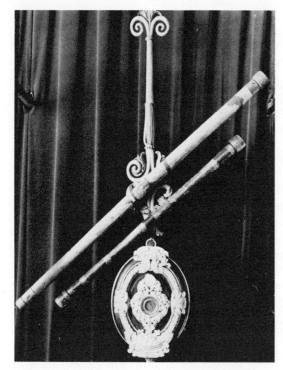

FIGURE 4-10 Galileo's telescope (Courtesy Florence Museum of the History of Science and Technology).

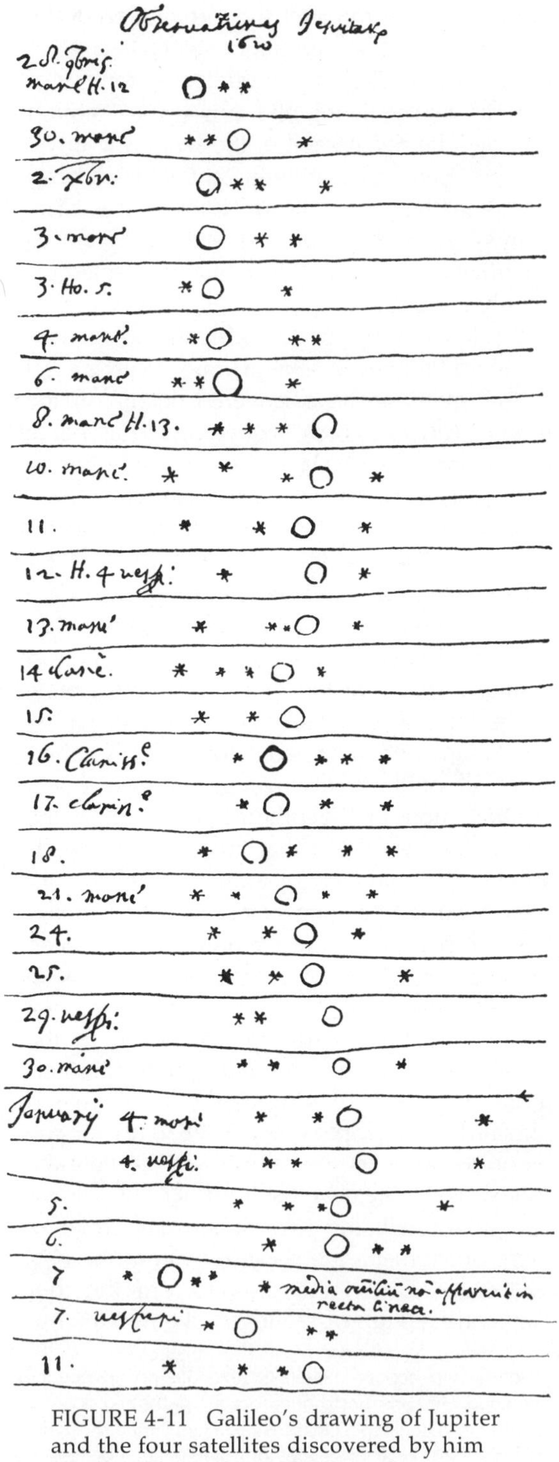

as it related to motion was grossly in error. For example, they concluded that the preferred state for earthly objects was not necessarily one of rest, a cornerstone of Aristotle's physics. To be sure, an object placed in horizontal motion by hand eventually comes to rest once the hand is removed. But Aristotle's physics could not explain why changing the shape of the object, or bringing it into contact with different kinds of surfaces, affected the duration of the motion.

Aristotle's writings influenced almost every area of intellectual inquiry. The failings of his teaching in mechanics concerned only a few scholars. Hence Aristotle's influence in other areas such as philosophy, political science, and astronomy remained undiminished while his physics was undergoing major revision by such scientists as Copernicus, Galileo, William

FIGURE 4-11 Galileo's drawing of Jupiter and the four satellites discovered by him (Yerkes Observatory Photograph).

Gilbert, and Francis Bacon. This revision reached its peak during the 100 years between 1540 and 1640. By the time Isaac Newton turned his attention to the problems of defining forces and the quantitative (mathematical) description of motion, the role of forces in nature was becoming clear. Indeed, Newton hastened to acknowledge that if he did have a longer range of vision in these areas than others of his contemporaries, it was because he "stood on the shoulders of giants. . . ."

Newton's First Law. Newton's three laws of motion represent a distillation of this earlier progress, reduced to formal statements. His most important contribution here, aside from the mathematical rigor, was to focus scientific attention on the concept of force. The essence of his arguments, especially as stated in the first law, is that a *change* in motion, either through slowing down, or speeding up, or a change in direction of an object, is necessarily the action of one or more forces. In the latter case multiple forces must be unbalanced:

Every material body persists in its state of rest, or of uniform speed in a straight line, unless acted on by an external force.

This statement is in direct conflict with Aristotle's concepts of force. Among these was the notion that for earthly objects a force was required to produce uniform motion. The first law implies that *either* no forces are acting on a body, *or* all of the acting forces exactly neutralize each other whether or not the object is at rest. Vectors are used to advantage in depicting forces graphically since the latter have both magnitude and direction. Figure 4-12 illustrates a few of the vector concepts. A more detailed treatment can be found in Appendix 1.

Newton's Second Law. Newton's second law is sometimes called the law of inertia. Inertia is a measure of the opposition, or resistance, of a material body to a change in motion when acted on by an impressed force. Inertia is a phenomenon common to our expe-

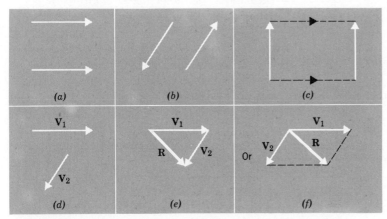

FIGURE 4-12 Some properties of vectors. The vectors in (*a*) are equal because their magnitudes (lengths) are equal and their directions are the same. In (*b*) the vectors are unequal because their directions are different. (*c*) Vectors can be moved in space; they remain unchanged if their original magnitudes and directions are preserved. Vectors can be added as in (*e*) and (*f*); the sum of two or more vectors is a *resultant*. Conversely, a vector can be resolved into components (see Appendix 1).

rience. Recall the tendency of our bodies to be thrown forward when a fast moving vehicle slows down suddenly. Observe how the occupants of an automobile, as well as the automobile itself, sway sidewise toward the outside of a sharp curve negotiated at too high a speed. These examples illustrate the inertia of the human body or vehicle to a change in motion.

Not all bodies have the same inertia, even when they are of the same size. A competent softball pitcher can deliver the ball to the plate with speeds in excess of 75 miles per hour. He couldn't come close to duplicating the feat with a 16 pound shot used in track and field competition.

Newton's second law can be stated in several ways. The following is a useful form:

The net external force acting on a material body will accelerate it in the direction of the force. The magnitude of the acceleration is directly proportional to the magnitude of the force and inversely proportional to the mass of the object acted upon.

We can write these relationships symbolically.

$$\text{net } \mathbf{F} \propto m\mathbf{a} \qquad (4.1)$$

$$\frac{\text{net } \mathbf{F}}{\mathbf{a}} = m \text{ (a constant)} \qquad (4.2)$$

where $\mathbf{F}$ is force, m is mass, and $\mathbf{a}$ is acceleration.

Equation 4.2 says that for a given object whose physical quantity of matter remains constant (or stated more economically, if the mass of the object remains constant), a change in the magnitude of a net force, or a change in the direction of the force, or *both,* produces an identical change in the acceleration of the object. Newton's second law is more commonly written,

$$\text{net } \mathbf{F} = m\mathbf{a}. \qquad (4.3)$$

We have here several extremely important concepts. One of them is *acceleration.* Accel-

eration is the rate of change of velocity which in turn is a quantity of speed having *both* magnitude *and* direction. Hence velocities can be conveniently represented by vectors as was the case with forces. The concept of velocity is also highly significant in astronomy and physics. The velocity of an object is said to change if its speed changes, if its direction changes, or if both change together. Note that if the velocity change is one of direction, the object's speed may, and often does, remain constant. Similarly a change in an object's speed does not necessarily involve a change in direction. And since in general velocity changes occur in finite time intervals or increments, the phenomenon is by definition an acceleration. The units of acceleration consist of a length unit divided by the product of two time units. Thus, an acceleration in metric units could be written *centimeters per second per second,* abbreviated as cm/sec².

Another concept arising from the relationship between the magnitude of a given acceleration and the force which produces it is *mass.* Mass is an intrinsic property of matter. The atomic or molecular structure of a body determines its mass. Strangely, the size (volume) of the body is not directly involved. To be sure, mass in the Newtonian frame of reference is associated with only finite bodies (including, of course, parcels of gas or liquid), and all such bodies occupy space. Hence the relationship between a quantity of mass and the space it occupies is treated separately as the *density* of the body. Density is defined as mass per unit volume. From this point of view density is a descriptive term for the substance of which a body is composed.

We should note carefully that except when an object is moving at high speeds, speeds near the speed of light, the mass of an object is independent of where or how it is measured. There are relativistic changes in the mass of high speed particles and objects that need not concern us here. Newtonian mechanics, in reality a special case of the theory of general relativity applicable to ordinary speeds, is completely adequate for the present.

Newton's second law states that for a given force, a measure of the acceleration it imparts to a body uniquely determines the body's mass. Conversely, for an object whose mass is known, observing the magnitude of the acceleration it exhibits determines the magnitude of the force causing it. In fact if we know any two of the three quantities in Equation 4.3, the third quantity can be at once determined. The perceptive reader should be able to recognize that Newton's first law is just a special case of the second law.

Newton's Third Law. Newton's third law is much more subtle. Its validity often escapes an observer unless he is conditioned to recognize it. The third law is a statement concerning pairs of forces: all forces exist in oppositely directed pairs:

If one body exerts a force on another body, the latter exerts a reaction force on the first body, equal in magnitude, but oppositely directed.

We shall gain more insight to the third law after we investigate the nature of central and gravitational forces. For the present, consider a book lying on a table. The book pushes down on the table. Since in general it does not move downward, the table must be pushing upward on the book by an exactly equal but oppositely directed force. Again, a person pushes on a wall; the wall pushes back on the person. Provided that the wall does not move, this backward force acting on the pusher is again exactly equal to but oppositely directed from the applied force. Once more: a space vehicle pushes on the ejected exhaust gases of its rocket engine. The exhaust gases push on the rocket. In this case the design of the system is such that both exhaust gases and the vehicle are free to move, hence the vehicle accelerates at a rate exactly equal to, but in the opposite direction from, the exhaust gases. The fact that

space vehicles are free to accelerate and move about in space without, in the Aristotelian frame of reference, having the atmosphere to push on is concrete evidence of the validity of Newton's third law.

Universal Gravitation

The concept of gravity as an attracting force, meaning that unrestrained objects would "gravitate" toward the center of the attracting body, was well established by the time of Copernicus. But many of the fine details eluded the early investigators. For example, it was commonly accepted that if a light and a heavy object were simultaneously released from the same height, the heavier object would reach the ground first. The rate of fall was presumed to be proportional to the object's mass. Not until Galileo conducted a variety of tests of this hypothesis did the truth appear, that the rate of fall was independent of the object's mass. What had confused the early philosophers was the presence of an undetected frictional force offered by the air.

But gravity was conceived as merely a phenomenon of the earth. Until Newton boldly set forth his stunning theories, no marriage had occurred between terrestrial and celestial mechanics. It was Newton who expanded on Galileo's experimental treatment, showing mathematically that gravity was a force of attraction between *pairs* of masses. This does not imply that if more than two bodies are involved, the additional bodies are not affected by the attracting force. Consider three bodies, A, B, and C. There are three pairs of force-attractions: A and B, A and C, B and C. Where more than two bodies are involved, analysis of the mutual attractions becomes very complicated. For reasons which will become clear later, we shall for the present consider only the simplest case, the two-body problem.

Newton went so far as to theorize that the gravitational phenomenon involved all masses in the universe. In so doing he extended his terrestrial mechanics — exemplified in the three laws of motion — to objects beyond the earth. The laws of motion became universal laws. In the First Book of the *Principia,* Newton's magnificent treatise on the mathematical principles of natural philosophy (an archaic term for physics), he showed that the net force acting on a planet or satellite at any instant is directed toward the *center of motion* of the orbital path. Whether or not the path was circular, elliptic, or one of free escape depended upon the initial motion of the body at the instant the central force became operative. Newton argued thusly. Suppose that an object is moving with constant speed in a straight path. The net forces on the body are zero according to the first law. The positions of the body at the end of each successive, equal-time interval will be equally spaced along the path (Figure 4-13). Suppose further that the object is connected by a taut string to some central point not on the path (*O* in the figure). Then the triangles formed by the line connecting successive path positions and the string all have same area. A theorem in elementary geometry states that the area of a triangle is equal to one-half the prod-

uct of the base and the altitude. The altitude of a triangle is the perpendicular distance from the base to the triangle's *apex.* Note that it is often necessary to extend the base line well outside of the triangle in order to see this relationship. Now in the figure each base segment is the same length, and the vertical distance to the central point *O* (the apex) is the same for all of the triangles. Hence the triangles all have the same area. We observe that as the object moves from point to point along the path, the string sweeps over equal areas in equal intervals of time. This begins to look like Kepler's second law.

Newton's next step proceeded like this. Suppose the time interval between the successive space positions *M, N, P, R, S* of the body along its path is some small quantity Δt, pronounced *delta* t. Suppose further that on the arrival of the body at *N* a force in the form of a sharp blow directed toward *O* is delivered to the object (Figure 4-14). The blow is presumed to be of short duration, but powerful. Hence the body accelerates very quickly up to some speed determined by the force of the blow and the duration of its effect. After this time, the

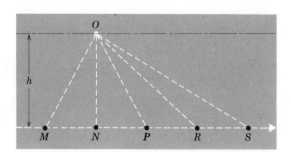

FIGURE 4-13 An object moving along a straight line according to Newton's first law will successively occupy the equally spaced points *M, N, P, R, S* in equal time intervals. A line connecting the object to an arbitrary reference point *O not* on the line will sweep over equal areas in equal time intervals. All of the four triangles shown have the same area.

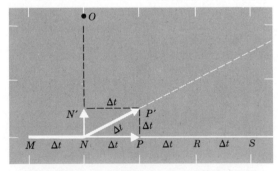

FIGURE 4-14 According to Newton's first law, if an object traveling along the line *PS* is acted upon at point *N* by a force which has direction *NO*, the resultant path will be along a line such as *NP'*. The velocity in the direction of *NP* is unchanged by the force acting at right angles to it; hence the travel time from *N* to *P* is the same as from *N* to *P'*.

body has a velocity directed toward O. If it were stationary initially, the resultant motion toward O would be all that the body would have. But the object is moving toward P from N when the blow is delivered. After the blow, the body has one component of motion still toward P; it also has another component directed toward O. The resultant motion is the *vector sum* of the two components. The vector sum can be graphically determined by constructing a parallelogram having the two component velocities as adjacent sides. The forward diagonal of the parallelogram is the new velocity, both as to speed and direction.

If the force does not act again, the object will continue along a new path in the direction NP'. Each new increment of distance will be of the same magnitude as that of the resultant velocity vector, **NP'**. A possible change in magnitude of the speed along the new path in general results from the fact that the displacement blow will not be perpendicular to the object's path at the instant the blow is delivered. Thus there may be a component of the deflection in the direction the object is initially moving which, depending upon the initial conditions, may instantaneously slow the object down, or even speed it up.

A succession of blows delivered at intervals Δt apart, each directed toward O, will result in successive displacements, and in general, in successively different speeds of the object along its resultant path (Figure 4-15). The more curious reader may wish to continue the construction. Depending on the magnitude of the deflecting force and the initial speed of the object, if the time intervals are made vanishingly small but *not* zero, the path may turn out to be a circle, an ellipse, one of the open curves, a parabola, or a hyperbola (Appendix 1). The point O is the center of force; it will be at the center of a circle, at one of the foci of an ellipse, or at the focus of the other two curves. *In other words if the body is continuously acted on by a central force it must move in accordance with Kepler's second law.*

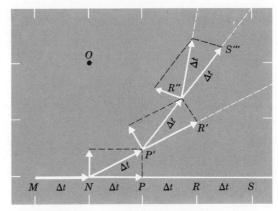

FIGURE 4-15   A succession of blows, each delivered with the same force after equal time intervals, and all of which are directed toward O will deflect an object originally traveling with constant speed along the line MS into a path around O. Depending upon the initial conditions, if the blows follow each other after infinitely short time intervals, the result is as if a continuously acting force is deflecting the object, and the path will be one of the conic sections (ellipse, circle, hyperbola, or parabola).

By reversing the argument, Newton was able to prove by a similar line of reasoning that *if a body is moving in accordance with Kepler's second law it must be acted on by a central force.*

Thus, Newton placed Kepler's first two laws on a sound mathematical basis; all that remained was to show the nature of the central force. Since the planets orbit around the sun, the central force in this case must be related to the sun. With respect to satellites, the central force must be related to the parent planet. We shall not pursue the development of the argument and its proof. We shall merely accept and apply Newton's results. Newton showed that the gravitational attraction between two bodies acted along a line connecting their centers of mass (Figure 4-16). Hence the central force in planetary motion is a gravitational force! By the third law of motion there are two equal and opposite forces of attraction, each acting on

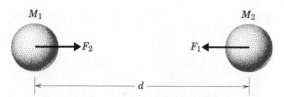

FIGURE 4-16 Gravitational force between two bodies. The force F_1 is an attractive force on M_2 directed toward M_1; the force F_2 is an attractive force on M_1 directed toward M_2. The forces act along a line containing the centers of mass of the two objects, and they are proportional to the product of the two masses. The magnitudes of the forces decrease inversely as the square of the increase in distance between the centers of mass.

By selecting a constant of proportionality, that is, a quantity which is constant for *all* cases for a given system of units of force, mass, and distance, we arrive at a mathematical statement of Newton's universal law of gravitation,

$$F_{\text{gravity}} = G \frac{m_1 m_2}{d^2} \qquad (4.7)$$

where G is a known universal constant found by experiment. Interpreted in prose, Equation 4.7 states: *the mutual force of attraction between two bodies is directly proportional to the product of their masses and inversely pro-*

one of the bodies. The force on each is proportional to the product of their masses,

$$F_{\text{gravity}} \propto m_1 m_2 \qquad (4.4)$$

where m_1 and m_2 are the two masses in question, and F_{gravity} is the force of attraction. Newton also showed that the force of attraction decreases with increasing distance of separation of the masses. The force varies inversely as the square of the distance of separation.

$$F_{\text{gravity}} \propto \frac{1}{d^2} \qquad (4.5)$$

This relationship shows that as the distance d becomes larger, the ratio $1/d^2$ becomes smaller. In a given situation, if d is doubled the force of attraction becomes 1/4 the original magnitude; if d is tripled, the force decreases to 1/9 of its original value, and so on. This is the significance of an inverse ratio.

The Universal Law of Gravitation. The relationships of Equations 4.4 and 4.5 can be combined since both apply in any given situation. We have

$$F_{\text{gravity}} \propto \frac{m_1 m_2}{d^2}. \qquad (4.6)$$

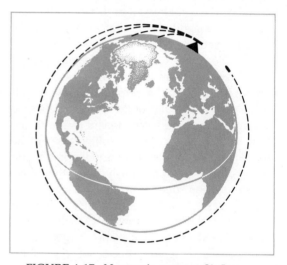

FIGURE 4-17 Newton's cannon. Sir Isaac Newton hypothesized that the greater the muzzle velocity of a horizontally fired projectile, the farther it would travel before gravity pulled it to the earth's surface. If the cannon were placed on the summit of a high mountain and had sufficient muzzle velocity, the projectile would fall not only toward the center of the earth, but it would also deviate from the initial horizontal motion at the same rate that the earth's surface curves away from the flight path. The net result would be a projectile in orbit around the earth. Newton's hypothesis was verified completely with the successful orbit of Russia's *Sputnik I* in October, 1957.

portional to the square of the distance between their centers of mass.

Newton went on to show that if gravity is the central force acting on an orbiting body, the square of the time it takes to complete one revolution is directly proportional to the cube of mean distance of the body from the center of force, and that this relationship is valid for any orbiting body. The precise relationship often is influenced by minor disturbances from a third (or more) body. There is no exact solution for the so-called n-*body problems* where *n* is an integer larger than 2. Hence problems in celestial mechanics most generally have to be solved by numerical approximation methods using high speed computers. This does not, however, detract from Newton's universal law of gravitation in any way. To the contrary, the *n*-body problem provides supporting evidence for its truth.

Here is a precise, quantitative verification of Kepler's third law. Since the three laws of motion are also involved, the Newtonian synthesis of terrestrial and celestial mechanics is now complete. In passing, it is interesting to note that Newton predicted the orbital velocity of an earth satellite (in his case a projectile) at a given height above the earth's surface (Figure 4-17). His prediction was completely verified with the advent of the space age a little over 300 years later.

Apparent Motions of the Planets

It now appears propitious to relinquish our position as observers on a stationary earth platform at the center of the universe. We do this *not* because this frame of reference is wrong, but because the alternative theories, especially that of Kepler, are exceedingly more simple, more complete, and in general more tenable. We shall still consider the heliocentric solar system, even with elliptical planetary orbits, a hypothesis, not a fact, until we are able to prove the motions of the earth. These proofs must be independent of the data used to derive

the hypothesis. We shall attack and solve the problem in the next chapter.

We can, however, take note of some persuasive factors which in the interim lend support for our new position. Galileo's telescope displayed magnified images of the planets and moon and in so doing disclosed several unsuspected details. The images of the planets were disks, not points; Venus displayed phases similar to those of the moon. Jupiter's surface appeared to be banded, and Saturn had a beautiful ring system which defied explanation; both planets appeared flattened, not spherical. In addition Jupiter had at least four satellites. The moon had surface features resembling terrestrial plains or seas, and mountains. But no amount of magnification changed the apparent point images of the stars. Presumably they are extremely distant from the earth.

If the latter case is true, we are compelled to discard the concept of a real celestial sphere as being a feature of the universe. We do this on the grounds that the required surface speed at the celestial equator is totally irrational. Even if the celestial sphere were only as far distant as Saturn, and the classical investigators knew it was much more distant than that, the stars on the celestial equator would be traveling in excess of 60,000 miles per second!

The variable speed of the planets in their orbits is consistent with the ellipse hypothesis and the central force theory. The closer a planet is to the sun, the stronger is the gravitational force acting on it. And since the orbits are not circular, there is generally a component of the gravity force which alternately speeds the planet up or slows it down. A similar argument in connection with the hypothesis of an orbiting earth accounts for the apparent variation in the speed of the sun along the ecliptic. This is just a reflection of the earth's variable speed in its elliptic orbit.

Retrograde Motion. The retrograde motion of the planets can be explained as a perspective effect caused by the changing positions of both

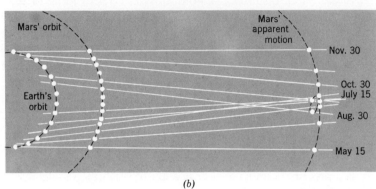

FIGURE 4-18 (*a*) The retrograde loop of Mars for 1971. Vertical coordinates are hours of right ascension, horizontal coordinates are declination in degrees. (*b*) Plan view of Earth-Mars configuration for the 1971 retrograde loop. Mars' apparent motion is projected on the right. (Some of the projected positions have been omitted for clarity.)

the planet and the earth relative to a stationary backdrop of stars. The variation in the shape and size of the retrograde loops is consistent with variations in the respective planet-earth positions with orbits which are not quite in the same plane. An understanding of the retrograde motion concept is made easier with the aid of an analogy drawn from our experience.

Suppose we are traveling in the fast lane of traffic, overtaking a slower moving vehicle at a steady pace. When viewed against any convenient reference backdrop, a range of hills, for example, the slower moving car appears to slacken its pace past the landscape. As we draw near it finally appears to move backward against the backdrop, and continues to do so until we are some distance ahead of the other vehicle. Eventually it appears to resume its forward motion against the background.

Figure 4-18 illustrates the corresponding planetary hypothesis as viewed from earth, and as it would appear when viewed from above

the plane of the orbit. As soon as we can successfully prove that the earth is revolving around the sun, we can accept this hypothesis as fact.

SUMMARY

The early models of the universe—those of Plato and Aristotle and Ptolemy as *well* as those of Copernicus and Tycho Brahe—all had serious defects. In every case some feature of the observed behavior of the system was left unaccounted for. The concept of concentric spheres gave way to a more acceptable geocentric theory. But the artificiality of epicycles traveling on deferent circles, of equants, and occasional tampering with the data, in time took its toll.

Copernicus came closer to the truth. But man often finds himself reluctant to relinquish long-held, comfortable notions. So it was with Copernicus who could not give up the philosophical necessity of circular orbits. Neither could he dispense with epicycles, nor could he fix the center of his system exactly on the sun. In truth, the system was not quite heliocentric.

Kepler made the great transistion in thinking from the long-held classical notions to those of modern astronomy, aided substantially by geometric analyses. From the earliest times, man's speculation about his universe was dominated by qualitative, philosophical evaluations. Mathematics was slow in entering the scene; when it did make its presence felt, the change in thinking became dramatic and relatively rapid. Based on his analysis of the extraordinarily good data left to him by Tycho Brahe, Kepler proposed a heliocentric, elliptical orbit system, and he described it by the three empirical laws of celestial mechanics which bear his name.

Newton provided the capstone to 21 centuries of evolving astronomical thought, marred, it is true, by 10 centuries of inertia during the Dark Ages. Based upon the work of many eminent scientists, especially that of Copernicus, Kepler, and Galileo, Newton successfully promulgated a new mechanics, which he described in terms of his three famous laws of motion. Newton completed the synthesis of terrestrial and celestial mechanics by extending his three laws of motion into a universal arena and adding the great law of universal gravitation. His total success in verifying mathematically Kepler's laws of motion has provided astronomy with a celestial mechanics which has remained essentially unchanged and unchallenged for over 300 years.

Questions for Review

1. Is a "stationary earth" a totally unacceptable frame of reference? What are the disadvantages of this concept?

2. What is meant by unsymmetrical periodic motion? Can you think of an example other than the one used in the text? (Hint: start with a dictionary definition of unsymmetrical.)

3. List some of the salient features of scholastic mechanics as related to astronomy.

4. What was Ptolemy's major contribution to astronomy? Explain.

5. What was Copernicus' major contribution to astronomy? Explain.

6. What role did the observations of Mars' positions play in the development of Kepler's laws?

7. List Kepler's laws of motion.

8. What observations made by Galileo lent strong support for a heliocentric theory as opposed to a geocentric universe?

9. State Newton's laws of motion.

10. What is Newton's universal law of gravitation?

11. What is the relationship between Newton's laws, including the universal law of gravitation, and Kepler's laws?

12. Discuss retrograde motion. Is it a real motion or is it a relative motion? Explain.

13. Given three spheres: Sphere *A*; mass 50 grams, radius 5 centimeters. Sphere *B*; mass 250 grams, radius 10 centimeters. Sphere *C*; mass 100 grams, radius 7 centimeters. Which of the two spheres when placed with their centers 25 centimeters apart will attract each other with the greatest force? Which pair of spheres would attract each other with the least force under the same conditions? Explain.

14. What is the approximate period in years (to the nearest whole number) of a planet that is 5.0 astronomical units from its sun?

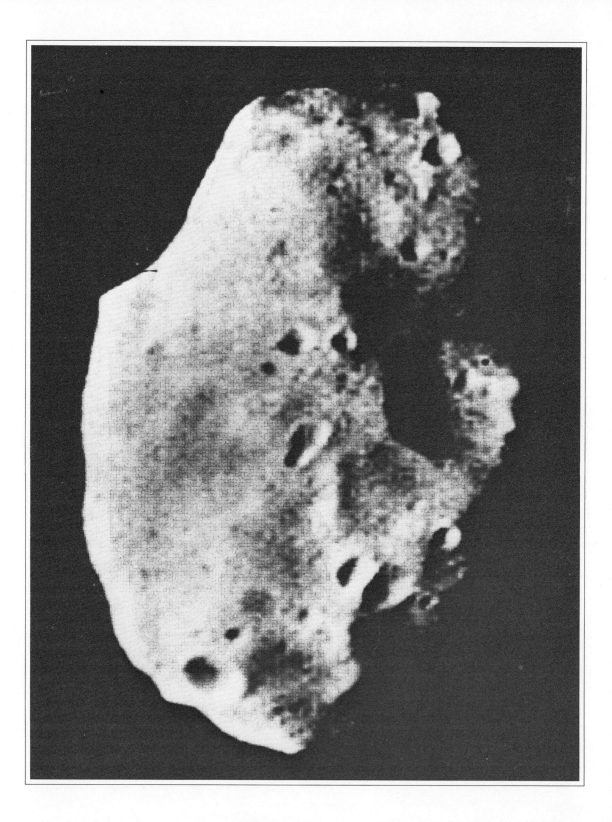

PART TWO
The Solar System

Prologue

If knowledge about the universe is our main concern, why should we
scrutinize the earth in great detail? Indeed, why bother with it at all? The
answer is that we do not wish to emulate the ancient astronomers—and
some not so ancient—who landed in the quagmire of indeterminate
answers to their astronomical problems because they failed to look
before leaping. To be sure, we can measure the angular size of the moon
with a suitable protractor; the ancients did likewise. But what does this
tell us about the real physical size of the moon? We need additional data,
the acquisition of which is intimately wrapped up in measurement techniques that we develop and learn to use right here on earth. Failure to
obtain this data was one of the shortcomings of the early astronomers.
We shall find that the earth is smallish, less than 8000 miles through, and
that it is flattened top and bottom as though it had been squeezed in a
vise. We shall also find the necessary techniques for determining the
earth's motions, one of which is its 67,000 mile per hour rush around the
sun. This is so that we may finally and with confidence escape from the
intellectual prison of a stationary earth concept.

To accomplish all this we must make two prominent departures from
methods that have been successful in the past, but which no longer suffice. One departure is to find and adopt an "external" frame of reference,
one that is not anchored to the earth. For how can we measure motion if
the yardstick moves with the moving body? The other departure is to
use the properties of light itself, instead of just the information it carries,
in a new and different way of acquiring data.

The Earth: Dimensions and Motions

5

We are inclined to take the earth more or less for granted. All of man's known history, and most of his experience, is intimately involved with this larger member of the earth-moon binary planet system (L., *binarius,* two at a time). Yet in many respects man knows more about the structure and behavior of the solar system, of the stars, even of the galaxy or of galaxies, than he does about his own planet. Much of what we know about earth has been the result of inductive logic; man has had to combine laboratory procedures in physics and chemistry with astronomical methods in order to discover the earth's physical nature. Then by a process of elimination the irrelevant, the mystical, the metaphysical, and the impossible notions are rejected. Only then does the earth, a microcosm of matter in the scale of the universe, take on added dignity and properly become an object of awe for the beholder.

Optical Tools: A New Requirement

Much of what we learn about the earth by direct measurement or by inductive logic forms the basis for deducing the physical characteristics of the remote objects of the universe, including other planets and satellites. Hence we need a somewhat intimate acquaintance with the instrumentation by which the necessary data are acquired. We must study the earth in rather complete detail, not as a geological specimen, but as an astronomical body.

We shall make use of some ordinary and familiar measuring devices: a weighing scale or balance, a clock, and a meter stick. The latter is the fundamental unit of length in the metric system of weights and measures. (We shall use the metric system from this point on; Table 5.1 lists some of the more commonly used quantities. A more complete tabulation is included in Appendix 1.) In addition to these devices, we must turn to a pair of specialized optical instruments together with their accessories: the telescope and the spectrograph. Much of the data which will provide us with proofs of the earth's motions, and most of the data by which we come to know the rest of the solar system and the universe, come to us in

Table 5.1   Selected Conversion Factors (Approximate Values Only)

1 centimeter (cm) = 0.4 in.
1 meter (m) = 3.28 ft
1 kilometer (km) = 0.62 mi
1 mile (mi) = 1.61 km
1 cubic centimeter (cm³) = 0.06 in³
1 kilogram (kg) *weighs* 2.2 lb at the earth's surface
1 gram (g) *weighs* 1/28 oz
1 gram per cubic centimeter (gm/cm³) = 62.4 lb/ft³
1 mile per hour (mph) = 1.61 km/hr
1 kilometer per hour (km/hr) = 0.62 mph
1 centimeter per second (cm/sec) = 0.033 ft/sec = 0.02 mph
1 gram (g) *weighs* 980 dynes (dyn) at the earth's surface
1 kilogram (kg) *weighs* 9.8 newtons (Nt) = 980,000 dyn
Acceleration of gravity (g) = 980 cm/sec²

See Appendix 1 for a more complete table.

the form of visible light. Let us set the stage for the utilization of this data.

Some Principles of Geometric Optics and Refraction.
Visible light is a small segment of the total array, or spectrum, of *electromagnetic radiation*. The latter is a form of energy propagation which needs no physical medium for support; thus it travels fastest in a vacuum, with a speed of very nearly 3×10^{10} cm/sec.† *Electromagnetic* refers to the electrical and magnetic characteristics of this type of energy propagation when it is being considered from the standpoint of a wave disturbance. We shall examine the wave characteristics of light in the next section. As detailed in Appendix 3, light can, for many purposes, be thought of as a narrow beam, or ray, which obeys definite geometric principles as it proceeds through transparent media. This is our present frame of reference.

With respect to human perception, visible light ranges in color from darkest red to blue-violet.†† In between are all shades of red, orange, yellow, green, and blue. White light is composed of all of the colors of the visible spectrum in exactly the same proportions as in light coming from the sun. Indeed, by definition pure sunlight is white.

Electromagnetic Waves and Photons.
The propagation of electromagnetic radiation in a vacuum (or a near vacuum, such as space) is accomplished by means of a progressive build up and decay of electrical and magnetic fields. The two fields interact in such a manner that as one decays, the other builds up, and so on. A

---

† This symbolism is called *scientific notation* and is a convenient way of representing and computing with very large and very small numbers. Thus, 3×10^{10} represents the number 30,000,000,000 (30 billion). See Appendix 2 for more details on this system.

†† Color is a physiological phenomenon in response to electromagnetic radiation of a discrete range of energy. See Chapter 10 and Appendix 5 for discussions of electromagnetic phenomena.

---

field is an abstraction that refers to a region in space where electrical or magnetic forces exist. This subject is treated in more detail in Appendix 5. For our present needs, we shall find it sufficient to accept the word of physicists that the progressive disturbance represented by the variation in the electromagnetic fields propagates at a constant velocity c (p. 422). The disturbance pulses travel in trains of *waves*, each with a finite wavelength, λ. Radio waves are quite long, being in the meter wavelength range. Light waves are much shorter, on the order of hundred-thousandths of a centimeter. The number of waves arriving at a detection site or mechanism in a specified time interval depends upon the wavelength; we define the event as *frequency*, or number per second, symbolically ν. The unit of frequency is reciprocal time, 1/sec. Since the speed of propagation in a given uniform medium is constant, the relationships among speed (or velocity), frequency, and wavelength are significant:

$$c = \nu\lambda, \qquad \lambda = \nu/c, \qquad \nu = c/\lambda.$$

Under certain circumstances, electromagnetic energy behaves as though it is propagated in discrete bundles, or packets, called *photons*, rather than as continuous emission. Nevertheless the wave characteristics are often also discernible, so it is convenient to think of the photons as being embedded on the wavefronts of electromagnetic waves. Because the waves oscillate as they travel through space, so do the photons. We must realize that these concepts are mere mental abstractions; while physicists know precisely the physical nature of electromagnetic radiation in terms of mathematical constructs, it is not possible to translate these into perceptual images. But the preceding crude analogies serve reasonably well. The importance of the photon concept becomes apparent later in the text.

The Telescope.
As is discussed in more detail in Appendix 3, two classes of optical instruments provide most of the information from

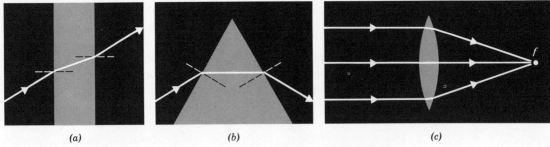

(a)   (b)   (c)

FIGURE 5-1   The basic principles of a refracting telescope. Light rays are deflected when they pass through media of different optical density. (*a*) Deflection through a glass block with parallel faces; (*b*) deflection through a prism, essentially a glass block with nonparallel faces; (*c*) deflection through a lens which may be thought of as a stack of many prisms with different angles between the faces. The lens (*c*) is capable of bringing parallel rays from a source into focus at a point *f*.

which astronomical data are developed. The telescope is the more familiar of the two. Contrary to popular belief, the chief utilization of a telescope in astronomy is in conjunction with a photographic plate or film holder. Thus the telescope is in reality a camera, and in this sense it is superior to the human eye. The eye is easily deceived, and it does not have the ability to retain images, whereas a photographic negative yields a permanent record of a given observation.

The principal advantage of a telescope is in its ability to gather light and bring it to a focus, yielding a small, relatively bright image of the object (Figure 5-1). The *light gathering power* of a telescope is a function of the *aperture* (the diameter) of the *objective* lens or mirror. The light gathering power increases in proportion to the increase in the area of the objective, which in turn is proportional to the *square* of the radius of the aperture (p. 408). Thus, a 40-inch telescope will gather 4 times as much light as a 20-inch instrument, and 16 times as much as a 10-inch telescope.

A second function of a telescope is to separate closely adjacent points, that is, to *resolve* detail in an image. The *resolving power* is directly proportional to the aperture of the telescope. Thus, a 40-inch telescope will resolve

twice the detail that a 20-inch telescope will accomplish.

A third and least important function of a telescope is to provide a magnified image of the object for direct visual observation. In general, images formed by a telescope are not magnified in photographic procedures. It is a far simpler procedure to enlarge the photograph, and better results are obtained that way. We should note at the outset that the images formed of all stars are just points of light. No amount of magnification will do any more than increase the apparent size of the image formed by the telescope objective; even so, the images of stars remain point images.

Magnification is accomplished with the aid of an auxiliary instrument, the eyepiece (Figure 5-2). An eyepiece is simply a short-focus, small-aperture telescope which focuses on the image formed by the objective. The magnification is proportional to the ratio of the focal length of the objective to the focal length of the eyepiece. Thus a telescope with an objective focal length of 100 cm will magnify an extended image such as a planet or portion of the moon 50 times when equipped with a 2 cm focal length eyepiece. It will magnify 500 times when the focal length of the eyepiece is 0.2 cm.

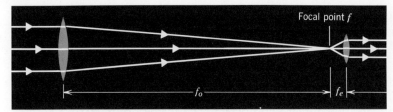

FIGURE 5-2   Magnification in a refracting telescope. Light from an image source is brought to focus at f; a supplementary lens then diverges the light so that the resultant image covers a large area on the retina of the eye, giving the sensation of enlargement. The amount of magnification is determined by the ratio of the focal length of the objective lens to the focal length of the eyepiece, or $M = f_o/f_e$.

The Spectrograph.   A second class of instruments includes the *spectrograph*. In practice it is an auxiliary instrument used in conjunction with a telescope. As detailed in Appendix 3 a spectrograph is an optical attachment which disperses, or spreads out, a beam of light into all of its component colors. In practice the light beam is passed through a very narrow slit, then rendered into parallel rays by a lens (Figure 5-3). The parallel rays then pass through a train of *prisms* which disperse the light into its spectrum. The spectrum is then brought to focus on a photographic plate. If one or more of the *wavelengths* of the light is missing or is somehow reduced in intensity, this will show up in the spectrum as a dark image of the slit. Because of the shape of the slit these images are often called *lines,* or better, *spectral lines.*†

For reasons which will become clear in Chapter 10, some information about an ob-

† The reader who is unfamiliar with spectroscopes and spectrographs should study carefully the discussion in Appendix 3. These instruments are of paramount necessity in providing information about most of the other members of the universe.

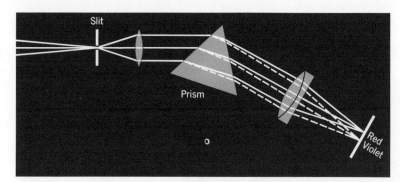

FIGURE 5-3   Schematic optical train for a single-prism spectrograph. If the light beam consists of several wavelengths, the shorter wavelength components are deflected through a greater angle by the prism than the longer wavelengths. The light from the source is then said to be dispersed; any missing wavelengths will appear as a dark shadow or image of the slit.

ject's motion can be determined from the position of the known spectral lines for a given object with respect to a *comparison spectrum* which is superimposed on the negative or plate at the time the photo is taken. If an object is moving *toward* the earth, the spectral lines will be shifted toward the blue end of the spectrum, in proportion to the speed of the object. Similarly, the spectral lines are shifted toward the red end of the spectrum if the object has a component of velocity directly away from the earth. No displacement of the lines occurs if the object is moving across the line of sight. Hence we speak of the velocities which produce the displacements as *radial velocities*. For the present this is all we need to know about telescopes and spectrographs.

Shape and Size of the Earth

The average literate person is so sophisticated concerning some of the facts about the earth, such as its shape, its rotation, and its revolution about the sun, that it seems almost superficial to dwell on these points in a contemporary course in astronomy. On the other hand, much can be learned from the tenacity and ingenuity of astronomers in their assault upon the unknown before the advent of sounding rockets, before artificial earth satellites, and before

space travel. Some of the early Greek philosophers knew that the earth was spherical, or at least approximately so, yet as late as the early 16th century many persons still were secure in their belief that the earth was flat. What are some of the arguments which lead indisputably to the fact that the earth is round?

To Pythagoras of Samos (6th century B.C.) is credited the first guess of record that the earth is spherical, although the records contain no evidence that he was able to prove it. Anaximenes (585–525 B.C.) correctly deduced that eclipses of the moon were caused by the shadow of the earth as cast on the moon by the sun. Aristarchus of Samos (3rd century B.C.) asserted that because the shadow of the earth on the moon always appears circular, the earth must be round. Only a sphere casts a circular shadow in every configuration.

Standing alone, this phenomenon is insufficient proof; it shows only that the earth is roughly circular in *section*. A disk-shaped flat earth, if oriented at right angles to and on the line connecting the sun and the moon, would cast a circular shadow (Figure 5-4).

The early Greek mariners discovered, and later travelers verified, that when traveling at right angles to the diurnal motion of the celestial sphere, that is, north and south, the stars shift position, higher or lower in the sky in

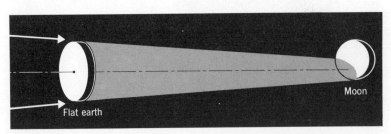

FIGURE 5-4   A flat, disk-shaped earth would cast a circular shadow on the moon if oriented as shown. Unlike a sphere which casts a circular shadow in any orientation, a flat earth would cast a very much different shadow, perhaps none at all if oriented edge-on to the moon.

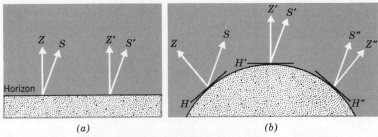

FIGURE 5-5 (*a*) The angular direction to a very distant reference star will be the same from every point on a flat earth. This does not hold true on a spherical earth because angular directions are referred to the local zenith which has a different space direction at every point on the surface (*b*).

proportion to the length and direction of the journey. Thus, the pole star sinks lower and lower for a southbound traveler, finally disappearing below the horizon when the observer crosses the equator. This phenomenon cannot occur on a flat earth. The explanation is that a traveler moving over a curved surface will generate a new zenith *direction* and a new horizon at every point on his journey (Figure 5-5).

Since 1940 photographs taken by cameras borne aloft by high altitude rockets plainly show the curvature of the horizon. An orbiting aritifical satellite is convincing proof in itself of the sphericity of the earth. The Apollo missions to the moon have clinched the case (Figures 5-6 through 5-8) for all except for small isolated groups of religious zealots such as one in England that fanatically holds fast to the flat-earth concept.

Size of the Earth. The next question, now that the approximate sphericity of the earth is confirmed, is how large is it? The problem of determining the earth's circumference or diameter from observations made on its surface is more difficult than it might appear. Eratosthenes of Greece, a 3rd century B.C. geographer and astronomer, is generally credited with making the first accurate determination of the circumference of the earth. He observed that at noon the vertical rays of the sun illuminated the bottom of a deep well at Syene (now *Aswan*), Egypt. Simultaneously at Alexandria, due north of Syene, the sun's rays were 7.2° from the vertical (Figure 5-9). Now, reasoned Eratosthenes, 7.2° represents 1/50, of a complete circle. Therefore the distance between Syene and Alexandria represents 1/50 of the circumference of the earth. If we are correct in our estimates of the length of the basic unit (the stadia) in which Eratosthenes reported his results, he came within 80 km of the modern value of 40,054 km, an extraordinary achievement. From the relationship, circumference = π × diameter, we can calculate the latter: diameter = 40,054 ÷ 3.1416, or 12,756 km. Compared to the presently accepted value, this is an accuracy of within 1%.

Eratosthenes' *method* was sound; the principal weakness in its application was the assumption that the sun's rays are parallel at the earth. This is a requirement if the angular difference in the direction of the sun's rays at different points on the surface of the earth is due solely to its curvature. We have no record of how Eratosthenes came to the conclusion that the sun is so distant that its rays are indeed essentially parallel at the earth.

The invention of the telescope provided astronomers with a tool for obtaining the precise

FIGURE 5-6 Curvature of the horizon as photographed by Apollo 9 astronauts. The clouds are associated with a storm system 1200 mi north of Hawaii (Courtesy NASA).

FIGURE 5-7 Crescent earth, photographed by Apollo 4 astronauts (Courtesy NASA).

angular direction of a selected reference star. Thus, reliance on the more crude determination of the direction of the sun's rays could be dispensed with. The one remaining necessity was a method for accurately determining large distances on the earth's surface. The breakthrough came in 1671, nearly 2000 years after Eratosthenes, when a French astronomer, Jean Picard, devised a wheel equipped with a mechanical distance counter. The wheel was rolled along the ground, and the distance traversed was read off directly.

We recall from Chapter 2 that the angular altitude of the celestial pole above the horizon corresponds to the latitude of the point where the observation is made. In particular, there is a one-to-one correspondence between the angular change in an observer's latitude and the angular change in the position of a given reference star. Thus, let two points in a north-south direction be marked on the surface of the earth. Let their separation be such that the angular direction of a reference star differs by 1° when observed from each of the points. Then the latter are 1° of latitude apart. The distance between the points represents the

number of kilometers per degree of latitude. If the earth were truly spherical, 360 times this distance would equal the circumference of the earth.

In practice the total distance itself is not directly measured. A convenient base line a kilometer or so in length is marked and carefully measured, not with Picard's wheel but with a steel surveyor's chain. The accuracy of the measurement normally exceeds one part in a million. Then a system of triangles is carefully laid out and the angles between the sides are measured with a high precision telescope called a *theodolite* (Figure 5-10). With the aid of trigonometry the distance between any two points in the triangulation network is easily determined. The network is designed to include a pair of points 1° in latitude apart. The distance between them can then be rather easily calculated.

Oblateness of the Earth. It became apparent even in the early determinations of the length of a degree of latitude that the distance is not the same at all places on the earth. In general there are more kilometers per degree near the

FIGURE 5-8 The earth, photographed by the Apollo 16 astronauts on April 16, 1972 about 1 hour after leaving earth orbit on their way to the moon (Courtesy NASA).

poles than there are near the equator. The significance of this fact becomes apparent when we tabulate several examples of the length of 1° of latitude as in Table 5.2. If we then draw a construction of the curvature of the earth incorporating this data as in Figure 5-11, we see that the earth is not perfectly spherical, it is *oblate,* or bulged around the equator.

Since the earth is not perfectly spherical there is no *single* value for its diameter. However, two of the quantities are important in determining the figure, or shape, of the earth. Let

us see how we can turn the information concerning the number of kilometers per degree of latitude to our advantage in solving this problem. First, we need the definition of a special unit of angular measure. In Figure 5-12 a portion of a circle including its center, two radii, and the included arc is shown. By definition the central angle between the radii which *subtend* an arc on the circle exactly 1 radius in length, is a *radian.* We know from geometry that there are 2π radii in the circumference of a circle. Hence a radius which sweeps out a

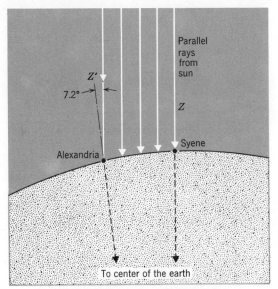

FIGURE 5-9 Eratosthenes' method for the determination of the size of the earth. Examination of the figure shows that the arc length of the distance between Syene and Alexandria is 7.2/360 of the whole circumference.

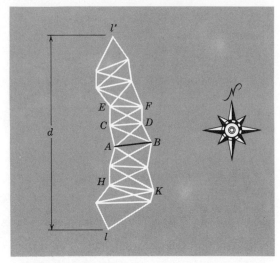

FIGURE 5-10 A triangulation network erected on a carefully measured base line *AB*. Knowledge of the length of *AB* and the size of all angles permits routine calculation of the distance *d* by elementary geometry and trigonometry.

TABLE 5.2 Length of Arc of the Earth's Surface at Various Latitudes	
Latitude	Length of 1° of arc, km
0°	111.3
15°	111.46
30°	111.62
45°	111.94
60°	112.1
75°	112.27
90°	112.43

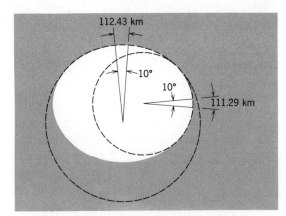

FIGURE 5-11 The curvatures of various portions of the surface of an ellipsoidal figure, shown here in section, are very different as is shown by the diameters of the reference circles. An angle of any value will subtend a longer arc on the larger circle.

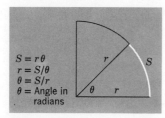

$$S = r\theta$$
$$r = S/\theta$$
$$\theta = S/r$$
θ = Angle in radians

FIGURE 5-12 Radian measure. An angle of 1 *radian* is subtended at the center of a reference circle by an arc 1 radius in length. If the arc length S is less than one radius r, the angle θ, is less than 1 radian in the ratio S/r.

circle turns through 360° and also through 2π (6.283) radians. In general, again referring to the figure,

$$S = r\theta. \qquad (5.1)$$

To determine r we solve equation 5-1 thus,

$$r = \frac{S}{\theta}. \qquad (5.2)$$

Now S on the earth can be measured (or calculated by triangulation), and the angle subtended at the center of the earth by S can be determined in degrees by direct astronomical observations as Figure 5-13 shows. An angle measured in degrees can be converted into radian measure by the relationship, $1° = 6.283/360$ radian. Then the radius r can be calculated from Equation 5.2 since S and θ are known. The earth's diameter is twice this quantity.

The best present values of the polar and equatorial diameters, subject to minor corrections as measurements are refined, are 12,713.5 km for the polar diameter, and 12,756.3 km for the equatorial diameter.

The *oblateness* of the earth is defined as the ratio of the *difference* between the equatorial and polar diameters to the equatorial diameter. Since the difference in question is 42.8 km, the earth's oblateness is 42.8/12,756.3, or 1/298. This ratio is of interest in comparing the oblateness of other planets with that of the earth. It is also exceedingly important in computations

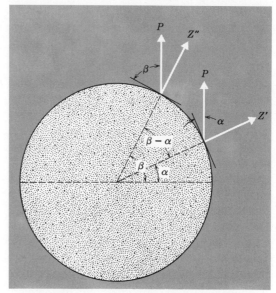

FIGURE 5-13 From elementary geometry the *altitude* of a reference star P above the local horizon is numerically equal to the latitude of the place of observation. Thus if the observation points are chosen so that the latitude difference is 1°, the arc length between the points is the number of miles per degree latitude. This relationship is local only, unless the curvature of the entire surface is uniform.

for the prediction of artificial satellite orbits and trajectories.

The Real Motions of the Earth

The notions that the earth revolves around the sun and spins about its own axis have a great appeal to reason in the light of Kepler's laws. But in order to banish any danger of accepting just another hypothesis, tenable or not, we must marshal the evidence which provides indisputable *proof* of these motions. Let us attack the rotation problem first.

If the earth is spinning, the diurnal rotation of the celestial sphere must be just a reflection of this movement. Moreover, the celestial poles will be just the extension of the earth's rotational axis. An observer at either of the earth's poles of rotation will observe the corresponding celestial pole exactly on his zenith.

Falling Body Proof for Rotation. To see the significance of the foregoing, let us employ as an analogy a phonograph record spinning on its turntable. It is easy to see that every point on the record completes its rotation about the center in the same time interval. Therefore the points on the rim must be traveling fastest because they have farther to go. Now imagine a straight line drawn on the record from rim to center. Each point on this line travels at a different speed than the adjacent points, with the highest speeds nearest the rim. If somehow a small particle could be propelled from one of these points, in the absence of frictional and gravitational forces, it would travel in a straight line, tangent to the circular path at the instant of release (Figure 5-14). This is in complete agreement with Newton's first law. We can apply this argument as follows.

Suppose, for simplicity, that we are on the equator on the roof of a tall building, perhaps 20 stories high. We release a marble from rest exactly 1.0 m from the smooth vertical wall of the building. The question arises: In the absence of air friction and wind currents will the marble strike the earth exactly 1.0 m from the wall, to the east of this point, or to the west of it?

Experimental evidence shows that the marble will strike the ground to the *east* of the point directly under the point of release. On the west side of the building the marble strikes closer to the wall, on the east side farther from it. For the case in question the point of impact is displaced approximately 1.2 cm east. This proves the earth is rotating. How? The base of the building is traveling eastward at exactly the linear speed of the earth's surface. But the top of the building *and the marble before release* are traveling a little faster toward the east (recall that east was defined in Chapter 2 as the direction of the rising celestial objects) because they are a little farther from the center of the earth. When the restraint on the marble is removed it begins to fall under the influence of

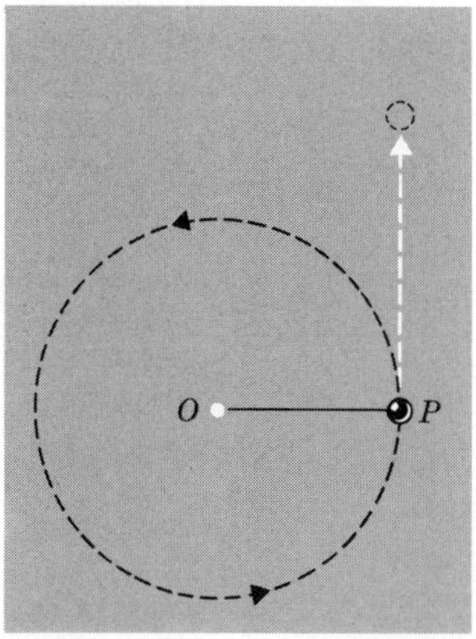

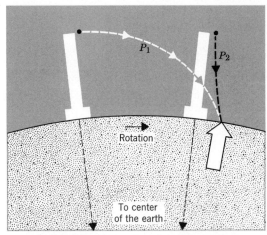

FIGURE 5-15 A marble dropped from the roof of a tall building in the absence of atmospheric friction and wind effects will describe a path P_1 with reference to the stars, and a path P_2 with reference to the building. The point of impact with the ground will be to the east of the point of release.

FIGURE 5-14 A particle rotating about O will, if the restraint is removed, fly off at a tangent from point of release P according to Newton's first law.

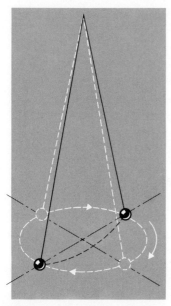

FIGURE 5-16 A Foucault pendulum initially swinging through an arc (black circles) with reference to the surface will in time appear to swing at right angles (dashed circles) to the original arc. If viewed from the stars, the direction of the swing does not change; the surface is seen to rotate under it.

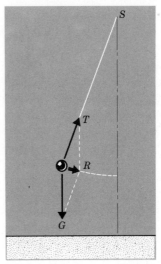

FIGURE 5-17 Two forces acting on a pendulum: gravity G and tension T in the cord from the suspension S. Since both forces act in a plane the *resultant* force which determines the swing of the pendulum is necessarily *in the same plane.*

gravity. But there is no longer any force propelling the marble eastward. In accordance with Newton's first law the marble's initial horizontal motion is preserved; it strikes the ground east of the expected point of impact (Figure 5-15). This behavior can only occur if the earth is rotating. The reader should try to ascertain if the same phenomenon would be apparent for a similar experiment performed at one of the poles.

Foucault Pendulum. The falling body experient, while satisfactorily proving that the earth spins, lacks something in performance. Because we cannot exclude atmospheric friction and the effect of wind, we need statistical treatment of a large number of repetitions in order to demonstrate the proof. There are other, more convincing demonstrations available. In 1851 a French physicist, Jean Foucault (1819–1868), suspended a heavy metal ball by a fine wire some 65 m long from the dome of the Pantheon in Paris. He very carefully set the ball swinging in a vertical plane over a reference pattern marked on the floor. After several minutes the plane of the oscillations appeared to have rotated slightly clockwise relative to the initial motion (Figure 5-16). If the suspension is assumed frictionless and the wire is free from twist during a short period, a not unreasonable assumption, the only possible explanation for the apparent rotation of the plane of the oscillations is that the earth is turning counterclockwise under it.

The analysis of the motion is at once provocative and convincing. When the pendulum is at one extremity of its swing and is momentarily stopped, two forces impel its subsequent motion (Figure 5-17). Gravity acts vertically on the ball, and tension in the wire is directed toward the suspension. The two forces are in the same plane, thus horizontal rotational motion is impossible. Moreover, the two forces are not directly opposed, thus there is a *resultant* force acting in the plane of the oscillation. In fact, the resultant force produces the motion. According to Newton's second law the pendulum cannot deviate from its original plane of oscillation which is fixed relative to the stars. Therefore, the earth must be spinning under the pendulum. If the experiment is performed at either pole, the apparent rotation of the plane of swing is 24 hours. At the equator the

plane of oscillation is fixed, not only relative to the stars, but also relative to the surface of the earth. This is easier to understand when we realize that at the equator the earth's axis of rotation is at all times entirely within the plane of the Foucault pendulum's swing. Thus the swing of the pendulum, as well as all stationary points on the earth's surface, is fixed relative to this axis (Figure 5-18).

Thus the Foucault pendulum demonstration appears to fail at the equator; in reality this apparent failure lends additional support to the proof of the earth's rotation. It can be shown

FIGURE 5-18 Foucault pendulum (Courtesy California Academy of Sciences).

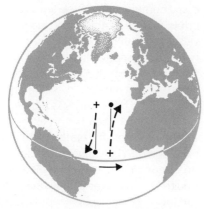

FIGURE 5-19 The Coriolis effect. A projectile appears to deflect to the right of a target when the event occurs in the Northern Hemisphere because of the relative difference in linear speeds to the east between the gun and target. Ocean and atmospheric currents also display the Coriolis effect.

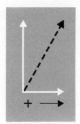

FIGURE 5-20 A projectile follows a path whose horizontal component is a straight line when measured with respect to a fixed reference system such as the stars. Upon leaving the muzzle the projectile retains the eastward motion of the gun while it acquires a muzzle velocity in the direction in which the gun is pointed. The resultant path is shown by the dashed vector.

that the period of oscillation of the pendulum is a function of the latitude. As the latter becomes smaller, that is, tends toward zero, the period of a Foucault pendulum becomes longer. At the latitude of Washington, D.C., the period is approximately 39 hours; at Baton Rouge, Louisianna, the period increases to about 48 hours.

The Coriolis Effect. Let us now engage in a classical scientific activity, the *gedänken* (thought) experiment. For physicists and astronomers this has been a most fruitful means of coming to grips with any problem which requires a technological approach to its solution. We should note that we cannot stop after thinking our way to a solution. This would put us in a class with Aristotle, and his methods have long since been refuted. We must subject our notions to controlled, valid tests.

Imagine a rifle pointed *north* in the Northern Hemisphere, aimed exactly at a thin vertical wire a few hundred meters distant. Assume that a bullet is fired from the gun in a frictionless,

windless atmosphere. It may come as a surprise to see the projectile apparently deflect to the right in a curved path and miss the target to the east. We reverse the positions of the gun and target. The bullet again appears to deflect to the right in a curved path, but this time it misses the target to the west (Figure 5-19). What is the explanation?

Both the gun and the bullet in its chamber are being carried to the east with the linear velocity of the earth at that point. The target also has a velocity to the east, but it is *less* than that of the gun. This is understandable because the target is fixed to a point on the earth's surface which is *closer* to the axis of rotation than the gun. When the projectile leaves the gun barrel it continues eastward with its initial speed according to Newton's first law. But it has acquired a velocity toward the north which is also preserved after the bullet is in free flight. The resultant path with reference to the fixed stars is shown in Figure 5-20. However, with reference to the earth, each point over which the bullet passes in its northward flight is moving eastward a little slower than the point immediately to the south. The cumulative description of the bullet's flight is the observed curved path.

The Real Motions of the Earth   87

The phenomenon is called the *Coriolis effect* after G. G. Coriolis, a French mathematical physicist, who first demonstrated its character quantitatively in 1844. We can better appreciate the Coriolis effect when it is viewed against a fairly modern background. In World War II, capital ships carrying 12- and 16-inch rifles were still employed in naval warfare. In target practice, for example, a projectile was fired at a target 9000 m away. Even after allowances were made for wind effects, the projectile struck the surface nearly 100 m to the right of the target when the practice took place in middle-northern latitudes. Hence a correction for the Coriolis effect was necessary.†

Atmospheric circulation patterns are decisively influenced by the Coriolis effect. In treating such phenomena quantitatively, Newton's laws must be modified to include the Coriolis and other curvature terms. The reason for this is that Newton's is a stationary, or *inertial,* frame of reference. But a rotating earth constitutes a *noninertial* frame of reference for objects in free flight, that is, not in contact with the earth's surface. In these situations the inertial form of Newton's laws is inadequate. We should note in passing that the Coriolis effect is independent of gravitational accelerations because the latter are at right angles to the horizontal components of motion which show the Coriolis deflection. The reader should be able to reason that the Coriolis deflections are to the *left* in the Southern Hemisphere, and he should also be able to deduce that the Coriolis effect is zero at the equator and becomes a maximum at the poles.

The Revolution of the Earth About the Sun

We term the earth's motion around the sun its *revolution* to distinguish it from the earth's *ro-*

† This passage is from Cole, F. W., *Introduction to Meterology,* John Wiley & Sons, New York, 1970. This reference also contains an easy to understand, qualitative discussion on inertial and noninertial coordinate systems.

tation. Astronomers employ the term revolution to describe rotational (circular) motion about an axis *external* to the body; they use the term rotation in connection with motion about an axis *within* the object. If Kepler was right, the axis of the earth's revolution around the sun passes through the latter and it is perpendicular to the ecliptic, which then becomes the projection, or reflection, of the earth's orbit as seen against the stars. What arguments can we advance in proof of the earth's orbital motion? Even though we have now rejected the concept that the celestial sphere spins around a stationary earth, this does not constitute a basis for rejecting the sun's possible annual motion around the earth. We must turn both the telescope and the spectrograph to our use in attacking this new problem.

<u>Annual Parallax of the Near Stars.</u> Prior to the invention of the telescope, and for some time afterwards, the absence of an observable parallax of the near, or bright, stars was a strong argument against a heliocentric system (p. 9). It was not until 1838 that F. W. Bessel (1784–1846), a German astronomer, first successfully measured the distance of a near star, *61 Cygni,* relative to the more distant stars. He accomplished this task by determining the parallax of the star, the significance of which is illustrated in Figure 5-21. Since then the parallax of some 5000 stars has been determined, but the displacement is so minute that the parallax of less than 2000 stars is known with confidence. For the present we are less interested in the application of stellar parallax in determining distances than we are in using it as a proof of the earth's orbital motion.

An examination of the parallax of a number of stars in random directions on the celestial sphere discloses an interesting result. The near stars (the only ones displaying parallax) in a direction perpendicular to the plane of the ecliptic describe annual *parallactic ellipses.* Other near stars in the direction of the plane of

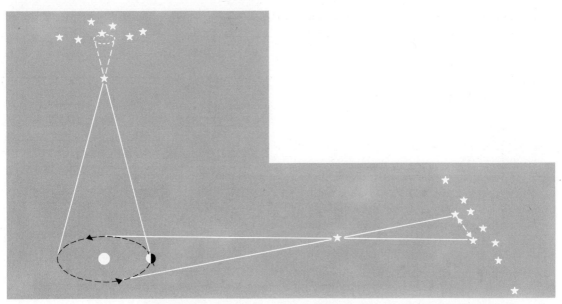

FIGURE 5-21 Parallactic displacement of the near stars. The near stars appear to oscillate in position relative to more distant background stars; the displacement path is a projection of the orbital motion of the earth on the celestial sphere.

the ecliptic describe annual oscillations along a line. In between, the near stars describe parallactic ellipses which are more nearly closed than in the first case. It is as though the ellipses are being viewed obliquely. These factors—the annual periodicity of the phenomenon, open ellipses above and below the plane of the ecliptic, and the linear oscillation in the ecliptic plane—all point to the fact that it is the observer and the earth to which he is fixed that are in motion. The earth is revolving about the sun, not the reverse.

Semiannual Doppler Shift. Another equally convincing proof of the earth's revolution is disclosed by spectrographic analysis of many stars at random. Here the phenomenon is not restricted to near stars. But it is restricted to stars near the plane of the ecliptic. An analysis of the spectrograms of a large number of such stars shows that each one displays a semiannual radial velocity of approach toward and recession from the earth. Called the semiannual Doppler shift (or Doppler effect) after its discoverer Christian Doppler (1803–1853), a noted German physicist, the phenomenon is very interesting as well as an exceedingly valuable observational aid. Let us here examine it carefully.

It is probably safe to say that everyone has observed the change in pitch of a sound as either the observer approaches, passes, then recedes from the source, or the source does the same with respect to a stationary observer. In either case the pitch is higher on approach and lower on recession. The greater the relative speed between observer and source, the larger is the change in pitch. The effect is analogous to the behavior of light and other electromagnetic radiation, except that in this case there is no variation in the speed of light, relative or otherwise. We shall here anticipate a more detailed discussion of the properties of electromagnetic radiation in Chapter 10.

The Revolution of the Earth About the Sun   **89**

We note that if a train of identical electromagnetic waves (p. 75) passes an observer at constant speed, the frequency of the passage of the waves depends upon the length of the waves; if the waves are shorter, the frequency is higher, and vice versa. Now the color (or position) of a spectral line in the spectrum (p. 77) depends upon the frequency of the light waves, since all travel at the same constant speed. If somehow a given spectral line could be made to change its real *or its apparent* wavelength, the frequency would also change. Thus, the longer the wavelength, the lower the frequency, and the redder the light. Similarly, the shorter the wavelength, the higher the frequency, and the bluer the light. The key idea here is that the speed of propagation remains constant.

Suppose a point object (such as a star image) emits a particular light frequency. We examine the relative position of the leading point (*P*) of a given individual wave with respect to the emitter (*S*) at the instant that the last point in the wave is emitted (Figure 5-22). Now suppose the object is moving radially toward an observer. The leading point on a given wave is examined relative to its position with respect to the emitter as before. But by the time the last point of the wave is emitted, the emitter has moved through a finite distance (from *S* to *S'*) toward the observer. Since all parts of the wave travel at the same speed independently of any motion of the source (Appendix 5, p. 422), the effect is to foreshorten the wave; the amount is proportional to the speed of the emitter. Sometimes this foreshortening is referred to as the "accordian" effect. But a train of such foreshortened waves reaches the observer with the speed of light, and it is interpreted as being light of a higher frequency. Thus the light is shifted toward the blue end of the spectrum. A similar argument holds for the Doppler shift toward the red end of the spectrum for a receding object. In practice it matters not whether the object is stationary and the observer is moving, whether the object is moving

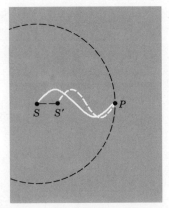

FIGURE 5-22 The Doppler effect. A transverse wave is emitted by a stationary source *S*; the form of the wave is illustrated by the curve *SP*. If during the emission of the disturbance the source moves say from *S* to *S'* the wave appears to be shortened.

and the observer is stationary, or whether both are moving relative to each other. Now let us return to our argument.

For a given star having initially, say, a relative radial velocity in the direction of the observer, the Doppler shift of the spectral lines toward the blue decreases in magnitude until after about 3 months it disappears. Shortly thereafter the same star displays an increasing Doppler shift toward the red which reaches a maximum in another 3 months. The cycle continues through a decrease in the Doppler shift toward the red, and a final increase to a maximum Doppler shift toward the blue at the end of an elapsed year. Since the phenomenon is observed for all stars near the plane of the ecliptic, but not for stars in a direction perpendicular to this plane, and since all stars in a given direction show a relative Doppler shift of the same magnitude, we must conclude that it is a relative effect caused by the motion of the earth in orbit around the sun.† The calculated

† Some stars may have a *real* motion toward or away from the earth, but this intrinsic motion can be averaged out in a series of observations, thus yielding the residual Doppler effect related to the earth's motion.

(a)   (b)   (c)

FIGURE 5-23 The aberration of rainfall, analogous to the aberration of starlight effect. To the stationary observer in (*a*) vertical rain appears to fall vertically. (*b*) If the observer begins to move he must adjust the angle of his umbrella to compensate for the apparent direction change of the falling rain. (*c*) The angle change is proportional to the speed of the observer.

magnitude of the earth's average orbital speed is approximately 30 km/sec or 108,000 km/hr.

<u>Aberration of Starlight.</u> A final, but not the sole remaining, proof that the earth revolves around the sun is shown by the *aberration of starlight* effect. We can best understand this phenomenon through the mechanism of another familiar analogy. Suppose a pedestrian is standing on a street corner waiting for the light to change. It is raining moderately, with no wind or breeze; thus the raindrops are descending vertically (Figure 5-23*a*). When the light changes the pedestrian starts across the street, and with his forward motion the rain appears (to him) to slant in from a position somewhere in advance of his motion. The pedestrian instinctively tips his umbrella forward to

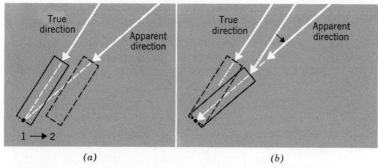

(a)   (b)

FIGURE 5-24 If a telescope is pointed in the true direction of a star while the earth carries it from point 1 to 2 as in (*a*), light rays will appear to hit the side of the telescope tube. (*b*) The telescope must be inclined as shown with respect to the *apparent* position of a star. The angle of inclination is the aberration angle for starlight.

provide compensating protection (Figure 5-23b). But a taxi is coming up fast in the inside lane. Fearing that it will not stop before reaching the crosswalk, the pedestrian smartly increases his pace; he tips his umbrella even farther forward to compensate for the apparently increased angle of the falling rain (Figure 5-23c). Fortunately the cab stops in time. The pedestrian stops on reaching the far curb, and scowls back at the cab driver. As he does so the pedestrian automatically restores his umbrella to a vertical position because the rain again appears to be falling vertically. With respect to the ground, the rain never deviated from the vertical. It was the combined motion of the pedestrian *and* the falling rain which produced a relative, *resultant* slant path for the raindrops.

A similar phenomenon occurs with respect to light energy. Light is propagated with a very great, but still finite, speed. Thus a telescope, which is fixed to an earth which is hurtling along in its path at 30 km/sec, must be tilted forward slightly in the direction of motion. This prevents an incoming light beam from being intercepted by the sides of the telescope tube before it completes the desired optical path (Figure 5-24). The aberration of starlight effect is the apparent change of position of a star in a direction in advance of the earth's motion. All stars show the phenomenon, each describing an aberration ellipse or oscillation depending on the star's position relative to the plane of the earth's motion. These aberration ellipses are similar in shape to the parallactic ellipses discussed earlier. Since the aberration of starlight is produced only in connection with the relative motion of an observer with respect to a star, the earth carrying the observer must be moving. The aberration ellipses prove that the earth is in orbit around the sun.

Size of the Earth's Orbit. Once we have established the average speed of the earth in its orbit, it is a simple matter for us to estimate the dimensions of the latter. From our everyday

experience we have learned that the product of the *speed* of travel and the total elapsed *time* for a journey will be equal to the distance traversed. We apply this argument to the earth, using the speed of 30 km/sec and the elapsed time of 1 year for a complete circuit (31.6 million seconds). The product is the length of the earth's orbit, 948 million kilometers.

Now from geometry we know that the circumference of a circle is 3.14 . . . times the diameter, or $C = \pi D$ where the symbol π (pi) represents the constant ratio 3.14. . . . If we divide both sides of this equality by π we have

$$\frac{C}{\pi} = \frac{\pi D}{\pi}$$

or, since $\pi/\pi = 1$, $C/\pi = D$. Now since both quantities on the left side of the equation are known we can find D, the required diameter. Evaluation yields a mean diameter for the earth's orbit of 298 million kilometers in round figures. One-half this value is the *radius* of the earth's orbit, 149 million kilometers. But this quantity is just the average earth-sun distance. The assumption that the earth's orbit is circular is, of course, not strictly correct. As we shall see later it is an adequate first approximation for it tells us the order of magnitude of the true earth-sun distance. By definition this quantity is an *astronomical unit (AU)*.

Precession of the Equinoxes

In an earlier chapter we noted a puzzling fact; the sun in its annual apparent journey around the ecliptic reaches the vernal equinox some 20 minutes before it arrives at the place among the stars where the immediately preceding vernal equinox occurred. As we indicated, it is as though the vernal equinox is drifting westward among the stars so as to meet the eastward approaching sun ahead of schedule. We are now in a position to understand the cause of this phenomenon.

As we saw in Chapter 2, the ecliptic, which is just the projection of the earth's orbit on the

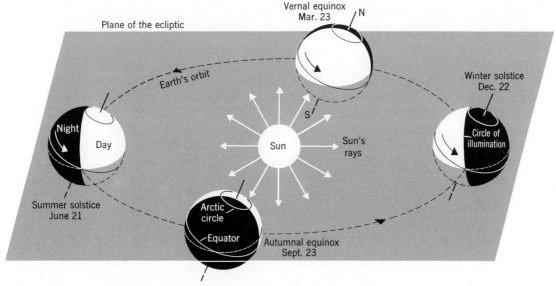

FIGURE 5-25 The seasons result from the angle of inclination of the earth's axis relative to its path around the sun. Except for precession effects, the earth's axis points in a fixed direction in space (after Cole, F. W., *Introduction to Meteorology,* John Wiley and Sons, New York, 1970).

celestial sphere, is inclined to the celestial equator with a fixed angle of $23\frac{1}{2}°$. Hence, the earth's axis must be inclined to the orbital axis by the same angle. Moreover, the earth's north pole, which by definition is that pole which is *above* the plane of a planet's *counterclockwise* revolution about the sun, is approximately fixed among the stars, at least during the course of a year or two. We accept this premise on the evidence of a fixed celestial pole which is very near a star named, for obvious reasons, *Polaris*. The geometry of the configuration is shown in Figure 5-25.

Because of the tilt of the earth its midriff bulge is displaced out of the plane of the orbit. The sun's gravity represents a force trying to bring the bulge into the line connecting the centers of the earth and the sun. As we shall see in Chapter 6, the sun gets a big assist from the moon. But here one of the strangest ''beasts'' in physics rears its fascinating head. Instead of responding to the righting force, or

torque, by tipping its axis upright, the earth tips at right angles to *both* its axis of spin *and* the righting force (Figure 5-26). Cycle riders are subject to this phenomenon whether they are conscious of it or not. If a horizontal change in

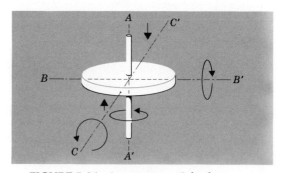

FIGURE 5-26 A gyroscope. A body rotating about axis *AA'* has a torque about axis *BB'* applied. Instead of tipping about axis *BB*, the body turns about axis *CC'* which is at right angles to both the applied torque and its initial rotation.

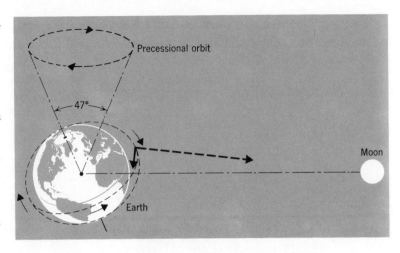

FIGURE 5-27 Precession. The moon's gravitational pull attempts to position the earth's equatorial bulge into the line connecting the centers of the earth and moon. This is equivalent to a rotation about an axis which points out of the page in the illustration. The earth gyroscopically responds with a rotation about a vertical axis, causing the earth's axis of rotation to describe a conical path as shown.

the cycle's direction is desired, the rider must tip the vehicle in a vertical plane in order to accomplish the change in direction without damaging consequences. A gyroscope behaves in a similar manner.

The net effect, called *precession,* results in the earth's axis tracing out a conical path about the ecliptic pole through a total angle of 47° (Figure 5-27). The motion is clockwise around the ecliptic pole. This effects a change in the plane of the celestial equator with respect to a fixed ecliptic. The two points of intersection of the two planes, the vernal and autumnal equinoxes, revolve clockwise around the celestial sphere with a period of about 25,800 years. For our purposes we can use an approximation of 26,000 years. It should be evident that Polaris will not remain the pole star for long; in 13,000 years it will be 47° from the celestial pole.

Precession is a disturbing complication for this reason: the primary references for the address of any star are the vernal equinox and the north celestial pole. Since both change position in the star field because of precession, the stellar addresses obviously change. The right ascension and declination of a great many stars are cataloged with reference to some *epoch.* Precession requires us to make correc-

tions in this data; eventually a complete up-dating of the catalogs will be necessary. The mean places of 1500 prominent stars are listed with their corrected addresses in each annual edition of *The American Ephemeris and Nautical Almanac.*

Nutations and Perturbations

The earth experiences a number of minor disturbances (called perturbations) in its regular motions which, except in very precise position astronomy, are not serious handicaps. The moon varies in its distance above and below the ecliptic (p. 123) with a cycle of about 19 years. Hence its contribution to the precessional motion of the earth is not constant. As a result the celestial pole describes a small wavy pattern along the precessional circle around the ecliptic pole. The resulting *nutation,* or wobble, also results in a slight variation in the inclination of the earth's axis; the *average* inclination is 23½°. Each of the other planets makes a perturbing contribution. Collectively the planetary disturbance is about ¹⁄₄₀ that of the lunar-solar precession.

The tides, more fully discussed in Chapter 6, exert a retarding effect on the earth's rotational rate, thus slowly lengthening the

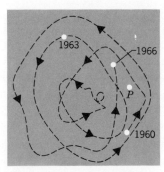

FIGURE 5-28 Wandering of the poles. *P* marks the relative mean position of the earth's north pole for 1900–1905. For uncertain reasons, perhaps alternate melting and rebuilding of the arctic snowpacks, the axis of rotation of the earth is not fixed on the surface of the earth. The path traced shows the wandering during the period from 1960–1966. This effect results in small but measurable variations in latitude on the earth. The entire path is confined to a square less than 35 m on a side.

day. The important effect here is that the earth's rotation is not an accurate clock. There are a few other small, erratic changes in the rotational rate which sometimes speed it up and at other times slow it down. These perturbations are not well understood, but it appears that they are related to changes in the distribution of the earth's atmosphere.

Finally, the axis of rotation itself wobbles within the earth, much like a stick through a candied apple. The probable cause is the uneven redistribution of a polar ice as it melts and freezes. The effect causes the earth's poles to wander around on the surface, but the disturbance is confined to an area about 2500 m² (Figure 5-28). Nevertheless it results in small but measurable changes in latitude on the earth.

Motions and the Seasons

What is there about the earth's motions and its orientation in space that causes the seasons?

The answer is not altogether obvious. We shall find it instructive to analyze the contributing factors.

The earth's orbit is an ellipse with the sun at one focus, completely in accordance with Kepler's model. The eccentricity of the orbit (Chapter 4) is not large, only 0.0167. This means that the ellipse differs from a circle by less than 1 part in 1000. Nevertheless, when the earth is at perihelion (nearest the sun) it is 5.2 million kilometers closer to the sun than when it is at aphelion (farthest from the sun). Both the perihelion and aphelion positions are on the major axis (long axis) of the ellipse at the points of greatest curvature. For this reason the major axis is frequently referred to as the *line of apsides* (L., *apsis,* a hoop or arch).

We have seen earlier that a planet is traveling fastest in its orbit when it is nearest the sun, and slowest when it is the most remote. These events occur around January 3 and July 4, respectively. Because of the earth's greater speed in the Northern Hemisphere winter months, the time interval between the sun's departure from the autumnal equinox and its arrival at the vernal equinox is 179 days, compared to the 186 days between the vernal and autumnal equinoxes. Interestingly enough, this difference was known to Athenian scholars as early as 432 B.C. Hipparchus in 120 B.C. suggested that this difference showed the earth was not exactly at the center of the sun's (supposed) circular orbit.

The sun exhibits a 3% variation in apparent angular size because of the annual variation in the earth-sun distance. This variation is too small to be detected except by direct precision measurement. When the earth is at perihelion it receives about 7% more radiant energy from the sun in a given time interval than it does when it is at aphelion. Moreover this greater energy input occurs during the southern hemisphere summer. Measurements show that the seasonal effects owing to the varying earth-sun distance are quite minor. The major contribution to seasonal temperature and weather variations comes from the inclination of the earth's axis of rotation. Except for precession effects, the direction of this inclination is fixed in space as shown in Figure 5-25. In either hemisphere the earth's axis is inclined toward the sun during that hemisphere's summer, and away from the sun during the corresponding winters. The variation in the relative heating efficiency of the sun's rays as a result of the earth's inclination is verified through observation as was discussed in Chapter 2.

SUMMARY

The motions of the earth are detectable only in terms of some external frame of reference that can be considered fixed relative to the earth. But the apparent motions could conceivably result from real motions relative to a fixed earth. Copernicus and Kepler hypothesized that it is the earth that is moving, and they presented powerfully persuasive arguments in favor of their position. Proofs are needed, however, in order to establish irrefutable evidence of the earth's motions.

The changing aspect of the celestial sphere to north-south travelers, and the fact that parallel rays from the sun strike widely separated points on the earth's surface at different angles, cannot be explained for a flat earth. Ergo, the earth is *not* flat; its sphericity is evidenced by the shadow it casts on the moon at lunar eclipses.

The size of the earth is determined by essentially the same method as was employed by Eratosthenes (Chapter 1). Refinements include measuring the angular direction of a reference star, rather than the sun, from the two observation points; the sun's center is difficult to observe directly. Variations in the number of kil-

ometers per degree latitude show that the earth is an oblate spheroid flattened at the poles by 42.8 km compared to the equatorial diameter.

That the earth is rotating is proved by the displacement of falling bodies toward the east, and by the apparent rotation of the plane of oscillation of a Foucault pendulum. Rotation is also proved by the Coriolis effect, an apparent horizontal curvature of the path of moving particles or objects.

The annual parallactic displacement of the near stars in a periodic cycle shows that the earth is in orbit around the sun. Additional support for this motion is furnished by the semiannual Doppler shift of the spectral lines of stars near the plane of the ecliptic. Further proof is shown by the aberration of starlight effect, an apparent displacement of all stars in the direction of the earth's motion about the sun.

The precession of the equinoxes, an annual westward migration of the places of occurrence of the equinoxes on the celestial sphere, is caused by a gyroscopic wobbling of the earth's axis of rotation as a result of the moon's gravitational pull on the equatorial bulge.

The earth's motion in its elliptical orbit results in a 7 day shorter period between the vernal and autumnal equinoxes than there is between the autumnal and vernal equinoxes. The reason for this is that the earth is closer to the sun, and therefore traveling faster, in the first instance. Yet the variation in earth-sun distance between perihelion and aphelion passage makes only a 7% contribution to the seasonal temperature and weather differences. The relatively fixed direction in space of the inclination of the earth's axis of rotation is responsible for the major portion of the seasonal variations.

Questions for Review

1. Give three functions of a telescope in the order of their importance.

2. Given two telescopes: Telescope A: Diameter of objective 20 cm; magnification 100×. Telescope B: Diameter of objective 10 cm; mangification 200×. (a) Which telescope objective will produce the brightest image of the moon? (b) Which telescope will produce the largest image of the moon? (c) Which telescope has the greatest resolving power? (d) What is the ratio of the light gathering power of telescope A relative to telescope B?

3. Give three modern proofs that the earth is round.

4. What is the diameter of a planet that has 64 km per degree of latitude?

5. Define oblateness as applied to a planet.

6. Give three modern proofs for the rotation of the earth.

7. Describe briefly three modern proofs for the revolution of the earth about the sun.

8. Discuss briefly the statement: "The semiannual parallax effect applies only to stars near the plane of the ecliptic." (Hint: first ascertain whether the statement is true or false.)

9. As a result of the precession of the equinoxes, will midsummer ever occur in February instead of August as at present? Explain.

10. How do the motions of the earth affect the seasons?

Prologue

How do we weigh an elephant while sitting on its back? How do we weigh the earth while riding as passengers on its endless excursions around the sun? An even more crucial question: How can we weigh other celestial bodies that are beyond our grasp? Why would we want to weigh them, anyhow? What justification suffices? Such purpose needs no justification, since it is intimately bound up with man's innate curiosity about himself and about the world in which he is captive.

Weighing an object in its simplest sense involves measuring the strength of a known gravitation pull on it. Once this quantity is determined, no matter what kind of derived or defined units we choose to report the results in, we have also determined how much mass — how much matter — the object contains. Lest we think weight or mass is without social or cultural implications, we should reflect on how the quantity weight enters into our everyday transactions.

In this chapter we shall review the steps made and the techniques employed in obtaining information about the weight and mass of several kinds of celestial objects, if for no other reason than to satisfy our curiosity about them. Curiously, it seems, the property of elasticity, similar to the properties of a stretched rubber band, enters into the methodology. In no other way can we get at such secrets as the inner structure of the earth's mass.

Then there is the problem of the earth's atmosphere, physiologically a blessing and a necessity, astronomically a curse. All of the information coming to us from the celestial universe beyond the earth rides along on electromagnetic radiation, three forms of which — light, heat, and radio waves — are astronomically indispensable. All of the information has to pass through the atmosphere if we are to receive it at the earth's surface. How can we *know* whether it all got through or not? In general, we need to know just what distortions of and omissions in the information received are the responsibility of the earth's atmosphere. Aside from that, knowledge about our own atmosphere yields insights to other atmospheres on other celestial bodies.

The Earth: Physical Properties

6

In our quest for knowledge about the universe the solution to another problem, the determination of the mass of the earth, has great significance. This information is of interest not solely because it concerns our planetary residence but also because at this stage in our quest the masses of all other astronomical bodies appear to be unknowable since they are beyond our reach. We shall discover, however, an avenue of attack on this problem in the process of finding the mass of the earth.

The Mass of the Earth

How shall we go about determining the mass of the matter which makes up the earth's vast bulk? The problem might be easier to solve if the earth were composed of just one kind of "stuff," but it is not. We know from experience that a given volume of water *weighs* less than the same volume of sand when both are weighed at the same location. Since Galileo demonstrated that under these conditions the acceleration of gravity is the same for all bodies regardless of their size or composition, we must conclude that the sand is more massive. We draw this conclusion from the definition of weight,

$$\text{net } \mathbf{F} = m\mathbf{g} = \mathbf{W} \qquad (6.1)$$

where $\mathbf{W}$ is the weight of an object whose mass is m when it is accelerated by gravity $\mathbf{g}$. Note the similarity of Equation 6.1 to Newton's second law, net $\mathbf{F} = m\mathbf{a}$. In the latter, $\mathbf{a}$ is *any*

acceleration; $\mathbf{g}$ is a particular one. Equation 6.1 shows that weight is defined as a force and it is proportional to the mass of the object whose weight is being considered. Because of the earth's rocky crust, we can now safely draw this conclusion: the mass of the earth is *at least* more than the mass of a sphere of water of the same dimensions as the earth. This is true provided a given volume of water has practically a constant mass. Experience shows that water is practically incompressible. Therefore, the *mass per unit volume* of water should be nearly constant, at least to a first approximation. This quantity, mass per unit volume, written *mass/volume* or in metric system units, grams per cubic centimeter (g/cm³), is defined as *density*. We usually symbolize density by the Greek letter ρ.

Now a little reflection on the application of Newton's second law, net $\mathbf{F} = m\mathbf{a}$ discloses a significant fact. If we apply a *known* force on an object and observe a particular acceleration, net $\mathbf{F}/\mathbf{a} = m$ defines the mass of the object which experienced the acceleration $\mathbf{a}$. But how do we know the magnitude of the net force? We observe the acceleration experienced by a known mass and define the product $m\mathbf{a}$ as the net force. But here we are trapped in an elliptical argument: we can determine the force if we know the mass, but to know the mass we must determine the force, provided of course, that the magnitude of the acceleration can be determined.

Of course, if we could ever determine a *first mass* we could use it as a standard of comparison for the determination of any other mass. How can we determine the first mass? We can't. We *define* it. The standard of mass (by definition) was selected in the late 18th century as the *inertia* offered by 1000 cm³ of distilled water at 4°C. For practical reasons the standard of mass has been redefined as the mass of a certain cylinder of platinum alloy deposited in the *Bureau Internationale des Poids et Mesures* at Sevres, just outside Paris. All other masses are determined relative to this primary standard. The kilogram, which was originally defined as the mass of 1000 cm³ of distilled water (*kilo* = 1000), is now defined as the mass of the metallic cylinder above. In astronomy such a small unit (at the earth's surface, 1 kg weighs approximately 2.2 lb) is impractical. In many situations a relative standard such as *one solar mass* ($M_\odot$) or *one earth mass* ($m_\oplus$) is used.

Thus we acknowledge that the concept of mass is a relative thing. Hence we are justified in evaluating the mass of the earth by a comparison method. The outline of the procedure is as follows:

Identical test masses m are placed in the pans of a very sensitive balance (Figure 6-1) and this system is allowed to come to equilib-

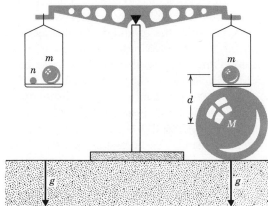

FIGURE 6-1 "Weighing" the earth. A suitable balance is initially in equilibrium with masses m, m in the pans. A large mass M is placed under one of the pans; the attractive forces between m and M disturb the equilibrium which is then restored by adding the small mass n to the opposite pan. Newton's universal law of gravitation is then used to compute the mass of the earth (see text).

rium. Next, a large disturbing mass M, perhaps 1000 times the magnitude of the test mass, is interposed between the earth and the right-hand pan; this upsets the equilibrium toward the right. Third, a small mass n found by trial and error is placed in the left-hand pan such that the original equilibrium is restored. We then have the gravitational forces between the earth and the two left-hand masses equal to the sum of the gravitational forces on the right: between the earth and the test mass, and between the disturbing mass and the test mass,

$$F_{(\text{left})} = F_{(\text{right})}.$$

But each of these forces is a gravitational force which can be represented in the form of Newton's universal law of gravitation, net $F = GmM/d^2$. The attractive forces can be symbolized as follows.

$$F_{(\text{left})} = G\,\frac{m \times m_\oplus}{r_\oplus^2} + G\,\frac{n \times m_\oplus}{r_\oplus^2},$$

$$F_{(\text{right})} = G\,\frac{m \times m_\oplus}{r_\oplus^2} + G\,\frac{m \times M}{d^2}$$

The symbols $\odot$ and $\oplus$ stand for the sun and earth, respectively. When they are used as subscripts they identify the particular quantity in question with the sun or earth. For example, $m_\oplus$ is the earth's mass and $D_\odot$ is the sun's diameter. We shall use this notation throughout the book. In the preceding equations, the earth's mass is represented by $m_\oplus$, the restoring mass by n, the earth's raidus by $r_\oplus$, and the distance between the centers of the test mass and disturbing mass M on the right side as d. But the two forces are equal since the balance is in equilibrium. Therefore, the two *sums* of the gravitational components are equal. Moreover we can subtract the common factor $Gmm_\oplus/r_\oplus^2$ from each side and divide out the gravitational constant G. The result is

$$\frac{n \times m_\oplus}{r_\oplus^2} = \frac{m \times M}{d^2}. \tag{6.2}$$

The Mass of the Earth   101

A simple algebraic rearrangement produces

$$m_{\oplus} = \frac{r_{\oplus}^2 m \times M}{nd^2} \qquad (6.3)$$

where all of the quantities on the right are previously known or are directly measured. The value of the earth's mass determined this way is 6×10^{24} kg, or 6.6×10^{21} tons of mass.

What would be the mass of the earth if it were composed entirely of water? The volume of a sphere of the earth's radius r can be computed from

$$V = \frac{4\pi r_{\oplus}^3}{3}. \qquad (6.4)$$

The magnitude is 1.09×10^{24} liters (1 liter = 1000 cm^3). By the early definition of standard of mass each liter of water contains 1 kg of mass. Thus, the mass of a sphere of water of the size of the earth is 1.09×10^{24} kg.

We see that the earth's mass is 5.5 times this value. Evidently the earth's mean density is 5.5 g/cm^3. But the density of rock samples from the earth's surface averages around 3.0 g/cm^3; therefore, the earth must be made up in a large part of matter that is substantially more dense than rock. As is so often the case in science the solution to one problem often raises another of even greater complexity: What is the internal structure of the earth?

Internal Structure

Strange as it may seem, knowledge of many of the physical properties of the earth is more inaccessible to our investigations than many of the properties of stars and of galaxies. A blanket of water averaging about 4.5 km in depth covers 71% of the surface of the earth. A significant portion of this *hydrosphere* is frozen over. Ice also covers most of the continents of Greenland and Antarctica to depths believed to exceed 3000 m in places. The deepest mines penetrate the earth's surface not much more than 2000–3000 m; the deepest wells reach down less than 10,000 m. How, then, can we

know the internal structure and composition of the earth? Indeed, *do* we know, or do we merely just surmise its structure? We cannot here pursue this problem in any detail; it is one for research and analysis by geologists and geophysicists. Yet seeking an explanation for the unexpectedly large mean density of the earth is a tempting challenge.

Physicists have learned that matter in general can transmit elastic vibrations. Certain plastics such as putty or modeling clay are excluded because the transmission of the deformations depends upon the ability of the matter to resist the distortions. Two broad categories of elastic vibrations are important here. One, called *compression waves* and usually designated *P*-waves, consists of successive compressions and rarefactions of the medium. These disturbances are propagated through the medium in a direction parallel to the successive compressions and rarefactions, that is, in the direction of the applied stress. The other, called *transverse waves* and designated as *S*-waves, consists of periodic oscillations which are propagated perpendicular to the direction of the applied stress. Accordingly *S*-waves are shear waves because the propagation produces a shearing effect between adjacent particles of the medium. That is, the particles slide back and forth past each other at right angles to the direction of propagation of the disturbance. Figure 6-2 illustrates examples of each type of vibration wave. With the exceptions noted, all matter transmits compression waves rather freely. The denser the medium, the more efficient is the propagation and the greater is the speed of travel of the disturbance. Thus, compression waves are propagated best in solids, least efficiently in gases and liquids, and not at all in the absence of a supporting medium. Liquids and gases transmit tranverse waves so feebly that for all practical considerations we may assert that these disturbances are transmitted only by solids.

How can we turn this information to our use? Earthquakes in the earth's crust or upper

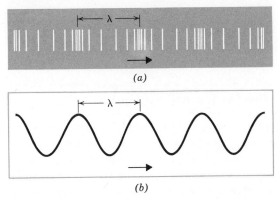

(a)

(b)

FIGURE 6-2 Graphical representation of
(*a*) compression waves and (*b*) transverse
waves. Sound waves are compression
waves; such disturbances require a
supporting medium for their propagation.
Electromagnetic waves are transverse
waves. Certain disturbances called *shear*
waves which have the form of transverse
waves can be propagated through solid
materials only. Electromagnetic waves do
not require a supporting medium. In the
figure the arrows show the direction of the
propagation of the disturbances; λ is the
symbol for wavelength.

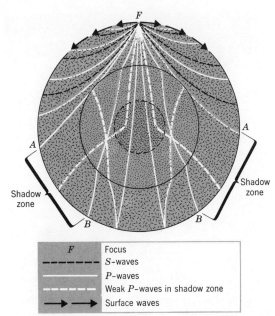

F	Focus
- - - -	*S*-waves
————	*P*-waves
- - - -	Weak *P*-waves in shadow zone
➤ ➤ ➤	Surface waves

FIGURE 6-3 Representative paths of seismic
(earthquake) waves through the earth.
S-waves (shear waves) cannot penetrate the
core; this suggests that the core is liquid
since liquids will not support
transverse-wave propagation. The shadow
zone is the region between the arrival
points of the most direct compression wave
from the focus F of the disturbance and the
nearest refracted compression wave. The
presence of very weak compression waves
(*P*-waves) in the shadow zone implies a
solid or very high density central region in
the core.

mantle (Figure 6-3), together with heavy explosive detonations and volcanic activity, generate both kinds of vibrations. Geophysicists have established elaborate networks of *seismic* observatories and have discovered *three* modes of propagation of elastic disturbances in the earth. These are *P*-waves (the letter *P* is used to indicate that compression waves are *pressure* disturbances), *S*-waves (*S* stands for *shear,* and identifies transverse, or distortional, waves), and complex surface waves, called Love-waves and Rayleigh-waves after their investigators. The latter are of no significant interest with respect to deducing the internal structure of the earth below the immediate surface, therefore we can safely ignore them here. Each of the other two types of waves under consideration has a distinctive range of velocities, governed only by the elastic constants and the density of the medium. Table 6.1 shows comparable val-

ues of the velocities with increasing depth in the earth. Laboratory experiments disclose the following properties of the wave propagation which puts seismology in the category of a practical art. Both wave forms can pass through solids. At interfaces between media of different densities, the waves are partly refracted, partly reflected. In one important case *S*-waves are created at an interface when a compression wave impinges upon it. Finally, *S*-waves cannot pass through gases or liquids.

From an exceedingly complex process of mapping the progress of hundreds of thousands

TABLE 6.1 Velocities of P and S Waves
at Different Depths in the Earth

Depth, km	Velocity of P-waves, km/sec	Velocity of S-waves, km/sec
33	7.75	4.35
100	7.95	4.45
300	8.58	4.76
413	8.97	4.96
600	10.25	5.66
800	11.00	6.13
1200	11.71	6.50
1600	12.26	6.73
2000	12.79	6.93
2400	13.27	7.12
2898	13.64	7.30
3000	8.22	
3400	8.76	
3800	9.28	
4200	9.70	
4600	10.06	
4892	10.44	
5121	(9.7)	
5121+	11.16	
5700	11.26	
6371	11.31	

of seismic disturbances that includes, in part, abrupt changes in the velocity of P-waves, the following hypothesis concerning the structure of the earth has emerged as the most probable.

1. A surface *crust* not uniform in thickness, averaging some 30 km, underlies the oceans and includes the exposed land masses. The average rock density is 2.7 g/cm³.
2. Under the crust, a *mantle* of so-called basement rocks extends to a depth of 3000 km. The mean density of the matter in this layer is 5.0 g/cm³.
3. Under the mantle is an *outer core* some 2100 km thick, with a mean density of 11 g/cm³. Available evidence indicates the outer core to be liquid, and probably a nickel-iron mixture.

4. Finally, an *inner core* supports the outer core and is centered on the earth's center. The mean density appears to be 14 g/cm³, and its composition is unknown.

The Earth's Atmosphere

There are two fundamental reasons why the earth's atmosphere is of interest to astronomers. One is that it serves as a model against which other planetary atmospheres can be measured and compared. The other is that the presence of the atmosphere is astronomically more adverse than beneficial; it hampers observation in a number of ways.

The atmosphere is a moving, turbulent fluid envelope composed of a mixture of gases. It is most dense in the lower levels; it decreases in density with height until it merges imperceptibly with the almost total vacuum of space. Almost all weather phenomena occur in the lowest layer which is called the *troposphere*. This layer reaches to a maximum height of less than 20 km. Yet some phenomena including the aurorae, the fiery trails of meteors, and certain radio wave reflections occur well above the troposphere. Atmospheric phenomena of one kind or another, excluding clouds and precipitation, can be detected up to 1000 km.

To a height of 80 km the mixture of gases is essentially homogeneous, that is, the percentage composition by volume is nearly constant. *Nitrogen,* a relatively inert gas, comprises some 78% of the mixture. *Oxygen,* so essential to almost all terrestrial life forms comprises 21%. The remaining 1% is divided unevenly between *argon* (0.9%), *carbon dioxide* (0.03%), and a whole host of other trace gases. The most important of the latter is *ozone,* a three-atom hybrid molecule of oxygen. The oxygen molecule normally consists of two atoms.

Water vapor is not considered a regular constituent of the atmospheric mixture. The water vapor content of the atmosphere is extremely variable in place and time, hence it is most conveniently treated as an atmospheric impurity, albeit a highly desirable one. Most of the water vapor is concentrated in the lowest 8 km because it enters the atmosphere by evaporation from the great expanses of water and snow or ice fields. Within this vertical restriction it is present everywhere over the earth, from slightly more than 0% by volume over arid regions, to a maximum of around 4% in the most humid locations.

Ozone is concentrated within a layer extending between approximately 15 and 25 km above sea level. The quantity of ozone in this 10 km layer is quite minute. We can get some idea of how rarified the upper atmosphere is by the following consideration. Atmospheric physicists have computed that if the ozone in this layer could be transported undisturbed and undiluted to sea level, yet undergo increased compression with decreasing elevation of that of the other gases, the 10-km layer would be compressed to a layer less than 0.6 cm thick. The importance of even this infinitesimal quantity of ozone will become apparent in the next section.

Atmospheric phenomena such as clouds, precipitation, turbulent motion, and absorption and scattering of certain fractions of the incoming electromagnetic radiation from all sources adversely affect astronomical observation. Clouds, fog, haze, smog, and precipitation, collectively referred to as *hydrometeors,* offer direct obscuration of incoming light; turbulence randomly distorts the images with time. Smoke particles, ice crystals, dusts, pollens, and similar materials, and even the gas molecules themselves selectively absorb a portion of the electromagnetic spectrum, or reflect it, or scatter the light. Yet in general, pure air is almost completely transparent to incoming radiation. The following are among the properties of the atmosphere considered most significant astronomically.

Atmospheric Absorption. As we shall see in considerable detail in Chapter 9, the radiation we receive from astronomical objects is com-
posed of other forms besides light. One classification scheme for radiation types is given in terms of the particular sorting or detecting devices which are more or less selectively responsive to particular wavelengths and energies. Thus, terms such as radio waves, microwaves (TV, FM radio, and radar), infrared waves (radiant heat), visible light (photosensitive devices and organs), and x radiation (the familiar x-ray photography) are all indicative of the way in which we interpret and utilize electromagnetic radiation.

The atmosphere selectively absorbs certain fractions of radiant energy. Incoming radiation from the sun, the principal source of electromagnetic energy for the earth, suffers some atmospheric depletion by absorption. X-ray energy is almost totally absorbed in the upper atmosphere; ultraviolet (uV) radiation is nearly all absorbed in the middle levels. There are mixed blessings in this. On the positive side is the benefit to terrestrial life-forms. Both of these kinds of electromagnetic radiation are lethal to most life-forms provided the concentration is sufficiently high. The biological community would be vastly different, if it existed at all, if this radiation were permitted to reach the earth in the concentrations presumably arriving from the sun at the top of the atmosphere.†

As previously mentioned, ozone provides the principal shield from ultraviolet radiation. Some gets through; sunburn and snowburn are symptoms of excess exposure to uV radiation. The variety and abundance of terrestrial life-forms is testimony to the efficacy of this radiation shield.

Carbon dioxide is a familiar product of volcanic emanations and of combustion processes, especially those involving fossil fuels such as oil and coal. There is some evidence,

† Since we are able to measure the concentration only *after* the radiation has passed through the atmosphere we are in a quandary about *how much* of each arrives from the sun, and correspondingly how much is absorbed in the atmosphere. We do make some progress in this area in Chapter 9.

as yet unconfirmed, that the carbon dioxide content of the atmosphere may be slowly increasing. But the gas is known to be fairly soluble in water. Meteorologists and oceanographers believe the oceans regulate the atmospheric content of carbon dioxide within very close limits.

Carbon dioxide readily absorbs and reradiates infrared radiation, principally in the range of the back radiation from earth to space. The complex absorption and reradiation acts as a blanket or shield against heat losses from the earth. Reliable estimates place the earth's mean temperature some 15°C warmer than would be the case in the absence of carbon dioxide.

Besides its obvious role in cloud and fog making, in precipitation processes, and in haze and smog formation, water vapor also absorbs infrared radiation, both incoming from the sun and outgoing from the earth. The exact amount is not known; the difficulty stems in part from the variability of the atmosphere's water vapor content, and in our inability to determine the exact amount of this radiation in the sun's emission spectrum. Hopefully current solar studies involving orbiting geophysical and solar laboratories outside the atmospheric envelope will assist in an early solution to this problem.

Atmospheric Diffusion. In physics *diffusion,* when applied to light, means random reflection from a rough or irregular surface or from particles whose size is of the order of the wavelength of the diffused light. Often the term *scatter* is applied to the latter form of diffusion. Atmospheric diffusion is familiar and commonplace although the explanation for this phenomenon may not be fully appreciated. Particles of dust and smoke, water droplets and ice crystals, and other similar matter reflect light rather brilliantly. Clouds are made visible by this process.

The largest-scale diffusion phenomenon in the atmosphere is the scattering of light by the air molecules themselves. These particles have radii of the order of one-tenth the wavelength of the scattered light. They are far more effective reflectors than absorbers of radiant energy. Lord Rayleigh (1842–1919), an English mathematician and physicist, demonstrated that the amount of scatter by gas molecules was inversely proportional to the fourth power of the wavelength of the scattered light,

$$S \propto \frac{1}{\lambda^4} \qquad (6.5)$$

where λ (lower case Greek letter "l," or *lambda*), represents the given wavelength, and S is the amount of scatter. Even without considering the effect of the exponent the ratio 1/λ is observed to become larger as λ gets smaller and smaller. Moreover if λ is a very small number raising it to the fourth power makes it vanishingly small in a hurry. For example $1/4^4 = 1/256$; $1/2^4 = 1/16$. Thus reducing the number which the exponent "feeds on" by a factor of 2 increases the magnitude of the whole expression sixteenfold.

The shortest wavelength in the visible light range is in the blue region of the spectrum. Thus, blue light is more effectively scattered than all other colors, especially more than the red components of white light from the sun. The scattering is in random directions; hence to an observer *within* the atmospheric envelope the blue light seems to originate from all directions. This is why the daylight sky appears to be blue.

The shade of blue depends upon the thickness of the atmosphere above the observer, discounting of course the diffuse reflection from larger particles which creates a milky cast to the blueness. Observers on high mountains or in high-flying aircraft are familiar with the noticeably bluer quality of the sky as compared to the sky color at sea level; astronauts have verified that the daytime sky is as black as the blackness of outer space when all of the atmosphere is below them.

A direct beam of light from the sun should

appear redder than it really is, since the blue component is scattered out of the path. This is truly the case, but the brilliance of the direct light from the sun and the brilliance of the sky dulls our senses to this effect. What, then is the explanation for the orange, or reddish color of the sun at sunrise and sunset? Part of the effect is the result of scatter. But the greatest contribution comes from selective absorption of the shorter wavelengths by the larger foreign particles in the atmosphere. The effect is most noticeable when the sun (or moon, which shines by reflected sunlight) is near the horizon. The absorbing path is much longer here than it is when the sun is nearer the zenith as Figure 6-4 illustrates.

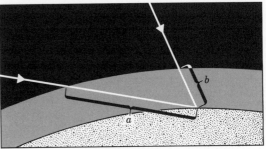

FIGURE 6-4 Graphical representation of the dependence of the optical path length (*a* and *b*) on the angle at which incident light rays impinge on the earth's atmosphere. The relative thickness of the atmospheric envelope is highly exaggerated in the figure.

Atmospheric Refraction. Suppose we were to examine the refraction of a ray of light entering the atmosphere as it passes through successively more dense layers of a transparent medium. The ray will be refracted away from the original path by an angle that increases in size with depth (Figure 6-5). Reference again to Figure 5-1a will also be helpful here. Now if we could "sight" back along the direction from whence the terminal end of the ray *seems* to come, the source which originated the ray would appear elevated above its true position. This is a somewhat crude analogy of what occurs in the atmosphere. We can think of the gaseous envelope as consisting of an infinitely large number of exceedingly thin layers. The optical path of an incident light ray will appear to curve as shown in Figure 6-6. While the refraction effect is noticeable, it is nevertheless quite small. Near the horizon an apparent point source of light such as a star appears to be displaced upward from its true position by as much as 1/2°. This is the angular diameter of the full moon. In fact, such stars will appear to rise while in reality they are still below the horizon. The refractive displacement is largest at the horizon; except in the most precise position astronomy the effect is negligible at angles exceeding 10° above the horizon.

Perhaps the most noticeable effect of the refractive capabilities of the atmosphere is observed in connection with the setting sun. As the sun sinks toward the horizon, the lower *limb* is refracted upward to a greater degree than the upper limb. (The limb is the circumferential edge of the apparent disk of a celestial body.) Hence the sun appears to be "flattened" at the bottom (Figure 6-7). Sometimes the extreme lower edge appears striated, or even detached from the main body of the sun. Atmospheric stratification, or layering, accounts for this appearance.

Twinkling Stars. If the atmosphere is turbulent, as it always is to some degree, light from an apparent point (dimensionless) source such as a star undergoes differential, or random, variable refraction with time. Occasionally the entire narrow beam is refracted out of the line of sight for an instant, causing apparent extinction of the source. More often the blue component is refracted away, only to reappear more intensely than ever as the red component is bent out of the line of sight. Thus the star appears to vary in brightness in a random manner, and to change apparent color in the process. The effect is termed twinkling.

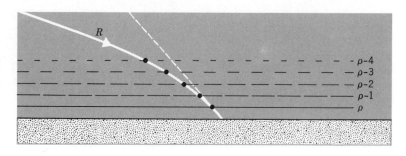

FIGURE 6-5 Cumulative refractive effects on incident light rays by multiple layers of increasing density. The density increases downward.

Extended bodies such as planets suffer the same refractive distortion. But when the source appears to be about the same diameter as the apparent turbulence cells, as is the case with planets, the effect averages out and the objects do not usually appear to twinkle. We must not accept this as a fool-proof test; severe turbulence will cause even the planets to twinkle somewhat.

The Earth-Moon System

Thus far in our discussions we have largely neglected the earth's companion planet, the moon. The reason for doing so is that the moon is sufficiently important an astronomical body as to warrant a starring role of her own. The moon takes her place, center stage, front in the next chapter. Here, we shall review the effects on the earth's surface of the gravitational interaction between the two bodies. This will provide us with a comfortable transition to the investigation of the lesser half of this celestial marriage, or celestial nativity, whichever it is.

Newton's universal law of gravitation and his second law of motion have important implications for the observed behavior of the earth and moon with respect to each other. If orbiting the sun were the only space motion the earth and moon have in common, the law of gravitation would have long since described the mechanics of their catastrophic collision. But we know the moon revolves around the

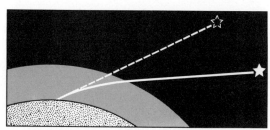

FIGURE 6-6 The image of a star appears higher than it really is because of the refractive properties of the earth's atmosphere. The atmosphere increases in density downward. In the figure the relative thickness of the atmosphere and the angular displacement are highly exaggerated for purposes of illustration.

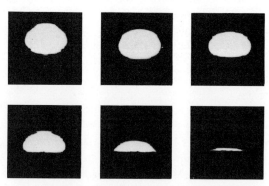

FIGURE 6-7 The apparent flattening of the sun as it sinks near the horizon as a result of atmospheric refraction. The lower edge of the disk is refracted upward more than the upper edge because the atmosphere is denser near the surface (Courtesy Lick Observatory).

earth at a rate which keeps it from falling to the surface. One important question yet to be answered is: Does the earth in turn revolve around the moon? Or should we ask, do they both revolve around some common center of mass (often referred to as a center of gravity)? The answer to this question comes with the determination of the shape of the moon's orbit (which we shall undertake in Chapter 7).

Tides. There is one group of terrestrial phenomena for which the moon bears the largest responsibility: the tides. *Tide* is an archaic word which has acquired a unique meaning in terrestrial physics. It is defined as an alternate rising and falling of the earth's surface—*both* water *and* land—under the combined gravitational attractions of the sun and moon. The sun contributes only one-third of the tide raising force in spite of its immensely greater mass. The moon's proximity to the earth, almost 400 times closer than the sun, gives it the gravitational advantage.

It is not too difficult to visulaize how the moon can raise a tidal bulge on the side of the earth facing it. The ocean masses are free to move; even the crust of the earth is sufficiently plastic that the continents rise slightly under the gravitational pull. However, there is a corresponding tidal bulge on the opposite side of the earth, and this bulge seems paradoxical at first. It is not, *per se,* caused by the gravitational pull of the sun; this opposite bulge appears even when the moon and sun are both on the same side of the earth. Let us analyze the problem of explaining the opposite bulge by taking one step at a time, as though the tides were indeed initially raised in steps (Figure 6-8).

1. The moon raises a tidal bulge on the nearest side of the earth. An analysis reveals that the near surface of the earth experiences a greater gravitational pull from the moon than does the center of the earth. The latter experiences a greater pull than does the far side of the earth.

2. The center of the earth moves toward the moon along the line of the tide raising force. It leaves behind the lesser attracted far surface material, again mainly water, until the gravitational pull of this bulging mass on the center of the earth just balances the lunar tidal force.
3. The result is two tidal bulges oppositely situated on the earth.

These bulges would be centered on the line connecting the centers of the earth and the moon, except for the earth's rotation. The great continental land masses continually slam into the tidal bulges of the hydrosphere. Friction displaces the centers of the bulges some distance in advance of the line of centers (Figure 6-9). We shall come back to this point in a moment.

The sun contributes a similar tidal effect. When the three bodies are approximately in line, the solar and lunar tidal forces add to produce the high, or *spring,* tides. These of course occur at the times of the new and full moons. When the sun and moon are in quadrature, that is, at right angles with respect to the earth, the tidal forces oppose each other; this produces the low, or *neap,* tides at the time of the moon's first and third quarters.

Locally observed tidal effects along a coast line are exceedingly complicated because of frictional variations, ocean currents, and prevailing winds among other causes. Local tide prediction is a necessary and difficult computational task.

The frictional collision of the continental land masses with the tidal bulges exerts a braking effect on the earth's rotation. The rotation rate of the earth is presently decreasing at a rate of 0.0016 second per day per century. This preposterous appearing unit is not as ungainly as it seems. Suppose the rate were stated in terms of so many revolutions per second per second. We would immediately recognize the quantity as an acceleration, or, if it decreased with time, a deceleration. Thus the peculiar ap-

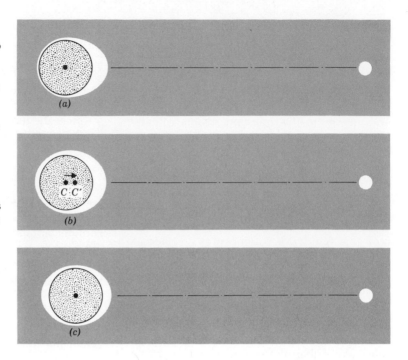

FIGURE 6-8 Step-by-step explanation for opposing tidal bulges on the earth. (*a*) The moon raises a tide on the earth's nearest surface. (*b*) The center of the earth moves from *C* to *C'* in the figure in response to the gravitational pull of the moon *and* the center of mass of the tidal bulge. The far side of the earth is attracted less because it is farther away. (*c*) When equilibrium is restored the pull of the tidal bulge on the far side balances the pull of the tidal bulge on the near side; the earth's center is then midway between the bulges.

pearing quantity should be recognized for what it is, a deceleration, in spite of the unfamiliar use of the given time units. Even though the deceleration rate is exceedingly small, the effect is cumulative. External events such as eclipses appear to have gained 3 hours on the clock during the past 2000 years. Perhaps some 4 billion years ago, the moon was only some 15,000 km from the earth; the earth's day was only 5 of our present hours long.

According to well established conservation principles in physics, we can say with confidence that in a closed system momentum is conserved. The earth and moon can be considered such a closed system. *Momentum* is a physical quantity obtained from the product of a moving body's mass and its velocity. Momentum is related to the energy of motion of the body, although it is *not* identical with it. The important consideration is that what momentum the earth might lose is gained by the moon. Since the earth is slowing down from

the tidal effects, it is losing rotational momentum. This momentum is gained by the moon in the following way. Recall that the tidal bulges are displaced away from the line of centers of the earth-moon system. This displacement is in the direction of the earth's rotation, which is *also* the direction of the moon's orbital motion (Figure 6-10). The presence of such a large mass of matter represented by the tidal bulges tends to speed up the moon in its orbit. According to Kepler's third law this will result in both a larger orbit and a longer interval for the month. As the earth slows down, the moon will spiral away, the days will become longer, and so will the month. Calculations indicate that after several hundred million years the day will become equal in length to the month. Both will be equivalent in length to about 47 of our present days. The earth will keep one face turned toward the moon, just as the moon keeps one face toward the earth at present.

DATE DUE

NOV 1 1 1994	
DEC 0 8 1994	
DEC 1 1 1994	
Feb 2 1994	
FEB 2 2 1995	
2-27-95	
MAR 1 9 1995	

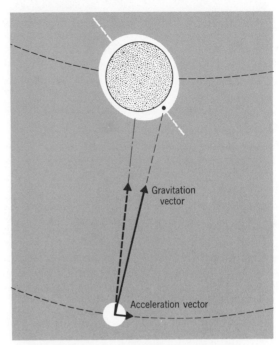

FIGURE 6-10 Acceleration of the moon in its orbit by the tidal bulge. The tidal bulge acts on the moon along a line represented in the figure by the gravitational vector. This vector can be resolved into components which show that part of the gravitational force acts toward the center of the earth, the remainder acts in a line along the moon's orbital path. But this is an accelerating force which tends to increase the orbital speed of the moon.

planetary radius and is called the *Roche limit*. The remnants will remain in orbit around the earth, in a form resembling the rings of Saturn. The time scale for this cycle is several tens of billions of years. Presumably other destructive forces will obliterate both the earth and moon long before this.

SUMMARY

The mass of the earth is determined from the differential attraction of the earth and a large test mass on a third mass. This test yields a value for the average density of the earth that is nearly twice that of the surface rocks. Seismic studies show how the earth's mass is distributed and offer strong evidence for a liquid core.

The earth's atmosphere hampers astronomical observation from the earth's surface because of its relative murkiness (obscuration and light scatter), its turbulence, and its refrac- tive properties. In addition the atmosphere conceals information about external radiating objects because of the way it absorbs certain fractions of the incoming radiation.

Finally, the earth-moon system is governed by mutual gravitational forces, one aspect of which is the earth's tides. The conservation of momentum in a closed system such as the earth and moon has definite implications for the ultimate fate of the moon.

Questions for Review

1. Outline briefly the procedure for determining the mass of the earth. Can this method be applied from earth to other planets?

2. What is the internal structure of the earth as revealed by seismic waves?

3. What is the main composition of the atmosphere up to about 80 km?

4. How is it that ozone, an irritating poison, is essential to terrestrial life-forms?

5. How does the presence of water in the atmosphere, in all of its various forms, affect astronomical observations?

6. Why is the sky blue?

7. Even though they are not discussed in the text, can you explain the phenomena of mirages?

8. What causes the twinkling of the stars?

9. Why is the moon more effective than the sun in producing tides?

10. Why do some locations on earth experience tides of 6–7 m, when the combined gravitational effect of the sun and moon can produce maximum tides averaging only 2 m?

11. What effect will tidal friction have on (a) the earth-moon distance. (b) the length of the day?

Prologue

The moon has always intrigued man, if for no other reason than that of her intrinsic beauty. Because he was so enraptured with this voluptuous princess of the night, man since time immemorial has ascribed anthropomorphic qualities to the moon and has included her among his menáge of immortals.

For some men, the moon became Diana, goddess of the hunt. For others it was the moon which controlled the fertility of women and arranged the destinies of nations. The moon was an object of *supreme* worship in ancient Ur. By contrast, South Pacific islanders and Indians of Canada thought that the moon was nothing more than a minor diety whose face was mottled by mud thrown by disgruntled earthlings.

The moon is prominent in the folklore of almost every nation. She has inspired poets the world over, from the unknown compilers of the Book of Genesis, to Homer and Theocrastus, to Virgil, to Shelley and Keats, and to Wordsworth and Coleridge and Shakespeare. These literary giants have all paid homage to the moon in their verse.

Superstition abounds in moon lore. The word lunacy attests to the supposed influence of the moon on the psyche of mortal man. The moon is also supposed to influence the growth of plants, as is attested to by the early almanacs that contain planting guides arranged according to the phases of the moon.

We have more pressing reasons for studying the moon. It is an astronomically important body because of its proximity to the earth; it has become accessible to direct exploration. Man for the first time has discovered what other celestial bodies may be like. Because of its nearness we use the moon as a target in developing optical investigation methods that we can then apply to more remote bodies with increased confidence.

So from the earth's distance of 400,000 km right down to direct laboratory inspection of moon rocks brought back by the astronauts, we can progressively investigate this companion planet of the earth: this dry, lifeless clod which is one-fourth of the earth's diameter, one-eighteenth of the earth's mass, yet which has a density only 60% as great. We shall examine the pock marks on its surface and the dark plains of its ancient volcanic flows; we shall chronicle its wiggles and wobbles as the moon figuratively squirms its way along its path to its ultimate destiny.

The Moon

7

The closing section of the preceding chapter hinted broadly at a problem that has intrigued astronomers and geoscientists for centuries. Was the moon formed with a separate identity at roughly the same time as the earth? Or was the moon wrenched violently from mother earth's womb in a cataclysmic birth? These questions have not yet been resolved and they may never be. On the other hand future Apollo mission explorations of the lunar surface, together with the Surveyor exploration of the surface of Mars scheduled for the last half of this decade, may yield hitherto unsuspected facts concerning the moon's origin.

There are a number of hypotheses concerning the event, none of which has been able to withstand completely the combined tests of observation, mathematical analysis, chemical analysis, or geophysical probability. In an attempt to gain a fuller appreciation of astronomy's position in the matter, let us review briefly some of the more prominent theories concerning the origin of the moon.

Origin of the Moon

According to some theorists the moon's mass was originally an integral part of the earth's mantle and crust. Presumably some passing astronomical body created enormous gravitational stresses on this matter which, in a sense, "floats" on the inner molten core of the earth. So great were these stresses that the lump of matter which is now the moon was torn loose.

It went into orbit along with what remained of the earth around a common center of gravity in accordance with Kepler's laws. The theorists point to a vast depression in the floor of the western Pacific Ocean as the place where the moon material came from. This is known as the tidal theory.

Another theory suggests that the moon was formed from an accretion of matter contained in the vast interstellar nebula which preceded all of the members of the solar system. Eventually, in accordance with gravitational laws, this material became organized into a spherical shape. The event probably occurred not too far from the earth, perhaps no more than 15,000 km distant, and the moon has since been receding from its original position in a manner outlined at the end of Chapter 5.

A third theory proposes that the earth acquired its companion planet by a capture process, sometime after the earth was well settled in its existence, but certainly more than 2 billion years ago.†

The arguments against each of these proposals are many and impressive. We can at most acknowledge qualitatively those objections which seem most plausible. The tidal theory, the first one mentioned, has serious defects. Any gravitational force that could succeed in wresting the moon material from the earth would have to greatly exceed the earth's gravitational hold upon it. Probably only a passing star could accomplish this feat. If this were the case, how did the organization of the solar system survive the intruder's presence? Furthermore, any such gravitational larceny could deprive the earth of all of its water and most of its

† The age of the universe and its members is argued in detail in Chapter 14, and to a lesser degree in Chapter 10.

FIGURE 7-1 Lunar surface features as viewed from Apollo 15 command module in orbit July 31, 1971 (Courtesy NASA).

atmosphere before the intruder succeeded in capturing a substantial mass of the earth's rocky mantle. At present the possibility of a gravitational cataclysm cannot be totally ruled out. But the question of how the earth acquired both a new ocean system and a new atmosphere poses formidable difficulties for the acceptance of this theory. Another objection is raised by the fact that analysis of earthquake waves discloses no scar tissue (which most assuredly would remain from such a violent event) in the earth's mantle. Such scar tissue, however, would not necessarily be present if the wrench occured either before the earth was consolidated or before the surface was molded by subsequent heating and melting of the mantle. Finally, how did this "liberated" material break the gravitational bonds with the intruder and remain in the immediate vicinity of the earth? It would seem an incredible coincidence that the passing body could tear the matter loose from the earth, yet be traveling so

fast that it got out of gravitational range before the moon matter could catch it.

The accretion theory is more plausible; the principle objection rests in the fact that the earth's larger mass should have captured the material first, instead of permitting a new gravitational center to form nearby. The capture process is an equally plausible theory. But where did the moon come from and how did it get close enough to the earth for the latter to capture it? Perhaps further clues as to the moon's origin will have to await laboratory analysis of additional moon material as it is brought back to earth.

The Moon's Motions, Real and Apparent

We have already noted that the moon periodically appears as a slender crescent just east of the sun and low in the western sky at sunset. As the days pass the moon is seen to be farther and farther east of the sun, and more and more of its apparent disk is illuminated (Figure 7-2). The moon progresses through its phases from crescent to first quarter in 6 or 7 days. At first quarter, half the disk is illuminated by the sun while half remains in darkness. A week after first quarter, the full moon, a fully illuminated disk, rises in the east as the sun sinks below the horizon in the west. The phases then regress through third or last quarter to a slender crescent which rises just before sunrise. For the next several days the moon disappears from view only to repeat the cycle.

Figure 7-3 illustrates the geometry of the earth-moon-sun positions and the relative phase appearance of the moon for representative positions. The moon is new when it is in the same direction as the sun. It is quite apparent why we cannot see the moon at this time. The overwhelming brilliance of the sun and the brightness of the surrounding sky, together with the fact that we are viewing an unilluminated hemisphere, render the moon invisible. As the moon swings eastward in its orbit we succes-

FIGURE 7-2   The moon at age 3 days, 5 days, 8 days, 20 days, 23 days, and 26 days. See Figure 7-12 for age 11 days, and Figure 7-21 for age 14 days (the full moon) (Courtesy Hale Observatories).

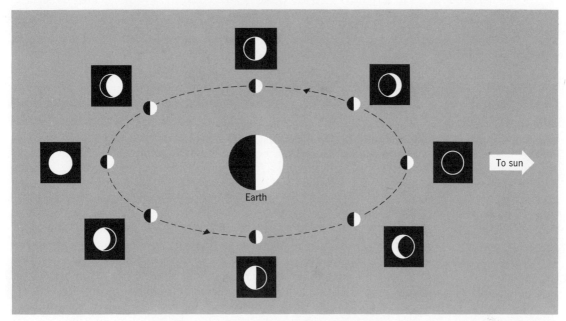

FIGURE 7-3 Relative space positions of the earth and sun, and the moon as it orbits the earth. Adjacent sketches show the phase appearance for each of the relative positions. Compare with the photographs in Figure 7-2.

sively view the moon at an increasing angular distance from the sun until full moon is reached. The pattern then reverses. The periodic nature of the lunar cycle, together with our knowledge of Kepler's laws, is conclusive evidence that the moon is nominally in orbit around the earth. We say nominally, because if the moon is in orbit around the earth, it is also in orbit around the sun. We have here the celebrated *n*-body problem of antiquity (Chapter 4) where, temporarily at least, *n* is taken to be 3. Because all three bodies—the sun, earth, and moon—are interacting gravitationally, a complete, long-term prediction of the moon's orbital motion is impossible.

Orbital Motion of the Moon.   Let us suppose that we were to make a very large number of observations of the moon's position on the celestial sphere. The observations extend over many complete lunar cycles. Among the com-

piled data are the moon's apparent right ascension, its apparent declination, and the times of each of the many observations. With this information we attempt to plot the moon's orbital path reference to the celestial sphere, which of course includes the sun's apparent path, the ecliptic. If the program of observations is sufficiently long, certain regular and periodic changes in the individual data become apparent. In order to study these phenomena in detail, the logical approach suggests finding average quantities and then using these as the reference coordinates or the reference values.

We discover that the moon's orbit is inclined to the ecliptic an average of 5°8' (Figure 7-4). The range is from 4°59' to 5°18'. Now if the moon's orbit is an ellipse with the earth at one focus as Kepler's laws predict, then the eccentricity of the ellipse should be detectable. Of course, the sun's gravitational disturbance will undoubtedly complicate the determination, but

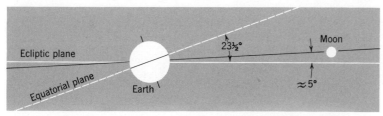

FIGURE 7-4 Linear projections of the orbital plane of the earth (the ecliptic), the earth's equator, and the moon's orbital plane. The moon's orbit is neither in the plane of the ecliptic nor in the equatorial plane of the earth.

at least a first approximation of the orbit should be within our capabilities. How do we proceed? According to the mathematical properties of an ellipse, the foci (Chapter 4) lie on the line joining the most widely separated points on the curve. If the earth is at a focus of the moon's elliptical orbit, the line joining the moon's closest approach to the earth and its most distant recession will pass through the earth. By definition the average of these two distances is the *mean earth-moon distance*. Clearly the earth will be displaced from this average distance which marks the center of the ellipse. If it were not, the focus containing the earth would be at the center. But under this circumstance the figure would be a circle, not an ellipse. Now the ratio of the distance between the center of the ellipse and the earth to the distance between the center and either extremity of the designated line (the semimajor axis) is the *eccentricity of the ellipse.*

Kepler's second law provides us with a method for determining the *least distant point* (*perigee*) and the *most distant point* (*apogee*) of the moon's orbit around the earth. At these points the moon will be traveling the fastest and the slowest, respectively, along its orbital path. At least this would be the case except for possible disturbances of the moon's speed by the sun (Figure 7-5).

Fortunately another line of attack is available to us, which is independent of the sun's potential interference. If the moon's orbit is mea-

surably eccentric, a detectable variation in the apparent diameter of the lunar disk should exist. Precise measurement discloses a 10% variation in the apparent lunar diameter. The largest diameter coincides with the moon's passage through the perigee point, the smallest diameter is associated with the passage

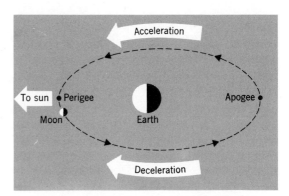

FIGURE 7-5 Kepler's laws, when applied to a two body system (earth and moon), predict the speed of the moon at various points in its orbit. But the presence of the sun's gravitational influence causes the moon to speed up when it is in relative approach toward the sun, and to slow down when in relative recession from the sun. Thus the moon departs from the two-body prediction because of the presence of a third body. The effect is present irrespective of the orientation of the moon's orbit (irrespective of the relative perigee-apogee positions).

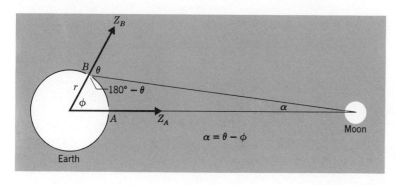

FIGURE 7-6
Determination of the
earth-moon distance by
the method of Ptolemy
(see text for explanation).

through apogee. With these two points thus determined, the next step is to compute the respective distances of these points from the earth.

One method used today is essentially the same as that devised by Ptolemy of Alexandria in 140 A.D. This is the same Ptolemy of Ptolemaic system fame. In Figure 7-6 simultaneous observations of the angular direction of the center of the moon with reference to each of the local meridians are made. Observation points are at A and B. Angle ϕ measured at the center of the earth corresponds to the difference in longitude between the two observers. For simplicity the observer at A records the time of the moon's transit across the local meridian. At this instant the radius from the center of the earth to the observer, the observer's local meridian circle, and the moon's center are all in the same plane. At this same instant the observer at B determines that the center of the moon is displaced to the west of his local meridian by the angle θ. With reference to the radius from the center of the earth to point B the direction of the line from observer to the center of the moon is $180° - \theta$. One of the theorems of plane trigonometry states that if two angles and the included side of any triangle are known, the remaining angle and the other two sides can be determined at once. We have the necessary information. The earth's radius from the center to B is the known side; the difference in latitude is one angle θ,

and the remaining angle is $180° - \theta$. The length of the side of the triangle that passes through point A and connects the centers of the earth and moon is just the quantity that we are after.

Since we are interested in the results of the computations, not with the computations themselves, we can pass directly to the computed values. The apogee distance in round figures is 403,000 km and the perigee distance is 360,000 km. The *average* of these quantities is the length of the semimajor axis of the ellipse; its length is 381,600 km. From these data we can determine the average eccentricity of the moon's orbit; it is 0.0549.†

The Moon's Sidereal and Synodic Periods.
When the moon is in the new phase, it is in line with, and between, the earth and sun. The moon is said to be in *conjunction* in this position. Similarly when the moon is full it is in line with, but on the opposite side of, the earth from the sun. The moon is said to be at *opposition* in this position. Strictly speaking this alignment is in a vertical plane containing all three bodies. Because of the inclination of the moon's orbit, the moon is usually slightly above or below the actual line connecting the earth and sun.

We now define the interval between two

---

† Perturbations by the sun and the earth cause the moon's orbital eccentricity to vary between 0.04 and 0.06.

The Moon's Motions, Real and Apparent **121**

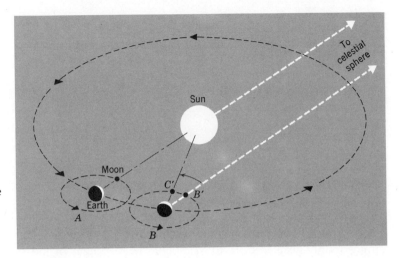

FIGURE 7-7 The difference between the moon's sidereal and synodic periods is just the time required for the moon to move in its orbit from point *B'* to point *C'* (see text).

successive conjunctions of the moon as its *synodic period* (or month). Similarly we define the interval between two successive arrivals of the moon at the same place on the celestial sphere as its *sidereal period* (or month). The synodic period depends on the relative positions of the sun, the moon, and the earth; it is the period of the cycle of the moon's phases, just over 29½ days long. The moon's sidereal period is 27⅓ days. Both periods are here expressed in units of earth-days. Figure 7-7 shows why the synodic period is longer than the sidereal period. At *A* both the sun and the moon as seen from the earth are in the same direction on the celestial sphere. At *B*, which is selected so that the moon as seen from earth occupies the same position on the celestial sphere as it did at *A*, the sun no longer is in the same line. The moon must move from *B'* in the figure to *C'* to effect alignment among all three bodies. It takes 2.2 days longer than the sidereal month to complete the synodic month.

We should keep this new terminology and its meaning conveniently at hand. The terms synodic period, sidereal period, conjunction, and opposition will be useful in subsequent discussions about other members of the solar system.

The Moon's Perturbations. In astronomical usage perturbation means a disturbance of the regular elliptic or other motion of a celestial body, produced by some force in addition to the one which causes its regular motion. If we were to attempt to predict the moon's future positions and motions as well as other astronomical phenomena concerning the moon on the basis of the kinds of data thus far indicated, we would experience embarrassing failure. Lunar periods, orbital eccentricity, times of perigee and apogee passage, inclination of the orbital plane, perigee and apogee distances, indeed almost every quantity shows differences from one set of observational data to another. It is only after an exhausting observational program extending over many lunar cycles that the patterns of the perturbations emerge. The sun is the principal offender in producing the disturbances, although Mars and Venus occasionally make their presence felt. Even though the sun is at a vastly greater distance from the moon than the earth — over 370 times farther — its immense mass makes the sun the dominant gravitational feature of the solar system. The sun's influence on the moon is twice that of the earth's.

Here are the principal perturbations: *evec-*

tion (a systematic change in the eccentricity of the moon's orbit, with a 32-day period); changes in the *inclination* of the moon's orbit; *regression of the nodes; progression of the line of apsides;* and *variation.* The interplay among the gravitational fields of the three bodies is so complex that the relative positions of all three are never exactly repeated. Indeed, the known components of the total lunar motion consist of more than 150 periodic variations along the ecliptic, nearly 150 at right angles to it, plus some 500 smaller variations. The chief interest in the latter variations is in connection with the study of complex motion in general.

The progression of the line of apsides refers to the turning, or rotation, of the moon's orbit around an axis perpendicular to the orbital plane; the period is approximately 8.85 years. The regression of the line of nodes is somewhat analogous to the precession of the earth's equatorial plane; this period is 18.6 years. Figure 7-8 can be of assistance in reaching an understanding of this phenomenon. The moon's orbital plane intersects the plane of the ecliptic in a line called the *line of nodes.* The point of intersection of the moon's path with the plane of the ecliptic when the moon ascends above the plane of the ecliptic is called the *ascending node* and is symbolized ☊ ; the corresponding point of descent below the plane of the ecliptic is the *descending node,* symbolized ☋. Knowledge of the direction of the line of nodes permits us to determine the orientation of the moon's orbital plane with respect to both the sun and earth. That the moon's orbit and the ecliptic do not coincide, and that the line of nodes regresses completely around the celestial sphere, lead to an interesting phenomenon. Depending upon the orientation of the orbit, the moon can be as much as 28½° north and south of the celestial equator, or as little as 18½° north and south. Figure 7-9 illustrates these extremes. Since the sun can never exceed a 23½° displacement north and south of the celestial equator, it is clearly possible for the moon to be seen higher in the sky, or

lower, than the sun ever is at any given location.

Because of the large perturbations in the moon's motion around the earth, we might well approach the problem from the point of view that it is the earth which disturbs the moon's motion around the sun rather than the reverse. Since the earth is in orbit around the sun, *so is the moon,* even though it is also in orbit around the earth. Precise measurements disclose that the moon's path is everywhere concave toward the sun, not "wavy" as we might at first think (Figure 7-10). Since the sun's gravitational pull on the moon is always twice that of the earth, the earth *cannot* deflect the moon away from the sun. The earth can and does periodically increase and decrease the curvature of the moon's path, just as the moon periodically increases and decreases the curvature of the earth's path around the sun. These resulting motions are in strict agreement with Newton's law of universal gravitation. If we recognize that it is the *common center of mass of the earth-moon system* (called the *barycenter*), and not the individual bodies, that follows an elliptical orbital path around the sun, then how it is that the moon and earth both follow a concave path around the sun becomes understandable.

By carefully comparing the sun's apparent angular size at new moon and at full moon we can compute how far the moon pulls the center of the earth away from the orbital path of the barycenter. The earth's center describes an elliptical path around the barycenter; the mean distance of radius of the path is 4730 km. This places the barycenter about 1640 km below the earth's surface! The elliptical path of the earth's center around the barycenter is a reflection of the moon's elliptical path around it, but it is reduced in size in inverse proportion to the respective masses of the two bodies. How this can be true we shall presently see in the discussion of the determination of the moon's mass.

With respect to the earth the moon travels

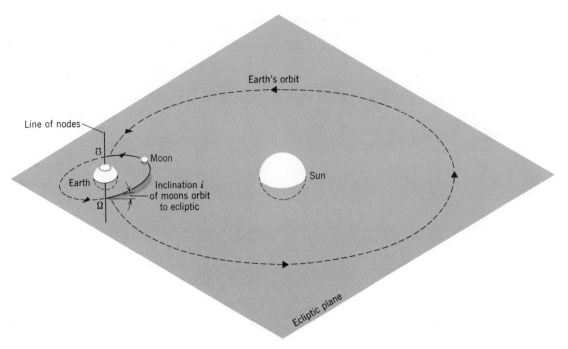

FIGURE 7-8 The *line of nodes* marks the intersection of the moon's orbital plane with the plane of the ecliptic. The moon rises above the ecliptic plane at the *ascending node* and descends below it at the *descending node*.

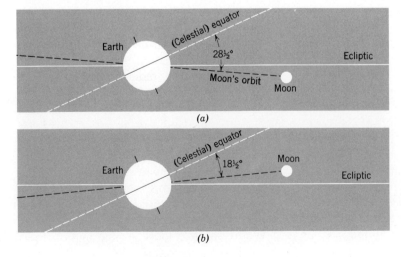

(a)

(b)

FIGURE 7-9 Depending upon the orientation of the moon's orbit as a result of regression, the moon can be as much as 28½ above or below the celestial equator, and as little as 18½ above or below it.

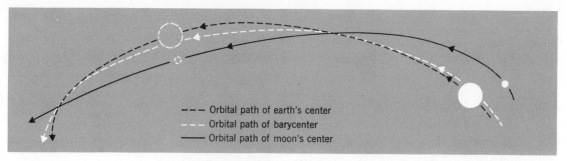

--- Orbital path of earth's center
--- Orbital path of barycenter
— Orbital path of moon's center

FIGURE 7-10 The center of mass of the earth-moon system follows an elliptical orbit in accordance with Kepler's and Newton's laws (see text).

eastward along its orbit approximately 13° on the average, per day. Also with respect to the earth, the sun apparently moves eastward along the ecliptic an average of approximately 1° per day. Thus with respect to the sun, the moon travels eastward 12° per day. This is easily apparent with respect to prominent fixed stars, and on occasion it is more apparent with respect to one or more of the planets. But the phenomenon is much more readily apparent when we measure the time interval between successive passages of the moon across the local meridian. The moon arrives on the local

meridian 50 minutes later on the average each day. The real interval between successive passages varies between 38 and 66 minutes. A number of factors operate to produce the observed variation in transit intervals. Among them are various combinations of the following: variations in the moon's orbital speed, variations in the sun's apparent speed along the ecliptic, and changes in the inclination of the moon's orbit with respect to the ecliptic. Included also are the precessional effects: regression of the nodes and the progression of the lines of apsides.

FIGURE 7-11 In autumn when the sun is near the autumnal equinox the full moon is near the vernal equinox (a). The rising of the moon is delayed only a few minutes each night. Six months later the positions of the sun and the full moon are reversed. At this time the moon's rising is delayed considerably on successive nights (b).

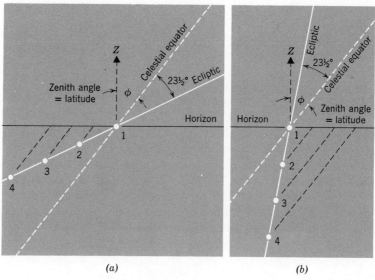

(a)

(b)

We might suppose that the time of moonrise would be as much later each day as is the time of transit across the local meridian. But observations show that such is not the case; the delay in rising time at 40° N latitude can be as little as 13 minutes or as much as 80 minutes. Compare this with the range of 38 to 66 minutes delay in transit times. For any given year the least delay occurs with the full moon nearest the time of the autumnal equinox; the greatest delay occurs with the full moon nearest the time of the vernal equinox. In 1971 these full moons occurred on October 4 and March 12, respectively. Figure 7-11 helps to explain the phenomenon. When the sun is at the autumnal equinox, the full moon is in the opposite direction on the celestial sphere near the vernal equinox. Since all objects rising above the horizon follow arcs of circles parallel to the celestial equator, we can see in Figure 7-11a that there is not much difference in the lengths of the paths the moon follows on successive nights in reaching the horizon when it is at the vernal equinox. Thus the rising times are not very much different. When the sun is at the vernal equinox the full moon is near the position of the autumnal equinox. Figure 7-11b shows that the angle between the ecliptic and the horizon is 47° greater at the autumnal equinox than at the vernal equinox. The moon, which is never very far from the ecliptic, has successively much longer paths to traverse in reaching the horizon when it is at the autumnal equinox. Hence the time of rising is much more delayed at this time. The reader should satisfy himself that at the time of the autumnal equinox the sun is descending into the southern celestial hemisphere, while the full moon is ascending into the northern celestial hemisphere. In the spring the situation is reversed.

At full moon nearest the time of the autumnal equinox there is a maximum number of days with full or nearly full moon in the early evening. The name *harvest moon* is given to this circumstance. In 1971 the delay in rising

FIGURE 7-12   The gibbous moon, age 11 days. The darker *maria* show up in excellent contrast t the lighter colored highlands (Courtesy Hale Observatories).

of the harvest moon was 32 minutes in the Northern Hemisphere at 40° N latitude; it was only 19 minutes in the Southern Hemisphere at 40° S latitude. It should be evident that while the harvest moon occurs near September 23 in the Northern Hemisphere, it occurs near March 21 in the Southern Hemisphere.

The Moon's Rotation.   Casual inspection, even without optical aid, discloses the same, mottled appearing surface of the moon always turned toward the earth. Figure 7-12 helps to verify this fact even though the fraction of this hemisphere facing the earth that is illuminated by the sun varies with the moon's phases. More exactly, as we saw in Chapter 2, for a brief period just after the first slender crescent of the moon becomes visible and just before the last slender crescent disappears, the rest of the moon's hemisphere can be seen very dimly lighted. Just as moonlight from the full moon dimly illuminates the night surface of the earth, so does the full "earthlight" dimly illuminate the night surface of the moon. A brief reexamination of Figure 7-3 will show that when the moon is between the sun and earth, the earth will appear full when seen from the moon.

We might, for an unthinking moment, conclude that the moon does not rotate about an axis as does the earth because of the unchanging appearance of its surface. We can, however, resort to a simple geometric analogy which quickly compels us to refute such a conclusion. We draw any convenient circle, perhaps 15 cm in diameter, and place a coin at its center. Another coin, a Jefferson nickel for example, is centered on the circumference of the circle so that the face on the coin is "looking" at the central coin. Now if we move the nickel along the circle, always keeping the face turned toward the central coin we find that the nickel must be rotated through 360° in making 1 revolution around the central coin.

The moon, then, must turn on its own axis with a rotational period that just matches its period of revolution around the earth. That these two periods are the same length can hardly be accidental. We know tidal action by the moon on the earth is slowing the latter's rotational rate. In all likelihood the earth has adjusted the moon's rotational rate by some similar tidal interaction.

If just a simple rotation of the moon were all that were involved, we would be able to see no more than 50% of the moon's surface. But there are other phenomena resulting from the geometry of the observer's position with respect to the moon and from the geometry of the moon's motions which enable us to see more than a single hemisphere. First, the moon's axis of rotation is inclined about 6½° from the perpendicular to its orbital plane. Second, and less important, is the 5° inclination of

FIGURE 7-13 Libration of the moon in latitude. South is at the top. The libration is easily seen with reference to the crater *Tycho*, top right center (Yerkes Observatory Photograph).

The Moon's Motions, Real and Apparent   127

the orbit to the plane of the ecliptic. These inclinations make the moon appear to first tip one pole toward the earth, and then tip the other pole one-half revolution later. As a result we are able to see somewhat beyond each pole during each lunar period (Figure 7-13). This is one of the three most prominent of the moon's *librations* (literally, a balancing, such as the oscillation of a chemist's beam balance). It is referred to as the *libration in latitude*.

A second type of libration, the *diurnal libration,* results from the geometry of our observational position as the earth's rotation carries us halfway around the earth in each 12 hours. This circumstance enables us to alternately see a little beyond each limb of the moon. A third type of libration, the *libration in longitude,* arises strictly in accordance with Kepler's laws. The moon rotates on its axis at very nearly a uniform rate, while at the same time it is moving with variable speed in its orbit. The net effect is an inequality during a given time interval between the angular displacement of the whole moon in its orbit. As a result we are able to see about 16° in excess of the hemisphere in a lunar east-west direction.

Because of the inequality of the periods of each of these librations, the total effect is at times additive. We can see as much as 59% of the moon's surface although never more than 50% at any one time. For the same reasons 41% of the surface is never visible

Physical Properties

Thus far we have considered the moon as an adversary of the earth (tides and perturbations) and as a member of the sun's captive family (motions). Earlier we utilized a classical technique for determining the moon's distance because we needed this information in the motions studies. Now we shall examine some techniques for determining the moon's size, its mass, its density and surface temperatures, its atmosphere (or lack of it), and its surface grav-

ity. Our methods will be general enough so that they can be employed in the studies of more remote objects. Lunar surface exploration has rapidly evolved from telescopic observation to direct sampling by man; therefore, let us defer any discussion about the physical appearance and structure of the surface to a later section on exploration techniques.

At this point we can reexamine the distance determination method and modify it slightly so that a general formula applicable to more distant bodies can be derived. The moon is a good starting place because its accessibility permits verification of and refinements in our methods.

Distance. Figure 7-14 illustrates the geometry on which the distance and size formulas are based. It resembles Figure 7-6 except that here we avoid reference to the center of the visible disk of the moon. An observer at *A* on the earth measures the apparent angular diameter *α* by taking a series of alternate sightings first on one limb and then on the other and averaging the differences. Simultaneously an observer at *B* measures the angular direction of one limb *C*, an arbitrary choice previously agreed upon between the observers. It is not essential for two observers to join in the effort; one observer can secure all the data. Suppose the observer is initially at *A*. He makes the required determinations there; 12 hours later he makes the measurement at *B* where the earth's rotation has carried him. In either case the distance *b* separating the observation points *A* and *B* is computed from the known latitudes and longitudes and the known dimensions of the earth. In the event one observer makes both sightings, the change in the moon's position in its orbit must be corrected for.

The procedure here differs from the former in that instead of measuring the angular directions with reference to the horizon and zenith distances on earth, we measure the moon's apparent displacement, or parallax, on the celestial sphere. Because the moon is so distant from the earth, the parallactic displacement is

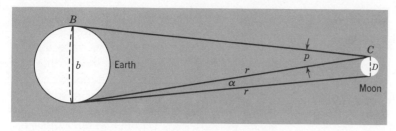

FIGURE 7-14 The distance from the earth to the moon (r in the figure) is calculated from the trigonometric relationship between the parallax angle p and the earth's diameter. Once the value of r is determined, the moon's diameter D can be calculated from the angular size α and r (see text).

less than 1°; for all other astronomical bodies it is even much less than this.

But even for a 1° parallax the arc of a circle subtended by the earth at the earth's distance, and passing through two points such as A and B, will be essentially the same length as the chord (straight line) connecting A and B. This fact permits us to make this simplification: the ratio of the arc AB to the angle $p°$ it subtends is equal to the ratio of the circumference of the whole circle (of which the arc is a part) to the 360° central angle of the circle. We let b represent the length of the arc AB. Then,

$$\frac{b}{p°} = \frac{C}{360°}. \qquad (7.1)$$

The circumference C of any circle is $2\pi r$ where r is the radius. In Figure 7-14 radius r for the arc segment AB is just the distance from A to C. This distance is not essentially different from the center-to-center distance of the earth and moon. Substituting $2\pi r$ for C and rearranging,

$$\frac{2\pi r}{360°} = \frac{b}{p°}. \qquad (7.2)$$

A mathematical equality is not altered if we perform an identical arithmetical operation on all terms of the equality. Therefore, we can divide both sides of the equation by 2π and multiply both sides by 360°. We obtain

$$r = \frac{360° \times b}{2\pi \times p°}. \qquad (7.3)$$

The distance r which is AC in the figure can be determined from the known distance b between A and B and the measured angle $p°$. In practice $360°/2\pi$ can be combined into a single number. Furthermore, practically every parallax is measured in seconds (''), not degrees ($1'' = 1/3600$ of 1°), so we shall incorporate this transformation,

$$r = \frac{360° \times 3600 \text{ sec/deg} \times b}{2\pi \times p°}$$
$$= \frac{206,265 \times b}{p°} \qquad (7.4)$$

where p is in seconds of arc and r will have the units of b. For practical reasons in connection with uniform procedures, observations taken at various latitudes are translated into equivalent units of the mean radius of the earth. The parallax of the moon is then called the *mean equatorial horizontal parallax* when it is used to obtain the moon's mean distance. The mean equatorial horizontal parallax for the moon is 57'3'' or 3423''. The mean equatorial radius of the earth is 6378.16 km. When we insert these values into Equation 7.4 we obtain

$$AC = r = \frac{206,265 \times 6378.16 \ km}{3423}$$

$$= 384,300 \ km.$$

In applying this formula in the determination of the distance to any celestial body, all that is necessary is to select a baseline b sufficiently long that a parallax angle $p°$ can be measured.

<u>Orbital Speed.</u> Now that we know the mean earth-moon distance we can find the mean orbital path length; the orbit is very nearly a circle whose radius is 384,300 km. From $C = 2\pi r = 2\pi \times 384,300$ we find the orbital path is very nearly 2.41 million kilometers long. If we divide this distance by the time it takes the moon to complete one round trip, we find the moon's average orbital speed is approximately 4700 km/hr or 1.1 km/sec. The earth's orbital speed averages 30 km/sec by comparison.

<u>The Moon's Size.</u> We can extend the use of the relationships of Figure 7-14 to compute the diameter of any celestial object *provided* the diameter is sufficiently large so that its angular size can be measured. In the case of the moon, the diameter subtends an arc of a circle, which is again proportional to the whole circumference in the same way that the angluar size is proportional to 360°. We have

$$\frac{D}{2\pi r} = \frac{\alpha}{360°} \qquad (7.5)$$

where D is the diameter of the moon, r is the mean earth-moon distance just determined, and α is the angular diameter of the moon. Again converting degrees to seconds of arc and rearranging the equation we obtain

$$D = \frac{\alpha \times r}{206,265} = 3476 \ km.$$

The ratio of the moon's diameter to the earth's diameter expressed as a percentage is 3476 km/12,742 km $\times$ 100 = 27.3%. Since the square of the radius of a sphere is proportional

to its surface area, and since the radius is 1/2 the diameter, the interested reader should be able to verify that the moon's surface area is $r_\mathbb{C}^2 / r_\oplus^2 \times 100 = 7.4\%$ that of the earth. The subscripts $\mathbb{C}$ and $\oplus$ refer to the moon and the earth, respectively. (The area of a sphere is found from $A = 4\pi r^2$.) The moon's surface area is thus just over 1/13 that of the earth. In a similar manner we can compute the volume of the moon to be only 2% or 1/50 the volume of the earth. We shall make use of this information presently.

<u>Mass and Surface Gravity.</u> Almost everyone is familiar with the old-fashioned schoolyard teeterboard. Most of us remember how we balanced it when the riders had different weights. The heaviest child sat closer to the center of balance (Figure 7-15a). Had the occasion arisen we could have verified that the product of the weight of each rider and his distance from the balance point (fulcrum) were equal.

The same principle prevails in the gravitational balance of two astronomical bodies such as the earth and the moon in revolving around their common center of gravity (center of balance) (Figure 7-15b). These two products are equal: the mass of the moon times the moon's distance from the balance point, and the earth's mass times the earth's distance from the balance point. In both cases the distances refer to the centers of the bodies, or

$$m_\mathbb{C} \times 384,405 \ km = m_\oplus \times 4730 \ km$$

where as before the subscripts $\mathbb{C}$ and $\oplus$ refer to the moon and earth, respectively. If we divide both sides of the equation by the moon's distance, we find that the moon's mass is about 1/81.3 of the earth's mass; in round figures the moon's mass is 1/81 that of the earth.

From the above facts we would expect that an object on the moon would weigh correspondingly less on the moon than it does on earth. It would not be 81 times less. We recall from Newton's law of universal gravitation that the weight of an object is proportional to the

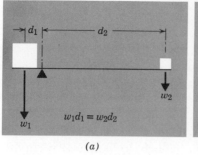
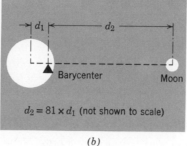

$w_1 d_1 = w_2 d_2$

(a)

$d_2 = 81 \times d_1$ (not shown to scale)

(b)

FIGURE 7-15 Mechanical analogy (a) shows the relative rotations of the earth and moon around their common center of mass (b). The latter is known as the *barycenter*.

mass of the attracting body, and *inversely proportional* to the square of the distance from that body's center of mass. An object on the moon's surface is only 27% as far from the moon's center as a corresponding object on earth would be from the earth's center. We find upon direct comparison with the situation on earth that the gravitational advantage with respect to distance is 13.7 times as great on the moon as it is on the earth. Similarly, the gravitational disadvantage, because of its smaller mass, is 81 times on the moon. The net gravitational disadvantage is 13.7/81 or approximately 1/6. This means that an object on the moon will weigh 1/6 as much as it would on earth, even though the object's mass is the same in both places.

Finally, now that we have the data on both mass and volume of the moon relative to the earth, we can compute the relative density of the lunar material. We recall that the average density of an object is defined as the mass of the object divided by the volume the mass occupies. For the moon

$$\frac{1/81\ m_\oplus}{1/50\ V_\oplus} = \frac{5}{8}\ \rho_\oplus$$

where $m_\oplus$ is the earth's mass, $V_\oplus$ is the earth's volume, and $\rho_\oplus$ is the earth's density. We found that the earth's density averaged 5.5 times that of water; the moon's average density is therefore only $5/8 \times 5.5 = 3.4$ times that of

water. Evidently the moon materially differs from the earth in composition.

Atmosphere and Temperature. Telescopic observation of the moon's limbs fails to disclose evidence of the scattering or diffusion of light that, if present, would produce at least a small twilight zone. Moreover, when the moon's disk passes in front of (*occults*) a star or planet, the light from the distant body is extinguished abruptly. The presence of a lunar atmosphere would cause the extinction to occur gradually as the light from the object penetrated through successively deeper layers of a lunar atmosphere. On this basis we can conclude that the moon has no measurable long-term gaseous envelope. Again, the surface gravity on the moon is so small that at the existing daytime temperatures, all ordinary gases could escape into space. We shall pursue this topic in somewhat more depth in the next chapter in our discussion of Mercury.

Surface water in either liquid or solid form on the moon appears to be an impossibility for the following reasons. Water or ice continually evaporates into the space above it at a rate that depends upon the temperature of the water. Evaporation ceases only when such a space is saturated, that is, it already contains as much water vapor as can physically exist under the given conditions. But on the moon, once the water evaporates it is a gas that can escape the gravitational attraction; thus, any surface water would have long since disappeared into space. This argument does *not* preclude the existence of water in the form of permafrost deep under impervious layers of lunar soil. Water could also be locked up in the crystalline structure of some minerals in a form chemists call *water of crystallization.*

Without an atmosphere to reflect and disperse some of the solar radiation and to trap some of the outgoing radiated heat, the lunar temperature range is much greater than that on earth. During the long lunar day (equivalent to two earth-weeks) the surface absorbs radiant heat energy from the sun at a faster rate than it can radiate the energy back into space. During the lunar night, the heat energy is all going out. Sensitive temperature measuring devices that interpret the rate of heat gain or loss and the wavelength of the energy involved are routinely placed at the focus of large telescopes. The data so obtained show that the sunlit surface of the moon reaches temperatures in excess of 145°C; the temperatures during the lunar night plunge as low as −130°C. The soft-landing Surveyor lunar probes of 1966 and 1967 carried instrumentation that corroborated the earth-based determinations. The more recent Apollo mission landings on the moon gave additional supporting evidence for the preceding conclusions.

Eclipses

In astronomy *eclipse* means the obscuration of light from one celestial body by another. Eclipses are not uncommon phenomena in that part of the universe accessible to our scrutiny. Notwithstanding that eclipses are interesting phenomena in their own right, they serve a more useful and scientific purpose. Out beyond the solar system the duration and frequency of certain eclipses, and the degree of obscuration, become almost the entire source of information concerning otherwise unseen celestial objects. The groundwork we lay here

will be advantageous later on in our investigations.

The most dramatic, and certainly the largest-scale eclipses measured in our local frame of reference involve the earth, moon, and sun. Eclipse terminology is almost entirely in reference to what an earth-bound observer witnesses. When the light from the sun is obscured by the moon, the event is called a *solar eclipse.* When the light to the moon is cut off by the earth we term this event a *lunar eclipse.* Clearly the moon passes between the sun and earth in the first case; in the second case the earth passes between the sun and the moon. Interestingly enough, when we see the shadow of the earth eclipsing the moon, an observer on the moon would see the earth eclipsing the sun.

Eclipses are termed *total* when the eclipsed body is wholly obscured; they are termed *partial* when only a fraction of the body is obscured.

Solar Eclipses.   When an observer sees an eclipse he is necessarily within the shadow cast by the eclipsing body. When the source of illumination is larger than the other celestial objects involved in the eclipse, as is the case with the sun, the geometry of the shadow has the form shown in Figure 7-16. There is a primary cone of shadow called the *umbra* (L., *umbra,* shade) based on the eclipsing body with the apex of the cone pointing away from the sun. An observer within this cone will see the sun totally obscured.

Light is radiated in all directions from every point on the sun. Thus we can visualize two pairs of rays emanating from the extremities of the sun's limbs which just graze the surface of the eclipsing body; in the case of solar eclipses this is the moon. All of the shadow lies within these boundaries. As Figure 7-16 shows there is a truncated conical shadow zone originating at the moon and expanding outward with an indeterminate base. This shadow zone encircles the umbra; it is called the *penumbra* (L., *pene + umbra,* almost shade). An observer within the penumbra will see only part of the solar disk obscured; how much depends upon whether he is near the position of the umbra (near totality) or near the outer limit of the penumbra (near-zero eclipse).

Another zone called the *annulus* (L., *annularis,* ring) extends beyond the apex of the umbra as though it were a mirror image of the latter. An eclipse observed from within this zone is not total because the eclipsing body does not completely cover the sun's disk. As a result an intensely brilliant ring surrounds the obscured central portion of the sun's disk. The name *annular eclipse* is given to this phenomenon.

At the risk of being repetitious we should again note that the term solar eclipse denotes that the sun is obscured by the eclipsing body. For earth-bound observers the moon eclipses the sun. For observers on the moon the earth would eclipse the sun. In Figure 7-17*a* observers positioned in the umbra would see a total eclipse; all others would see a partial eclipse. For observers positioned in the annulus cone as in Figure 7-17*b* the eclipse would be annular. This is the nearest to totality that any observer would see; everywhere else the eclipse, if visible at all, would be partial.

The geometry attending the various possible eclipse phenomena is somewhat complex. Variations in the earth-sun distance, the earth-moon distance, the sun-moon distance, and the inclination of the moon's orbit relative to the ecliptic all separately and jointly contribute to the nature of the observed solar eclipse. The moon's umbra averages 375,800 km in length; this means that on the average it does not reach the earth. Hence total solar eclipses are a less-than-average phenomenon. At the most favorable configuration between the earth and moon the umbra extends 28,350 km past the earth. Under this condition the greatest diameter possible for the umbra shadow on the earth is 275 km. On the other hand, if the umbra strikes the earth at a large angle, the shadow is

an ellipse. The least diameter will have a maximum value of 275 km as before, but the greatest diameter may be considerably larger than this.

What are the most favorable and unfavorable configurations for solar eclipses? The inclination of the moon's orbit plays the leading role in the determination. As Figure 7-18 demonstrates, the moon is either below the sun or above it except when it is very near one of the nodes of its orbit. For solar eclipses both the sun and moon must be within approximately 18° east or west of a given node. Otherwise the apparent disks will not make contact and no eclipse of any kind is possible. For a total eclipse to occur it is clear that both bodies must be essentially *at* the node. Since there are two nodes, and since the sun makes one circuit of the celestial sphere each year, solar eclipses can occur at most twice in any given calendar year except for the following circumstance. Because of the regression of the nodes of the lunar orbit of 20° per year, the sun can arrive back at a given node in as little as 347 days. Thus it is possible for the sun to be at one of the nodes twice and the other node once provided the first occurrence takes place in the first half of January. The odds are against such a sequence. On the other hand, the sun is in the vicinity of each node for about 36 days. Thus two partial eclipses of the sun can occur during this interval. If we sum up all of the possibilities, providing one eclipse occurs early in January, a maximum of five solar eclipses in any 1 year is possible.

A second condition for most-favorable configuration involves the apparent angular diameters of both the sun and moon. On the average they appear to be very nearly the same size, about 31' of arc. The moon's apparent diameter is largest when the moon is at perigee; the sun will appear smallest when the earth is at aphelion. When these conditions are met simultaneously, the moon's disk is large enough to cover the sun for a maximum of 7½ minutes in any one locality.

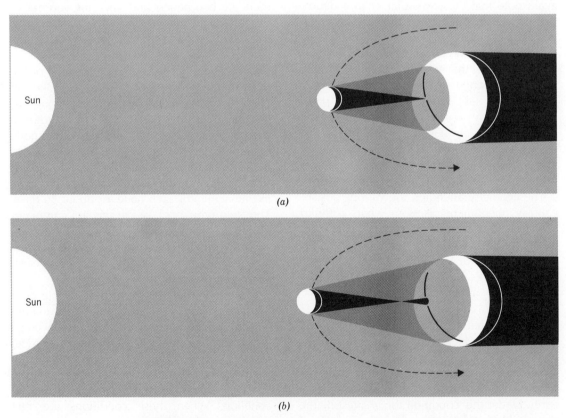

FIGURE 7-16 Geometry of the several shadow cones produced by an illuminating source larger than the object casting the shadows.

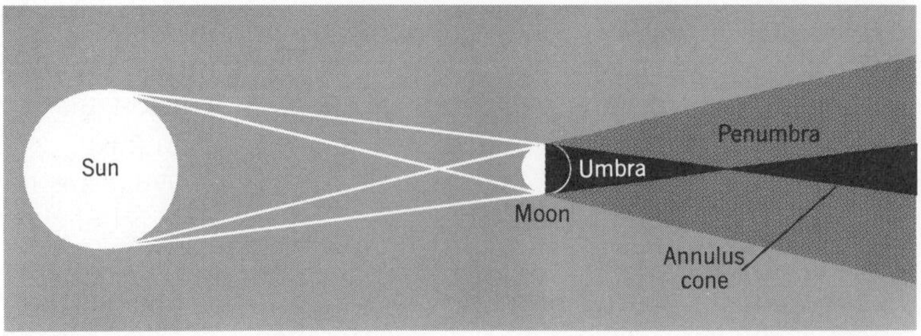

The most unfavorable condition for a solar eclipse finds the moon at apogee and the earth at perihelion. Under this circumstance the sun's apparent disk is larger than the moon's. The moon's umbra fails to reach the earth by over 32,400 km and at best an annular eclipse is all that is visible. Even then the sun and moon must both be within 15° east or west of one of the nodes for eclipse to occur.

Partial eclipses, other than the annular variety, are visible over a much larger area because the penumbra of the moon's shadow is

(a)

(b)

FIGURE 7-17 When the moon's umbra reaches the earth, a path of totality is cast within the penumbra (*a*). When the umbra falls short of the earth the annulus cone creates an annular eclipse within the penumbra.

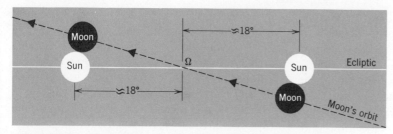

FIGURE 7-18 The ecliptic limit. If the moon crosses the ecliptic when the sun is within the indicated angular distance of the nodal point, an eclipse, partial or total, will occur. Otherwise the moon passes above or below the apparent solar disk and no eclipse is visible.

much more extensive than the umbra. It covers a circle exceeding 8100 km in diameter on the average. Thus, for most observations, partial rather than total eclipses are the rule (Table 7.1).

Lunar Eclipses. The circumstances surrounding lunar eclipses differ in several respects from those of solar eclipses. Eclipses of the moon are usually of much longer duration than solar eclipses. Whereas the edge of

TABLE 7-1   Total Solar Eclipses Through 1999

Date	Where visible	Maximum duration (minutes)
July 10, 1972	Bering Sea, Northwest Territory, North Atlantic Ocean	3
June 30, 1973	Guiana, Central Africa, Indian Ocean	7
June 20, 1974	Southern Indian Ocean	5
Oct. 23, 1976	Congo, Indian Ocean, New Zealand	5
Oct. 12, 1977	North Pacific Ocean, Hawaii, Columbia	3
Feb. 26, 1979	Pacific Northwest	3
Feb. 16, 1980	Central Africa, India	4
July 31, 1981	Caspian Sea, Central Asia, Japan	1
June 11, 1983	Indian Ocean, Java, Coral Sea	5
Nov. 22, 1984	East Indies, South Pacific Ocean	2
Nov. 12, 1985	South Pacific Ocean, Antarctic Ocean	
Oct. 3, 1986	North Atlantic Ocean	1
Mar. 9, 1987	Siberia, Arctic Ocean	3
Mar. 18, 1988	Sumatra, South China Sea, North Pacific Ocean	3
July 22, 1990	Finland, Arctic Ocean, Aleutian Islands	3
July 11, 1991	Marshall Islands, Central Mexico, Brazil	6
June 30, 1992	South Atlantic Ocean	
Nov. 3, 1994	East Indies, Australia, Argentina	4
Oct. 24, 1995	Iran, India, Malaysia, Central Pacific Ocean	3
Feb. 26, 1988	Central Pacific Ocean, Venezuela, Atlantic Ocean	4
Aug. 11, 1999	North Atlantic Ocean, Balkans, India	2

the shadow of the moon cast on the earth is sharp and distinct, the edge of the shadow of the earth cast on the moon is quite diffuse. Moreover the eclipse shadow on the moon's surface is dimly suffused with light ranging in color from dark orange to coppery red. Even when the moon is totally within the earth's umbra this illumination renders the more conspicuous features of the lunar surface plainly distinguishable. Often the moon's immersion into the earth's penumbra is imperceptible. Because of the diffuse character of the umbra's edge, the moon's entry into it is not an abrupt phenomenon. Finally, the instant of the onset of a lunar eclipse is the same for every observer on earth who can at that time see the moon.

Because of the earth's larger diameter its umbra is much longer than the moon's. At the average earth-moon distance the cross section of the diameter of the earth's umbra is 8500 km. If we take into consideration the average speed of the moon in its orbit and the speed of the earth's surface in the same direction as the moon's motion, we find that the moon can remain immersed in the earth's umbra for a maximum of 1 hour and 40 minutes. This is more than 13 times longer than the maximum time for a total solar eclipse.

Illumination of the earth's shadow on the moon is the result of refraction and diffusion of sunlight in the earth's atmosphere. In Figure 7-19 rays of sunlight are depicted as passing close enough to the earth's surface to encounter the earth's atmosphere. Atmospheric density increases along the path. As we saw in Chapter 4 the increase in density refracts the rays toward the earth's surface; those rays which miss contact with the surface pass on by with a relative change in direction as shown. The refracted light partially illuminates the geometrical shadow cone; at the same time it shortens the effective cone of totality. But it is the geometrical cone which the moon intercepts. The reddish hue is caused by dispersion and absorption of the blue fraction of sunlight in the earth's atmosphere; only the relatively red wavelengths get through.

Lunar eclipses are about 25% less frequent than solar eclipses. As illustrated in Figure 7-20 total eclipses of the sun are visible somewhere on earth whenever the moon is within the limits shown at $L_1 - L_1'$. Total eclipses of the moon occur only when the moon is entirely within the earth's umbra as at $L_2 - L_2'$. Since the moon is outside the ecliptic limits $L_2 - L_2'$ for a greater proportion of the time than it is within it, clearly the frequency of total lunar eclipses is less than that for total solar eclipses. A similar argument holds for partial eclipses.

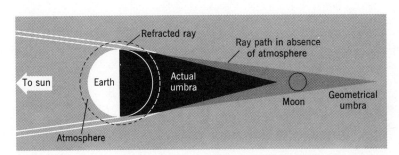

FIGURE 7-19 Refraction of the solar rays in the earth's atmosphere shortens the effective terrestrial umbra such that the moon never experiences a total eclipse. Blue light is refracted most; hence the geometrical umbra is illuminated with red light. This gives the moon its coppery color during totality.

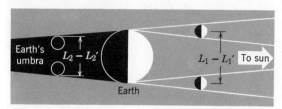

FIGURE 7-20 Eclipse limits. Lunar eclipses occur when the moon is within the limits L_2–L_2'; solar eclipses occur when the moon is within the limits L_1–L_1'. Both situations occur only when the moon is near either of its nodal points (see text).

Total solar eclipses are important for a reason other than those mentioned earlier (p. 132). When the sun's disk is obscured by the moon, its inner and outer atmospheres become visible. Normally the blinding brilliance of the sun's disk prevents observation of the much fainter atmosphere. Thus, conditions are optimal for studying the structure, extent, composition, and radiation characteristics of the sun's atmosphere during total solar eclipses. We shall return to this topic in Chapter 10.

Physical Characteristics and Exploration

Our knowledge concerning the physical characteristics of the moon's surface has advanced concomitantly with improvements in observational techniques. These have progressed from unaided visual inspection, to telescopic examination, to vehicles which at first crash landed and then subsequently soft landed on the moon's surface. The climax came in the form of visual examination following man's first footsteps on the lunar soil. Let us trace this progress and see both the limitations and the advantages in each giant step.

With the unaided eye only the gross features of the lunar surface are visible. The moon looks mottled by what appear to be some 30 large, smooth-textured, roughly circular areas, much darker than the rest of the surface, and

covering about one-half the moon's disk (Figure 7-21). Some fanciful viewer in the dim past thought the arrangement of the dark areas resembled facial features; the "man in the moon" has been with us ever since.

The first *systematic* study of the moon's surface was made by a handful of astronomers beginning with Galileo. These observers used low power telescopes because nothing better was available. To their eyes the lighter regions of the moon appeared extraordinarily rough, while the dark areas still seemed relatively smooth. Galileo assigned the general term *maria* (L., *mare*, sea) to the darker areas because of their resemblance to bodies of water (Figure 7-22). Very imaginative names for specific regions of the maria have come down to us from this era of exploration. Vivid examples are found in such names as *Mare Tranquilitatis* (Sea of Tranquility), *Mare Fecunditatis* (Sea of Fertility), *Mare Serenitatis* (Sea of Serenity), and *Mare Crisium* (Sea of Crises).

Continued improvement in telescopic instrumentation and in exploration techniques soon led to the discovery of more than ten separate mountain ranges or systems. A few mountain peaks stand in splendid isolation in the lunar maria. John Hevel of Danzig (1611–1687) named many of the ranges and separate mountains in his treatise on the moon entitled *Selenographia*. For the most part these features of the lunar landscape are named in honor of their earthly counterparts: the Pyrenees, the Alps, the Apennines, the Cordillera, and the Carpathians (Figure 7-23).

The most numerous and to many the most dramatic of the moon's surface features are the craters. They range in size from no seeming lower limit, up to nearly 240 km across (Clavius and Grimaldi). Improvements in resolution afforded by larger and larger telescopes disclose increasing numbers of smaller and smaller craters (Figure 7-24). Most of the 30,000 or so craters visible from earth are circular areas ringed with mountain walls. Some of these are clearly of advanced age;

FIGURE 7-21   The full moon, age 14 days (Courtesy Lick Observatory).

younger craters have been formed on top of the older features: on the mountain walls and in the crater floors. Relatively few craters are seen in the maria. Several of these have the appearance of being "drowned" in that only the extreme tops of the mountain walls are exposed. This suggests that the maria are of more recent vintage than the crater epoch; smaller craters may have been completely obliterated, and the larger ones appear to be filled with some sort of effluvium (Figure 7-25). That craters are a permanent feature in the maria almost certainly disproves the ocean hypotheses. Other evidence which we shall examine presently is even more convincing. The maria are better termed as plains.

Some of the craters have floors that are well above the level of the exterior terrain, although most of them are lower. In many the mountain

ring surrounding the crater floor does not rise abruptly from the exterior surface; the whole terrain slopes gradually upward to the mountain ring summit. The interior slopes are almost always quite steep (Figure 7-26). Besides having typical smaller craters in their floors, some of the larger craters have several mountain peaks of substantial height.

The measurement of the heights of the lunar features is not particularly difficult. The angular elevation of the sun above the horizon at any given time for any given point on the moon is easily calculated. The length of the shadow cast by the feature in question is scaled off with respect to the known dimensions of the moon. Then from elementary trigonometry the height of the feature can be determined (Figure 7-27). In this way some of the mountain peaks are determined to be over 6100 m high. Some of the

FIGURE 7-22 The *Mare Serenitatis* (the Sea of Serenity) (Courtesy Lick Observatory).

central peaks in the craters rise nearly 2150 m above the crater floor; many of the walls approach 6000 m in height above the surrounding terrain.

Crater diameters are scaled from the known lunar dimensions. Over 150 craters exceed 81 km in diameter. An interesting phenomenon awaits the first visitor who stands in the center of the floor of a crater such as Clavius, 240 km in diameter. Because of the sharp curvature of the moon's surface (Figure 7-28), the observer would not be able to see any of the crater's wall; it would be entirely below the observer's horizon. This makes the term *walled plain* very appropriate when it is applied to these large craters.

Before speculating on the origin of the craters, let us examine some of the other surface features which are accessible to telescopic investigation. These include the great *Alpine Valley*, a deep, almost straight gorge cutting through the Alps toward Mare Imbrium; numerous clefts, or crevasses, or fault depressions, collectively called *rilles*; straight, abrupt fault scarps; and last but not least, the *rays*.

The exact mechanisms responsible for the formation of the Alpine Valley or the rilles cannot be determined from telescopic observations, although tentative hypotheses about the processes can be made. The Alpine Valley (Figure 7-29) is so straight, and it slashes across the ridge of the Alps so deeply, that the depression appears to be artificial. Best estimates from telescopic observation place its length at 185 km and its width up to 21 km. One supposition which was dispelled by the lunar orbiting satellites was that a huge meteoric fragment, or perhaps a small planetoid, crashed into the

FIGURE 7-23   The *Apennines* mountain range.
Prominent are the craters *Archimedes* (top center)
and *Eratosthenes* (extreme lower left) (Courtesy
Lick Observatory).

FIGURE 7-24   The crater *Clavius* (Courtesy Hal
Observatories).

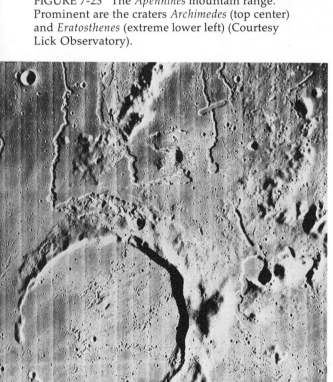

FIGURE 7-25   The crater *Prinz*, a drowned crater,
as photographed by Lunar Orbiter V (Courtesy
NASA).

FIGURE 7-26   The crater *Tycho*, photographed b
Lunar Orbiter V (Courtesy NASA).

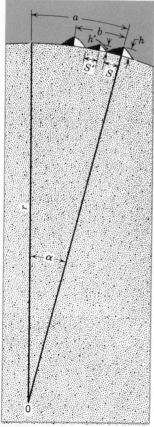

FIGURE 7-27 The crater *Theophilus*, with a diagram illustrating the method of measuring mountain heights (see text).

lunar surface at a low angle and smashed its way through all obstacles, stopping only when its energy was exhausted.

Rilles are for the most part narrow, shallow depressions that are occasionally straight, sometimes arcuate, and most frequently sinuous (Figure 7-30). The rilles, crevasses, and scarps all stand out best when the sun is near the lunar horizon and shadows are long. Thus they must be relief features. Based on evidence derived from the angle of illumination, some of the rilles have gently sloping sides; others are very steep, resembling cliffs. Considering telescopic evidence alone, the rilles bear striking resemblances to watercourses, or dry river beds. Careful examination, however, shows that some rilles change direction abruptly. Others wind around the base of topographic obstacles, notably at the edges of the maria. Others are offset as though the surface is displaced by what geologists term a strike-slip fault. Still other rilles extend up crater walls, down and across the crater floor, up and over the central peak, and up and out the other side of the crater. All of these variations almost surely rule out water erosion; they all suggest that the rilles are fault-zone features.

The rays and the maria stand out in sharp contrast for an entirely different reason. The maria appear darkest and the rays appear lightest at or near full moon. At this time the sun's rays are from directly *behind* the earth and the observer; therefore, the surface of the moon with respect to the observer is illuminated vertically and there are no shadows to accentuate relief features. We can therefore safely presume that the maria and rays are principally coloration phenomena, although the color in itself yields no clues as to how either was formed.

Lunar rays comprise an impressive array of bright (under full moon), light-colored material that diverges from a number of the more prominent craters like splashes of flour or chalk (Figure 7-31). Many of them extend for remarkable distances over the moon's surface. The rays as-

Physical Characteristics and Exploration   **141**

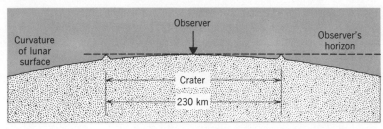

FIGURE 7-28   The curvature of the moon's surface is so great
that an observer standing on the floor of a crater such as *Clavius*
(Figure 7-24) would be unable to see the mountains of the rim;
they would be below his horizon.

sociated with Tycho extend more than 2100
km. Whether or not the craters were of impact
origin or were the result of volcanic activity,
the rays certainly appear in the telescope to be
some sort of ejected material.

Definite fault scarps, of which the *Straight
Wall* is the most prominent example, can be
plainly seen with the aid of a telescope (Figure
7-32). The Straight Wall is a cliff over 370 m
high and more than 120 km long. The face of

the cliff slopes upward toward the east at an
angle of 41°.

Origin of the Lunar Surface Features

Exploration of the moon made a giant leap
forward with the Ranger series of lunar impact
vehicles and the subsequent Surveyor soft-
landing instrument vehicles. As Rangers 7, 8,
and 9 sped toward self-destruction on the
moon with impact speeds of nearly 9700
km/hr, they sent back a total of over 17,000
truly magnificent television pictures. For the
first time man learned that craters and cra-
terlets as small as a few meters in diameter existed; also exposed to his view were rocks and
boulders as small as 1.5 m across (Figures 7-33
and 7-34).

The Surveyor explorations permitted transmission of televised pictures from the scene of
activity. For the first time a supposedly smooth
mare (*Oceanus Proscellarum*) was seen to resemble a rock-strewn terrestrial desert, a gently
rolling surface pocked with craters from a cm
across up to 455 m (Figures 7-35 through
7-37).

The Russian Luna vehicles and our own
lunar orbiters provided high resolution pictures
of the back side of the moon never before seen
by man. Surprisingly, this portion of the
moon's surface is markedly free of large maria.

FIGURE 7-29   Telescopic view from earth
of the *Alpine Valley*, extreme left center
(Courtesy Lick Observatory).

The surface is an extremely rough, cratered highland area (Figure 7-38). The lunar orbiters also provided close-up views from much more favorable angles than is possible by telescope of otherwise familiar topographic features (Figures 7-39 and 7-40). It became clear, for example, that the Alpine Valley was indeed a fault-zone depression, not an ejecta nor impact scar (Figure 7-41).

The Apollo missions which first placed man in orbit around the moon and then permitted him to finally set foot on our sister planet are still in progress, although the lunar-landing phase is completed. This spectacular scientific and engineering accomplishment has opened up new vistas for space exploration. Scientists have at last been able to subject material, known for certain to originate on another celestial body, to laboratory analysis. They have studied directly heat flow from below the moon's surface by means of instrumentation

FIGURE 7-30 A portion of the *Sirsalis Rille* on the extreme western portion of the visible lunar disk. The rille is clearly younger than the crater *De Vico A* (center), but older than the smaller crater *De Vico AA* in the floor of *De Vico A*, since it indents the rille (Courtesy NASA).

FIGURE 7-31 South central portion of the full moon showing the ray system associated with the crater *Tycho* (Courtesy Lick Observatory).

Origin of the Lunar Surface Features   143

(a)

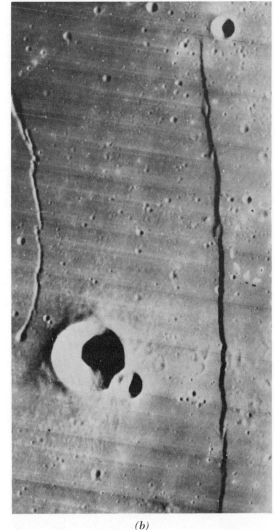

(b)

FIGURE 7-32 The Straight Wall (*white arrow*), a fault scarp in the southeast portion of *Mare Nubium* (Sea of Clouds). View (*a*) is from earth, view (*b*) is from a lunar orbiter (Courtesy Hale Observatories and NASA).

actually in place on the moon; they have examined seismic vibrations in the lunar crust triggered by natural and man-made phenomena.

It is too early to draw conclusions about the moon's origin and structure with any degree of finality. After all these are questions which, when applied to the earth, are still puzzling. But this much is tentatively established: lunar rock samples show similarities in chemical composition to their terrestrial counterparts.

There are, however, enough differences in crystal structure as to strongly suggest that the moon had a similar but not identical origin to that of the earth. That is to say, the moon appears *not* to have been formed from the earth.

As for the craters, some appear to have been formed by the impact of large meteorites or planetoids and smaller missiles all the way down to the size of a grain of wheat. Other craters appear to be of volcanic origin; many of the larger craters resemble terrestrial *cal-*

FIGURE 7-33 Ranger 9 photograph 2 minutes 50 seconds before it crashed to destruction in the crater *Alphonsus* (center) (Courtesy NASA).

deras, volcanoes whose central portion has collapsed into the void left by ejected material. High resolution photographs taken by the lunar orbiters of the floors of the larger craters disclose a texture strongly suggesting lava flow (Figure 7-42). The maria are probably part lava or similar igneous extrusives and part volcanic dust and ash.

Details of the structure of some of the rilles include strings of small craters along the course of the rille. These lend great support to the hypothesis that the rilles are fault structures associated with vulcanism (Figure 7-43). Mare Imbrium (Sea of Rains) in many respects appears to be an impact zone and the adjacent mountain ranges, notably the Alps, conceivably are ejecta from some tremendous collision impact. Yet prominent lunar geologists point to some features which indicate that the mountain structures were developed *after* the floor of the mare was laid down. The answer to this problem still eludes us.

The Taurus-Littrow area, site of the Apollo 17 mission touchdown, last of the current series of lunar landings, revealed few geological surprises. Among them, however, is evi-

FIGURE 7-34 Ranger 9 photographs a few seconds before destruct. The small white circles denote the impact area. Resolution in the last frame, incomplete because of destruction on impact, is about 0.5 m (Courtesy NASA).

FIGURE 7-35 The *Oceanus Proscellarum* area as seen by Lunar Orbiter III from an altitude of 52 km. The distance along the horizon is about 282 km (Courtesy NASA).

FIGURE 7-36 The western edge of *Oceanus Proscellarum* (center) as photographed by Lunar Orbiter III. The large crater at right center is *Damoiseau*, about 65 km in diameter. View is looking south (Courtesy NASA).

dence that the valley where the lunar lander, *Challenger,* came to rest, was probably caused by faulting at the time of the impact-formation of the adjacent Mare Serenitatis. The valley is surrounded by mountains that appear to be composed of both uplifted crustal material and

ejecta from the impact. Subsequently the valley was filled with lava flows.

On top of the lava formation are two distinct layers of material resulting from more recent mantling. One is a light-colored material, thought to be debris from landslides from the

FIGURE 7-37 The *Marius Hills* near the center of *Oceanus Proscellarum,* photographed by Lunar Orbiter III. View is toward the east, with the crater *Marius* at the extreme top right center. The crater is unrelated to the volcanic field of the mare; ridges and domes in the volcanic material are the significant features (Courtesy NASA).

FIGURE 7-38 The far side of the moon as photographed by Lunar Orbiter III. The prominent feature is the crater *Tsialkovsky*, discovered by Russia's Luna III (Courtesy NASA).

south massif. (A massif is a large mountain mass.) The other is a darker material thought to be *pyroclastics*, debris originating from violent volcanic action.

The discovery by geologist-astronaut Harrison Schmitt of orange-colored soil, at one place forming radial stripes surrounding crater Short, almost certainly verifies that vulcanism contributed to the final form of the lunar landscape (Figure 7-46). Such orange-reddish soil is characteristic of volcanic *fumaroles*. Fumaroles on earth are deposits from the last gasps of dying vulcanism wherein outpourings of gases and water vapor oxidize the surrounding soil to the characteristic reddish color.

That the uppermost soil was not caused by spallation (chipping or crumbling) from rocks and boulders such as that shown in Figure 7-47 was deduced by geologist Schmitt from differences in color, as well as in grain size and structure, evidence of different chemical structure.

Perhaps one of the most significant results of the first direct examination of lunar material came from radio-active dating of the samples. One esoteric rock has a probable age of 4.6 billion years. An igneous sample from the Sea

FIGURE 7-39 The crater *Theophilus* as seen from Lunar Orbiter III looking south. Altitude at which the photo was taken was 55 km. A much older crater, *Cyrillus*, is partially visible at extreme right center (Courtesy NASA).

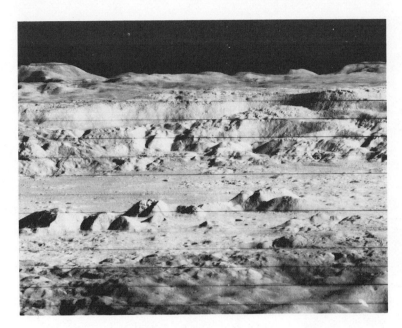

FIGURE 7-40 Close-up photograph of the floor of crater *Copernicus* as seen by Lunar Orbiter II. The mountains rising from the floor of the crater are about 600 m high (Courtesy NASA).

of Tranquility is dated at 3.7 billion years old. This is almost conclusive evidence that the basaltic material, in some of the maria at least, was extruded and crystallized at least 1 billion years *after* the moon was consolidated. (Basalt is a hard, dense, dark form of volcanic rock.) Corresponding age records on earth have been obliterated by wind and water erosion, mechanisms which definitely have *not* weathered the rocks on the moon. Perhaps the key to our further understanding of the evolution of the solar system will be found in future studies of the lunar surface. It is an intriguing possibility.

SUMMARY

The origin of the moon remains a matter of considerable conjecture. Among the several possibilities are: physical separation or division from the earth; separate and simultaneous accretion along with and in the immediate vicinity of the earth from the nebular dust and debris that spawned the entire solar system;

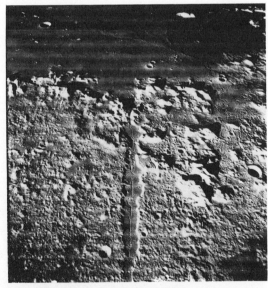

FIGURE 7-41 *Alpine Valley* as photographed by Lunar Orbiter V (compare with Figure 7-29) (Courtesy NASA).

FIGURE 7-42 Portion of the crater *Tycho* taken by the high resolution camera aboard Lunar Orbiter V. Compare with Figure 7-26 (Courtesy NASA).

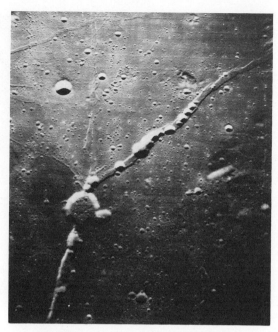

FIGURE 7-43 Crater chains along the *Huygens Rille.* Crater Huygens is at the junction of the two segments of the rille (Courtesy NASA).

FIGURE 7-44 Apollo 15 astronaut David Scott and the Lunar Rover on the edge of *Hadley Rille,* photographed by Lunar Module pilot James Irwin (Courtesy NASA).

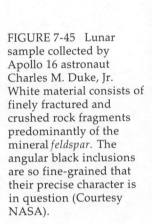

FIGURE 7-45 Lunar sample collected by Apollo 16 astronaut Charles M. Duke, Jr. White material consists of finely fractured and crushed rock fragments predominantly of the mineral *feldspar*. The angular black inclusions are so fine-grained that their precise character is in question (Courtesy NASA).

and a separate formation elsewhere in the solar system with ultimate capture by the earth. On purely logical grounds, the latter of these choices appears slightly more plausible than the others.

The changing phases of the visible surface of the moon result from the relative geometric configuration of the sun-earth-moon positions. Even though the moon revolves around the earth, in a strict sense the same can be said of the earth with respect to the moon. Neither orbit is convex toward the sun because the sun exerts the dominant gravitational force. Neither planet can recede from the sun as a result of the gravitational influence of the other body. As seen from earth, the moon's synodic period, the period of the moon's phases, is 29½ terrestrial days; her period relative to the stars is 27⅓ earth-days. The difference is a result of the earth's advancing motion along its orbital path during the lunar period.

Because of variations of the moon's speed in its elliptical orbit, of the regularity of its rotational rate, of the changing vantage point of an earth-bound observer resulting from the earth's

rotation, and of the precession of the moon's orbit, 59% of the lunar surface can be seen from earth at one time or another.

The moon is near enough to the earth that both its distance and its diameter can be determined quite accurately by elementary geometry and trigonometry. More important, the methods thus invented are readily applicable to distance determinations for very remote objects such as nearby stars, provided their parallactic displacement against background stars can be measured.

The mass of the moon is estimated from the magnitude of the displacement it imposes on the earth's geometric center relative to the orbital path around the sun. It is the center of the earth-moon system that obeys Kepler's laws; Newton's universal law of gravitation is applied to the observed motions to yield a value for the moon's mass of 1/81 that of the earth. The moon's volume is 1/50 that of the earth; its density is 5/8 the earth's mean density. The acceleration of gravity on the lunar surface is about 1/6 that of the earth.

Absence of appreciable atmosphere is at-

FIGURE 7-46   Apollo 17 scientist-astronaut Harrison Schmitt working near the Lunar Roving Vehicle at the *Taurus-Littrow* landing site. The orange-colored soil, discovered by Schmitt, a trained geologist, is clearly visible on either side of the rover. See text for discussion (Courtesy NASA).

FIGURE 7-47   Apollo 17 astronaut Schmitt shown standing near a huge split boulder at the *Taurus-Littrow* site. Geologist Schmitt is the only scientist among all of the Apollo mission astronauts (Courtesy NASA).

tested to by the lack of a visible twilight zone and the abrupt extinction of the light from occulted stars or planets. The presence of surface water is highly doubtful since it would evaporate freely in the absence of other gases and be lost to space. The surface gravity of the moon is so low that the planet could not retain an atmosphere under the existing temperatures.

Direct measurements by sensing instruments at the focus of large earth-based telescopes provide data on the rate of radiation heat losses and the wavelengths at which the radiation takes place. Theory then provides estimates of the temperature range on the moon from a midday +200°F to a lunar night time temperature of −173°F.

Eclipses of the moon and sun are interesting and predictable; again, they are purely geometric phenomena. Eclipses of the sun's disk by the moon provided the first means of examining the solar atmosphere which is nor-

mally invisible because of the sun's overwhelming brilliance.

Prominent features of the surface of the moon are the relatively smooth maria, the rugged mountainous highland areas, and the craters. Direct examination of the surface by the lunar orbiters and by laboratory examination of rocks brought back to earth by the Apollo astronauts show evidence that the maria, in part at least, resulted from lava flow or volcanic ash deposit during the succeeding billion years following the moon's consolidation. On the other hand some of the upland rocks and surface configurations suggest that both faulting and impace ejection were responsible. Thus the craters show dual characteristics of vulcanism on the one hand and of impact scars on the other. Most likely both kinds of events have occurred. Studies of these and other features, especially of the rilles, are being pursued intensely during the 1970s.

Origin of the Lunar Surface Features   **151**

Questions for Review

1. What are the principal theories concerning the origin of the moon?

2. As seen from the moon, does the earth go through phases similar to the moon's phases?

3. What was Ptolemy's method for determining the earth-moon distance?

4. Answer the following *true* or *false,* and explain your choice: If the moon circled the earth in the opposite direction but all else remained unchanged, (a) The moon would rise 50 minutes (on the average) earlier each day. (b) The moon would rise in the east and set in the west. (c) The moon would rise in the west and set in the east. (d) Tides on earth would be larger on the average than at present. (e) Eclipses of the sun would be more frequent than at present.

5. If an observer landed at precisely the center of the visible disk of the moon: (a) Would he always see the earth overhead? (b) Would the observer experience perpetual daylight? (c) If your answer to (b) is "no," then how many earth-days will elapse on the average between sunrise and sunset on the moon?

6. Answer the following either *true* or *false* and explain your choice: Neglecting atmospheric effects on earth, (a) A person on the moon could throw a baseball at least twice as fast as he could throw it on earth. (b) A person on the moon could throw a baseball at least twice as far as he could on earth. (c) An automobile on the moon traveling at the same speed and on the same kind of dry highway as on earth would skid at least twice as far in an emergency full-braking stop as it would on earth. (d) Providing the explosion was violent enough, the sound of a nuclear explosion on the moon could in theory be heard on earth.

7. Why hasn't the moon a detectable atmosphere?

8. Show the relative positions of the sun, moon, and earth for a total eclipse of the sun as seen from the earth.

9. What is the explanation for the occasional annular eclipses of the sun as seen from earth?

10. Why are lunar eclipses less frequent than solar eclipses?

Prologue

At this very moment man is marshalling his resources for the search for a suspected tenth planet far out beyond the orbit of Pluto which, until now, has been the absolute prince of darkness in the solar system. As recently as 1780, however, astronomers were unaware of the existence of the outer three of the solar system planets. Little Pluto escaped detection until 1930, and even then its discovery resulted from an incredible reliance on questionable, even doubtful data. The discovery of Pluto stands as one of the most remarkable coincidences of all time.

No other star has as large a known family of planets as the sun, although it is true that at least one star is now known to possess two attendant companions that probably qualify as planets. What are the odds favoring planetary systems as commonplace celestial phenomena? An even more significant question is the one of extraterrestrial life: Do we have living competition elsewhere in the universe? Does the possibility of the existence of life beyond the earth pose serious threats to our collective ego?

We cannot answer these questions yet. But study of the composition and physical behavior of the solar system is surely a logical step toward understanding the structure and disposition of matter elsewhere. Perhaps the enigmas of our own solar system family will help us reach this larger understanding. For example, why are the solar system planets sharply divided into two general classes according to size? Why are all of the planets traveling in the same direction around the sun? Especially, why do two planets spin in a direction opposite to their immediate neighbors? Finally, what similarities and what significant differences between the planets can we discover and turn to our advantage in understanding the larger scheme of creation? This chapter attempts to show how near we are to many of the answers, and also, how incredibly far we have to go for the others.

The Planets

8

The word planet (Gr., *planetes*, wandering) comes to us from ancient astronomy to denote moving celestial bodies as opposed to the so-called fixed stars. More particularly the reference was to those bodies that moved in regular paths around the celestial sphere: five starlike objects and the sun and the moon. Since to them the earth was a stationary space platform about which the universe rotated, the early astronomers included the sun and the moon in the family of planets. The five other objects were named Mercury, Venus, Mars, Jupiter, and Saturn. No other objects that qualified for the term planet were known to the ancients.

Modern astronomy, of course, has removed the sun and the moon from this category.† Added are the earth and the more recent planetary discoveries. In order outward from the sun are the *inferior* planets, Mercury and Venus. Inferior means, in the astronomical sense, closer to the sun than the earth. Beyond the earth are the *superior* planets: Mars, Jupiter, Saturn, Uranus, Neptune, and Pluto. Of these, Jupiter, Saturn, Uranus, and Neptune are often referred to as the major planets because of their great bulk. All of the others are known as the *terrestrial* planets because they more nearly approximate the earth in size.

Since their periods can be accurately determined, Kepler's harmonic law $P^2 \propto A^3$ enables

† The moon is still thought of as the earth's companion planet; together they form a double planet (binary) system. What is meant here is reference to the moon as an independent planet in the same category as the other planets

us to determine the planetary distances in terms of the mean earth-sun distance with reasonable accuracy. The mean earth-sun distance thus becomes useful as an astronomical yardstick; the assigned name, *astronomical unit (AU)*, attests to this utility.

The progression of the mean planetary distances from the sun in astronomical units is 0.39, 0.72, 1.0, 1.52, 5.2, and 9.52 for Mercury, Venus, Earth, Mars, Jupiter, and Saturn, respectively. The progression is broken only by the absence of a planet at 2.8 AU.

A rather extraordinary coincidence concerning this progression was first recognized by Titius of Wittenberg in 1766. It is classed as a coincidence because no theoretical explanation for the curiosity has yet been accepted. Publicity was given to Titius' discovery by W. E. Bode, Director of the Berlin Observatory, in 1772. It has since become known as Bode's law. Let us rediscover the law.

We first form a geometric progression — 0.0, 0.3, 0.6, 1.2, 2.4, 4.8, 9.6, 19.2, 38.4, 76.8 — where each term after the second is obtained by doubling the preceding term. Now if we add 0.4 to each term as in Table 8.1 we obtain a sequence that is remarkably close to the planetary distances (in astronomical units) known to antiquity as far out as Saturn, with one exception: a hole between Mars and Jupiter. The absence of a planet at 2.8 AU provided impetus for the search for the missing body. This was especially true after the discovery in 1781 of Uranus (q.v.) at almost exactly the distance predicted by the sequence. The circumstances surrounding the search are in part recounted in Chapter 9.

The strongest blow delivered against the acceptance of the Bode-Titius scheme as a statement of scientific truth was the discovery

TABLE 8.1 Bode's Law, the Bode-Titius Relationship.

	Planets known to the ancients							Post-antiquity planets		
	Mercury	Venus	Earth	Mars	?	Jupiter	Saturn	Uranus	Neptune	Pluto
	0.0 +0.4	0.3 +0.4	0.6 +0.4	1.2 +0.4	2.4 +0.4	4.8 +0.4	9.6 +0.4	19.2 +0.4	38.4 +0.4	76.8 +0.4
Bode's law distances in astronomical units	0.4	0.7	1.0	1.6	(2.8)	5.2	10.0	19.6	38.8	77.2
Actual average distances in astronomical units	0.39	0.72	1.0	1.52		5.2	9.52	19.2	30.0	39.4

in 1846 of Neptune (q.v.) at a mean distance of 30 AU from the sun, not the 38.8 AU of the Bode sequence. Its rejection as a scientific law was complete when Pluto was discovered at 39.4 AU, not 77.2 as Bode's law would have it. In fact, Pluto is approximately where Neptune should be in the Bode scheme; Neptune is almost precisely half way between the mean distances of Uranus and Pluto. But a planet at 30 AU is a major break in the sequence. Bode's law, to be valid, would have to hold for *all* observed planetary distances.

On the other hand, the *theory of dynamical relaxation* suggests that perhaps Bode's law does have elements of scientific validity. According to the theory, the original orbits of the planets of the solar system were very different from what they are now. In the same way that Jupiter perturbs the orbits of some of the lesser bodies in the solar system (Chapter 9), so did the various planets mutually perturb each other's orbits in a continuing process that ended in the essentially stable configuration we observe today.

Many astronomers and mathematicians have explored the problem. The best explanation so far for the planetary spacings in accordance with the Bode-Titius scheme resulted from a theoretical experiment performed by J. G. Hills at the University of Michigan in 1969. Hills devised 11 imaginary solar systems, with different sets of planetary masses and with elliptical orbits selected at random. He then employed a computer in calculating how the orbits would change as a consequence of the mutual perturbations, and how long it would take each system to achieve a stable configuration.

All 11 of the systems stabilized well within the period of time representing the age of our solar system. Moreover, all 11 imaginary systems obeyed mathematical expressions similar to Bode's law! The significance of this work still remains to be evaluated.

Planetary Configurations

An observer high above the plane of the ecliptic would map the relative space configuration of representative planets, the sun, and the earth as shown in Figure 8-1. Only two planets are shown: an inferior planet whose orbit is inside the earth's orbit, and a superior planet whose orbit is outside of the earth's orbit. An inferior planet is said to be in *conjunction* when it is seen from the earth in the direction of the sun. In this event both planet and sun have the same right ascension. Two possible situations exist. When the planet is between the earth and sun we say it is at *inferior conjunction*. When the sun is between the planet and the earth the planet in question is said to be at *superior conjunction*.

The line of sight from the earth which just grazes (is *tangent* to) the orbit of an inferior planet marks the maximum distance east or west of the sun that the planet can be seen. The tangent point on the orbit east of the sun marks the *maximum eastern elongation* for the planet.† A similar westward configuration is termed the planet's *maximum western elongation*. We shall establish units for expressing elongation presently. Clearly, an outer planet can never be at inferior conjunction although it can be and periodically is at superior conjunction. But there is a planetary position in the outer orbit which corresponds to inferior conjunction. It is on the same relative earth-sun line but is on the opposite side of the earth from the sun. Logically, this position is termed *opposition*. Just as logically, we observe that an inner planet can never be at opposition.

Outer planets have no maximum elongation in the sense that an inner planet does. Instead we frequently make use of the two points in an

† Eastward is defined as the forward direction of revolution about the sun.

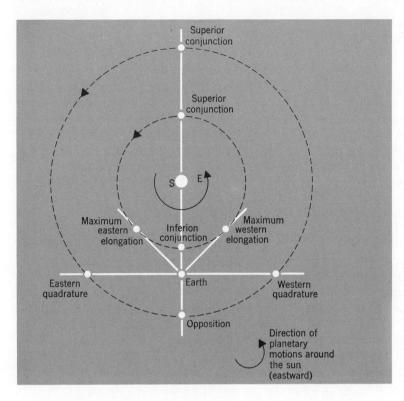

FIGURE 8-1
Configurations of representative inferior and superior planets relative to the earth (in plan view).

outer orbit which mark positions at 90° from the sun's direction. These points are called *eastern quadrature* and *western quadrature,* respectively.

A coordinate system we have not yet considered is of special utility in planetary-position astronomy. This system employs the plan of the ecliptic as the horizontal reference, and of two ecliptic poles as the extremities of a north-south reference. As usual, the line connecting the two ecliptic poles is perpendicular to the ecliptic. The vernal equinox is the zero point in the horizontal direction. A planet's position is then designated in *ecliptic longitude* and *ecliptic latitude*. The units are angular degrees. This coordinate system is analogous to, but of course not coincident with longitude and latitude on earth.

Orbital Elements

It is one thing to reject the notion that the planets orbit the earth and to accept instead that they orbit the sun. It is an entirely different matter to explain why we see them where they are, and to predict where we may expect to see them in the future. Success in this matter requires knowledge of the precise positions of the planets, including our terrestrial viewing platform, with respect to the sun. The planets, of course, occupy different positions in their orbits at different times. Each planet has a different period for its cyclic motion. Also, the planes of other planetary orbits are not quite in the same plane as the ecliptic. Thus the planets are periodically seen above and below the sun's apparent path around the celestial sphere. Therefore, we must be able to specify not only the position of the planets in their orbits, but also the unique shape and orientation of each orbit.

Except for perturbation effects which slightly distort them, each planetary orbit is an ellipse with the sun at one focus. Thus *all* of the orbits including the earth's orbit have at least one point in common: the position of the center of the sun. Since our vantage point is the earth, it is logical to select the plane of the ecliptic as the reference plane. Now since each planet travels in an orbit which lies in a plane that does not coincide with the ecliptic plane, each of the other orbital planes must intersect the orbital plane of the earth in a straight line. This notion follows at once from a theorem in plane geometry which states that the intersection of any two planes generates a straight line. In the solar system the straight line passes through the sun (Figure 8-2).

We have previously considered how the shape of an elliptical orbit can be specified by the length of its semimajor axis and its eccentricity (p. 55). For the present discussion let us symbolize these quantities as a and e, respectively. Let us further specify the angle between the orbital plane and the earth's orbital plane as the *inclination, i*. Next, we designate the line of intersection of the respective planes as the *line of nodes*. The nodes represent the two points of intersection of the line with the orbital path of the planet in question. In traversing its orbit the planet rises above the

plane of the ecliptic at the *ascending node,* and it descends below the ecliptic at the *descending node*. Clearly the nodes are at diametrically opposite points in the planet's orbit with respect to the sun.

A line of nodes can be oriented in the plane of the ecliptic at all angles through 360°. We must eliminate all of the mathematical possibilities save one, the correct one, for the given orbit. One way of doing this is to locate the ascending node with respect to some convenient reference point. The vernal equinox is a logical choice because its position is so well known. Thus the angle measured eastward from an imaginary line connecting the sun and the vernal equinox to the ascending node will establish a unique angular direction for the line of nodes. This angle is called the *heliocentric longitude of the ascending node,* and is symbolized ☊. We must be careful not to confuse this angle with the position of the ascending node of the moon even though both carry the same symbol. The inclination i and the longitude of the ascending node uniquely determine the position and the orientation of the given orbital plane with respect to the ecliptic.

But this is not all. Within its plane, the orbit can be oriented with the semimajor axis in any direction. We must specify the direction of the semimajor axis *in the plane* relative to some reference that is common to both the orbital plane in question and the ecliptic plane. The portion of the line of nodes connecting the sun

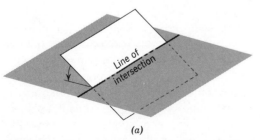

(a)

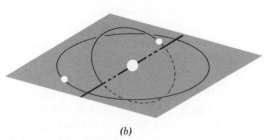

(b)

FIGURE 8-2 (*a*) A straight line is produced by the intersection of two planes; the line lies in both planes. (*b*) Intersection of any two solar system orbital planes; the straight line thus generated by the intersection passes through the sun.

and the ascending node will do very nicely. Since there are two semimajor axes let us arbitrarily select the one passing through the *perihelion point* (NL., fr. *peri*, nearest; Gr., *helios*, the sun hence the point on the planet's orbit nearest the sun). The angle measured from the line of nodes eastward to the semimajor axis will satisfy the requirement. The angle is called the *longitude of the perihelion point*, ω. The five quantities *a*, *e*, *i*, ☊, and ω, completely determine a given planetary orbit. These quantities are collectively referred to as the *orbital elements*. Table 8.2 contains the orbital elements of all known solar system planets. Figure 8-3 shows their relationships.

A sixth quantity is frequently referred to as an orbital element although it is not essential in fixing a planetary orbit. It does, however, make possible the determination of a planet in its orbit at any future time, provided, of course, that the planet's period is known. The element is defined as the *time of perihelion passage*, T. If we can determine just one time that a planet passes through the perihelion point of its orbit, then the application of Kepler's laws enables us to determine the exact position of a planet in its orbit at any other time. Unfortunately, the orbits of all of the planets change slightly and very slowly with time because of mutual perturbative interactions with one or more of the other planets. In practice long term predictions of planetary positions are not exact because the complexity of the perturbations makes mathematical treatment impossible. All orbital predictions must be revised and refined periodically. The mutual perturbations constitute *n*-body problems which, we should recall from Chapter 4, have no permanent *unique* solutions.

The orbital elements of a given planet can be determined from as few as three independent observations. But the subsequent mathematical analysis that yields the prediction values is long and tedious, even when more than three observations are available. The predictions are tabulated in readily available publications such

as *The American Ephemeris and Nautical Almanac.*†

Physical Characteristics of the Planets

With the coordinate systems and terminology now established, we can proceed to the discussion of the position astronomy of the planets. In the process we shall see what else there is about each planet that we can incorporate into the dossiers of our solar-system neighbors.

Mercury. Mercury, the planet nearest the sun, is aptly named after the messenger of the gods of Olympus in Greek mythology. It is the swiftest of the planets as Kepler's laws predict; Mercury's speed in its orbit varies from 37 km/sec when it is at aphelion, to 58 km/sec at perihelion passage. The variation in speed is employed in computing the rather large orbital eccentricity of 0.206. A synodic period of 116 earth-days and a sidereal period of 88 days are associated (through Kepler's harmonic law) with a mean Mercury-sun distance of 0.39 AU. Thus Mercury approaches as close as 46 million kilometers to the sun; it never recedes farther than 70.5 million km. Its tiny orbit restricts Mercury's excursions in maximum elongation to never more than 28° and never less than 18° from the sun. The larger value prevails when Mercury is at aphelion and the earth is so positioned that the line of sight from the earth is tangent to Mercury's orbit at the aphelion point. The smaller value prevails with a similar configuration with Mercury at the perihelion point. Variations from these values occur because of the changing vantage point as the earth circles Mercury's orbit, together with the fact that Mercury's synodic period of 116 days is not evenly divisible into an earth-year. This

† *The American Ephemeris and Nautical Almanac* is issued annually through the joint efforts of the United States Naval Observatory and H. M. Almanac Office, the Royal Greenwich Observatory. It is available through the Superintendent of Documents, United States Government Printing Office, Washington, D.C.

TABLE 8.2 Planetary Motion and Position Data†

	Mercury	Venus	Earth	Mars	Jupiter	Saturn	Uranus	Neptune	Pluto
Motion									
Sidereal period	88^d	225^d	365^d	687^d	11.8^y	29.5^y	84^y	164.8^y	248^y
Synodic period	115.9^d	583.9^d	—	779.9^d	398.9^d	378.1^d	369.7^d	367.5^d	366.7^d
Mean distance from sun (AU)	0.387	0.72	1.0	1.52	5.2	9.54	19.2	30.1	39.4
Mean distance from sun (km × 10^6)	57.9	108.	149.6	228	778	1426	2868	4490	5900
Precession (advance of perihelion)	$43''$	$8''.6$	$3''.8$	$1''.35$					
Orbital elements									
a — semimajor axis	(Equal to mean distance from the sun)								
e — eccentricity	0.21	0.007	0.017	0.09	0.077	0.056	0.045	0.0086	0.25
i — orbital inclination	7°	3°24′	—	1°51′	1°18′	2°29′	0°46′	1°46′	17°12′
☊ — heliocentric longitude of ascending node††	48°	76°.42		49°.34	100°.16	113°.43	73°.87	131°.5	110°
ω — longitude of perihelion point††	77°.4	131°.17		335°.54	13°.77	93°.30	172°.95	21°.16	223°.66

† Data from the Supplement to *The American Ephemeris and Nautical Almanac*, Epoch 1060.
†† These elements vary with time: values given are for the epoch of September 6, 1971.
(Notation such as 8″6 and 77″4 means seconds and tenths, and degrees and tenths, respectively.)

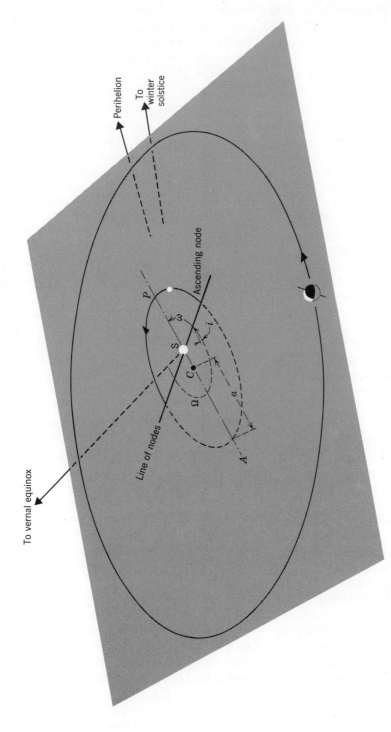

FIGURE 8-3 The orbital elements. Shown in perspective are the orbits of the earth and an inner planet (not to scale). Dashed portion of the inner orbit is below the ecliptic plane; the solid portion is above it. A, aphelion point; P, perihelion point; S, sun; C, center of the inner planet orbit; a, semimajor axis; i, inclination of orbit relative to ecliptic plane; Ω, heliocentric longitude of the ascending node; ω, longitude of the perihelion point; e (not shown), eccentricity — the ratio of the distance from orbit center to sun, and the distance from center to perihelion point.

means that Mercury is frequently seen at a maximum elongation that is intermediate in value to those associated with aphelion or perihelion passage.

Mercury is very difficult to see let alone study. It hides in the intense brilliance of the sun most of the time. Mercury usually sets so soon after the sun or rises so soon before the sun that twilight, rather than a dark sky, prevails. Moreover, the optical path to Mercury through the atmosphere is close to the horizon, thus the optical path length is so long that (the considerable) atmosphere turbulence and obscuration seriously inhibit adequate viewing.

Telescopic observations of Mercury are usually made during broad daylight when the planet is at its greatest heights above the horizon. That Mercury can be seen in broad daylight when it is most favorably located in orbital position is not a unique event. Several bright stars, Sirius and Vega for example, can be seen in daylight; so can the planet Venus when it is near the earth. In order to locate Mercury during daylight one has to know precisely where to look for it, information that can be obtained quickly from a published ephemeris.

Even at its most favorable aspect Mercury is a tiny object more than 81 million kilometers from the earth. The apparent diameter of its disk is usually less than 13″ of arc. This is 145 times smaller than the apparent diameter of the moon. Mercury's surface features are not pronounced; this fact plus the observational handicaps makes them extraordinarily difficult to see. Gross differences in shading constitute the only visible surface features (Figure 8-4).

The most reliable evidence indicates that Mercury's axis of rotation is nearly perpendicular to its orbit; the exact tilt is not yet known. Until recently (1965) the consensus among observers was that Mercury's rotational period was identical with its sidereal period, and hence it always presented one face toward the sun just as the moon presents one face to the earth. During 1965, however, Dr. Gordon Pettengill,

with the use of Cornell University's huge radio telescope in Puerto Rico, showed conclusively that Mercury rotates on its axis once every $58\frac{2}{3}$ days. This means the planet turns on its axis three complete revolutions for every two complete trips around the sun.

Two interesting phenomena arise from this behavior. First, astronomers on earth always see very nearly the same face and the same markings every time Mercury is in the most favorable position for viewing. Second, one complete diurnal period on Mercury is exactly twice as long as Mercury's year. This combination of rotation and revolution rates, together with the large eccentricity of Mercury's orbit, would present an unusual spectacle to a Mercurian observer at dawn when Mercury is at perhihelion. He would see the sun rise above the horizon, hang there briefly, then sink back down below the horizon only to burst upward again a little later and complete its apparent trip across Mercury's sky (Figure 8-5).

Because Mercury's axis of rotation is nearly perpendicular to its motion around the sun, whatever variations in seasonal climate it has arise from its rather large orbital eccentricity. In winter the climate is hot; in summer it is even hotter. When Mercury is at perihelion its sunlit surface intercepts nearly 7 times as much solar energy per square foot as does the earth. The absence of diffusion or refractive effects of light at Mercury's limbs suggests there is little or no atmosphere. Furthermore the reflecting power of its surface, only 6% of the light it receives, is characteristic of other solid-surface bodies (the moon and Mars) that have negligible atmospheres.

There are also theoretical grounds for assuming that Mercury can have no perceptible atmosphere. These are generally independent of direct observational evidence. This is a different situation from the one we experienced with regard to the lack of an atmosphere on the moon. There the observational evidence was almost conclusive in itself. Here the "seeing" at best is so bad that observational evidence is

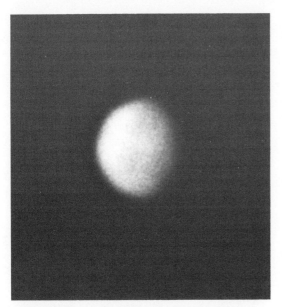

FIGURE 8-4 Mercury as photographed at the New Mexico State University Observatory. This photograph is among the best it is possible to obtain of this very difficult subject (Courtesy New Mexico State University Observatory).

FIGURE 8-5 Relation between the diurnal and orbital periods of Mercury as seen in plan view. Numbers 2 through 9 show Mercury's orbital position every 11 earth-days; position 1 is 5½ days before position 2. Index marks (*a*, *b*, etc.) extend above the surface and are fixed in place. Observe that the mark *a* in 1 is just at the sunrise terminator. In 2 it has risen into the sunlight zone, but in 3 it has sunk back to the terminator. Thereafter the sun rises, reaches the meridian (local noon) between positions 6 and 7, and sets at position 1*b*. Thus Mercury makes 1½ rotations with respect to the stars during which time it makes only ½ rotation with respect to the sun.

inconclusive; it needs other support. Fortunately that support is available. It was born out of exhaustive laboratory experimentation that led to reliable knowledge about the behavior of gases in general.

Atmospheric gas molecules move about, colliding with each other and thereby randomly acquiring or losing energy. In the atmosphere, however, there is no general restraint on motion except for the gravitational hold on these molecules. Their average speeds are governed in part by the manner and frequency of their random collisions with each other, but the temperature of the environment and the individual masses of the molecules are equally as important as the collisions. Kinetic (Gr., *kiñetikos*, to move) theory relates increasing speeds of motion of gas molecules with increasing temperatures.

Now the *velocity of escape* from any celes-

tial body is the speed an object must acquire in *free flight* (without additional propulsion or restraint) in order to get out of gravitational range before the retarding effect of gravity first stops, then reverses the object's motion. The magnitude of the gravitational field (a region in space where gravitational forces exist) at the surface of a celestial body can be computed with the aid of Newton's universal law of gravitation, provided, of course, that the mass and size of the celestial body are known. As we shall see presently, Mercury's surface gravity is low, only a fraction of that of the earth. Thus the escape velocity is also low. Because of the extreme surface temperature on Mercury, all known gases would have average molecular speeds in excess of that necessary to escape the gravitational bonds. Unquestionably whatever gases that were liberated during Mercury's evolvement escaped eons ago.

With respect to the surface temperature, the midday heat on Mercury's sunlit surface is ferocious by any standard. Astronomers have been able to measure the rate of radiation of heat from the surface of Mercury, and to divorce the data from the effects of reflected energy from the sun. The midday temperature at Mercury's equator exceeds 400°C, a temperature at which tin and lead are in the molten state. Some of the minerals could conceivably be near the plastic state. A description of Mercury's daytime surface as resembling a searing, seething cauldron may not be too unrealistic.

Corresponding data from the dark, or night, side of Mercury are not as valid because of the extreme weakness of the radiant energy. An upper limit appears to be on the order of −120°C; a temperature of −200°C appears more likely. The rapidity with which the surface material can plunge through a 600°C temperature range suggests that the heat-storage capacity of Mercury's surface rocks is low and that the conduction of heat through them is negligible.

Size and mass. From the known mean Earth–Mercury distance and the corresponding angular diameter of Mercury's disk it is a simple matter to compute the diameter as approximately 4880 km. Determining Mercury's mass is exceedingly difficult; the presently accepted value of 6% of the earth's mass is subject to revision at any time. The mass of any distant celestial body can be determined directly only by observing its gravitational effects on other bodies. Mercury has no satellites over which to exert gravitational domination. While it does

perturb Venus' orbit, the effect is very slight and the estimate of Mercury's mass from such observations is inconclusive. The best information comes from observing Mercury's effect on comets that infrequently come close to the planet, and from observing occasional encounters with planetoids (Chapter 9) whose eccentric orbits bring them fairly close to Mercury.

With a mass of 6%, a diameter of 38%, and therefore a volume of 6% of the corresponding quantities on Earth, Mercury's surface gravity is readily computed to be one-third of that of the earth; this is still twice the surface gravity of the moon. The velocity of escape from a celestial object can be shown mathematically to be directly proportional to the body's surface gravity. In Mercury's case it is one-third the velocity of escape from the earth, or slightly over 4 km/sec.

Since the ratio of Mercury's mass to its volume is the same as that for the earth, we conclude its average density to be the same, or about 5½ times the density of water. Since all known rocks, including those from the moon, have densities much less than this, Mercury's relatively high density suggests either a higher proportion of metallic substances in the rocks themselves or a metallic core resembling that of the earth.

Venus. There is only one object in the sky other than the sun and the moon that is bright enough on occasion to cast faint shadows on a dark night. This is the planet *Venus,* often referred to as the earth's "twin sister" because of similarities in size and mass; otherwise, they are completely unlike each other. Venus is named for the Roman goddess of love and beauty—the protectoress of gardens; Venus' symbol is appropriately a mirror, ♀, emblematic, perhaps, of the vanity that William Shakespeare called Woman. The early Greeks knew Venus as two planets: *Phosphorus,* seen in the early morning before sunrise, and *Hesperus,* an early evening, after-sunset object. They did not realize that Venus alternates from one side of the sun to the other and that the two objects were really only one. Pythagoras in the 6th century B.C. appears to have been the first among Western Civilization natural philosophers to recognize the truth.

Venus in its orbit alternates $3\frac{1}{2}°$ higher and lower in declination than the sun; this means its orbit is inclined by this amount to the ecliptic. Venus' sidereal period is 225 earth-days; when we express the period as a fraction of a year, we obtain from Kepler's laws a mean sun-Venus distance of 0.72 AU, or 108.1 million kilometers. Because the orbit of Venus is larger than that of Mercury, Venus swings out correspondingly farther in elongation east and west of the sun. As with Mercury, there is a variation in the magnitude of the displacement; the maximum is 48°, the minimum, 47°. This small variation yields an orbital eccentricity of only 0.007. Thus, Venus' orbit is nearly circular; her distance from the sun varies by only slightly over 1.6 million kilometers.

Venus is indeed a brilliant, beautiful early evening or early morning spectacle. It appears brightest near inferior conjunction. At this time Venus unfailingly attracts the attention and admiration of many spectators who otherwise are only casually aware of the splendors of this celestial sphere. Venus at its brightest is more than 15 times brighter than the brightest appearing star, *Sirius*. It is easily a naked-eye object in broad daylight if one knows where to look. Even at superior conjection when Venus is farthest from the earth, it is only $2\frac{1}{2}$ times fainter than at its brightest.

The distance between Earth and Venus varies from 41.9 million kilometers at inferior conjunction, to 258 million kilometers at superior conjunction, a sixfold variation in distance. We might expect that Venus' apparent brightness would also vary by a factor of 36 instead of a mere $2\frac{1}{2}$. Galileo was first to point out that Venus goes through phases similar to those of the moon. Also, Venus' apparent size does vary from 10" of arc to 64" of arc in diameter. But whereas its disk is fully illuminated

when Venus is at superior conjunction and its apparent size is smallest, only a thin crescent remains illuminated as it swings close to or away from inferior conjunction and the corresponding new phase (Figure 8-6). Venus' maximum brilliance coincides with the balance between her increasing apparent diameter and decreasing area of illumination. This equilibrium occurs 36 days before and after inferior conjunction.

Observers on earth will count 220 earth-days from the time Venus leaves superior conjunction until it arrives at greatest eastern elongation. But then it only takes Venus 72 days to move from there to inferior conjunction, or a total of 294 days for one-half of a synodic period. Thus the synodic period is $2 \times (220 + 72) = 584$ earth-days.

Why is the synodic period so much greater than the sidereal period? We can most easily answer this question after making an analysis of the geometry of two bodies moving relative to each other (Earth and Venus), with both moving relative to a third body which is taken as a stationary reference (the sun). In the process we may also discover why it takes Venus so much longer to move from superior conjunction to maximum eastern elongation than from the latter to inferior conjunction.

From superior conjunction to maximum eastern elongation Venus travels 220/225 of the length of her orbital path of 684 million kilometers, or 659 million kilometers. From Figure 8-7 we can see that this distance lacks being a complete orbital round trip by only 10° in angular measure. Meanwhile the slower moving earth covers 220/365 of its orbital path, an angular displacement of 223°. Thus at the end of 220 days both Earth and Venus have moved from their original positions 1:1' in the figure to position 2:2'. By definition Venus is now at greatest elongation relative to the sun as seen from Earth.

During the next 72 days Venus covers another 72/225 of her orbit, a distance of 217 million kilometers and is at position 3'. During

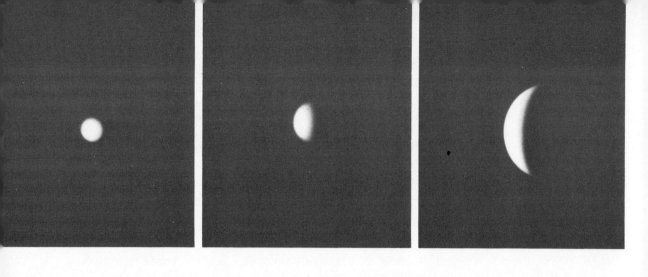

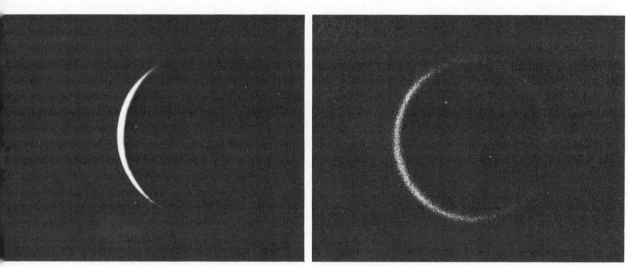

FIGURE 8-6   The phases of Venus as seen from earth. The differences in apparent size result from variations in the earth-Venus distance (Courtesy Lowell Observatory).

the same interval the earth moves to its position 3. By now, Venus has gained one-half lap on the earth, and both planets are in a straight line relative to the sun. By definition Venus is now at inferior conjunction. Since Venus' speed in her nearly circular orbit is essentially constant, the difference in time (220 versus 72 days) is simply the reflection of the difference in orbital distances needed to effect the desig-

nated alignments with the moving earth. The synodic period, then, is clearly the time it takes Venus to gain one full lap on the earth.

Physical characteristics of Venus.   Until the middle of the 1960s Venus was always something of an enigma. From appearances in any modest telescope it is quite obvious that a dense cloud cover completely obscures the

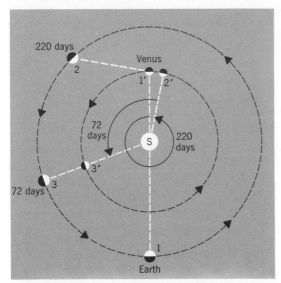

FIGURE 8-7 Elapsed time between superior conjunction, greatest eastern elongation, and inferior conjunction for the earth and Venus. At the initial moment ($t = 0$) the earth and Venus occupy the positions 1 and 1' respectively, relative to the sun. At the end of 220 earth-days the earth has moved to 2 and Venus to 2'; the position is that for greatest eastern elongation. After 72 additional earth-days the earth has advanced to position 3, Venus to position 3', and inferior conjunction occurs. Observe that superior conjunction and inferior conjunction occur in different parts of the respective orbits.

FIGURE 8-8 Venus. The surface is totally obscured by a dense cloud cover that in itself displays only vague and indistinct shadings (Courtesy Lick Observatory).

surface. This gives the planet an albedo of 0.49, considerably higher than the earth's 0.35.† The cloud cover is devoid of all markings save irregular and poorly defined shadings (Figure 8-8). Thus we wonder: How big is Venus? How massive is it? Does it rotate? Does Venus have seasons? How would the outside world appear to a hypothetical Venusian?

† *Albedo* is the fraction of incident radiation, principally visible light, that is reflected from the surface.

The original answers to these questions have given way to more precise measurements provided by modern technology. Thus, for example, measurements made with the aid of earth-based optical instruments yielded two values for Venus' diameter. Against a black sky, the extreme brilliance of the planet's disk causes observers to overestimate its size at 12,505 km in diameter. Again, when Venus transits (passes in front of) the sun's disk, the sun's overwhelming brightness causes an underestimate of the diameter at 12,182 km. Thus a compromise is necessary. The most acceptable value obtained from direct optical examination is 12,255 km.

During the early and middle 1960s radar was employed at the Lincoln Laboratory of the Massachussetts Institute of Technology and at the Arecibo (Puerto Rico) Ionospheric Observatory of Cornell University in an attempt to improve on optical capability. By measuring the round trip time of high-frequency radio waves which are reflected from Venus' surface, one can compute the distance to the latter with considerable accuracy. This is possible because radio waves travel at the speed of light, a constant value in a vacuum.† But the precise distance of Venus' center from the earth is known from the orbital constants at the time of observation. The difference between this quantity and the measured surface distance yields the radius of Venus, or, what is equivalent, the diameter. Venus' diameter from these data is 12,166 km with an uncertainty of 5 km.

In 1967 two spacecraft, Soviet Russia's Venera 4 and United States of America's Mariner 5 were launched on trajectories that took them to the vicinity of Venus. Mariner 5 passed within 49,000 km of Venus; Venera 4 curved in toward the planet and ejected a capsule which parachuted down toward the surface. Many kinds of measurements were made, and the data from both spacecraft correlated well. However, the diameter of Venus as deduced from Venera 4 was 12,213 km, a value incompatible with the high reliability of radar measurements. At present, scientists are in considerable agreement that Venera 4 ceased transmitting data before it reached the surface, not after it supposedly crash landed there. As a result, the radar determination still holds first place as the most reliable estimate for the diameter of Venus. Hopefully even better data will result from the Mariner-Mercury mission which includes a Venus fly-by, scheduled for launch in late October or early November, 1973.

Estimation of Venus' mass is a more difficult problem than determining its size. As we have seen, Newton's universal law of gravitation provides the only means of "weighing" a celestial body. But Venus has no natural satellites whose motions can be studied; only Venus' perturbations of the earth or occasional encounters with asteroids or comets normally

provide the necessary information. On the other hand Venus' encounters with Venera 4 and Mariner 5 in 1967 furnished first-hand information about Venus' mass, information that supported an estimate made from earth-based observations of 82% of the earth's mass. With both the mass and size of Venus thus known within reasonable limits, we can calculate its density to be 5.2 times that of water. Thus in the matter of size (96%), mass (82%), and density (95%) with respect to the corresponding terrestrial quantities, Venus is truly earth's twin. But here the similarities end.

Venus' cloud cover extends more than 56 km above the surface; earth's cloud cover rarely reaches 16 km. The Mariner 5 fly-by, Venera 4's 1967 penetration into the atmosphere, and Venera 7's successful transmission of data from the surface of Venus in 1970 verified many of the previous estimates of the composition of Venus' atmosphere. Carbon dioxide makes up more than 90% of the total, compared to 0.03% in the earth's atmosphere. The nitrogen content is 7% compared to 78% on earth; oxygen (found only in the upper levels) has a concentration of between 0.4 and 1.6% with a probably value of 1%, compared to earth's 21%. The concentration of water vapor appears to be less than 1/10,000 that of the earth's atmosphere.

Venera 7 also yielded values for the surface temperature of between 500 and 550°C, values that are consistent with estimates made by microwave radio measurements from earth. The Russian probe also determined a Cytherean (Gr., pertaining to *Aphrodite*, hence *Venus*) atmospheric density of 60 times, and an atmospheric pressure of 90 times, the earth's sea level values. These data concerning temperatures, pressures, and gaseous concentrations have interesting connotations for the history of the evolution of the earth's atmosphere. Considering the greater mass of Venus' atmosphere as reflected in its density and pressure at the surface, the masses of nitrogen and oxygen are about the same as in the earth's atmosphere. If

---

† In Chapter 14, the propagation of radio waves through space that is not a perfect vacuum is considered.

all the carbon dioxide on earth that is dissolved in the oceans, chemically combined in the carbonate rocks, and locked into the shells of marine shellfish were free in the atmosphere, the quantity would be comparable to that in the atmosphere of Venus.

Liquid water almost certainly cannot exist on Venus' surface because of the high temperatures. But large volumes of water are essential in chemical processes involving carbon dioxide, certainly for keeping it in solution. The presence of so much carbon dioxide in Venus' atmosphere lends plausibility to the argument against appreciable water on Venus.

The comparatively large percentage of oxygen is puzzling. At the existing temperatures oxygen combines freely with most surface materials. Exhalation of the original oxygen from the interior of Venus, and for that matter from the interior of the earth, seems improbable. Oxygen would have combined to form oxides of metals and silicates first. Most authorities agree that the bulk of the earth's oxygen was produced by plant life long after the earth was consolidated. The presence of corresponding plant life on Venus is precluded by the excessive temperatures. Thus, the question as to the source of Venus' oxygen remains unanswered.

An observer on Venus would find the landscape only dimly illuminated because of the great thickness of the cloud layer. His view would be through a murky haze, perhaps even a "whiteout" would prevail. If he could see for a moderate distance, his landscape would have a strange appearance. Because of the atmosphere's extreme density, light would be super-refracted, that is, bent upward in all directions from the observer. He would have to look upward at an angle estimated in excess of 15° to see his horizon. Even when standing on a flat, level surface, the observer would think he was standing in the center of a shallow bowl!

Until refinements in radar technology made pinpointing minute targets possible such as is needed in planetary astronomy the rotational period of Venus was a matter of wide conjecture. Published estimates of the period ranged from as little as 22 hours to as much as 365 days. All of the estimates were made in terms of supposedly observed rotation of the cloud cover. In the early 1960s investigators at Goldstone, California using the Jet Propulsion Laboratory's giant 210-ft radar tracking telescope discovered several localized areas on Venus' surface that had high reflective capabilities for radio waves. By tracking these features across an otherwise inscrutable surface, Venus was discovered, to everyone's great surprise, to rotate clockwise, or retrograde with respect to its orbital motion, with a period of 243 days. As we shall see later, only one other planet rotates about an axis in a retrograde sense.

As far as can be determined, Venus' axis of spin is not appreciably tilted from the vertical with respect to its orbital plane. Thus, with a circular orbit, no axial tilt, and an extremely dense and therefore presumably efficient heat-conducting atmosphere, Venus has no seasonal climate. It is hot, everywhere, all of the time. Because of the greenhouse effect of its atmosphere, Venus' surface is hotter than that of Mercury, even though Venus is on the average more than 48 million kilometers farther away from the sun.

The combination of Venus' slow retrograde rotation and a sidereal period shorter than the rotational period leads to an unexpected diurnal phenomenon. An observer on Venus would see the sun rise in the west and set in the east every 117 earth-days, even though Venus makes slightly less than one-half rotation during this time (Figure 8-9).

Mars. Mars is the nearest of the superior planets; it is the only one whose orbital configuration and atmosphere combine to make direct inspection of its surface by visual means possible. Mercury's surface is visible, but the planet is so small and so remote that little can be learned from such inspection. Mars is so distant and the earth's atmosphere is so turbu-

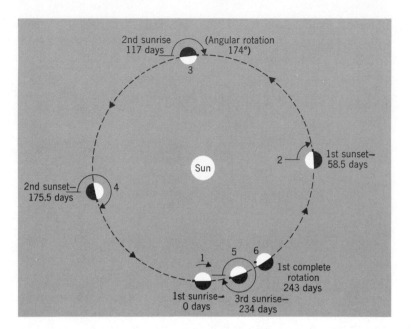

FIGURE 8-9 Plan view showing the relation between Venus' rotational and orbital periods. Small arrows show the amounts of Venus' rotation relative to initial reference position 1. Venus' orbital period is 220 earth-days.

lent, that even at the most favorable viewing periods, optical observation is ordinarily very poor (Figure 8-10).

Attempts to secure detailed photographs of Mars' surface features from earth are frustrated for other reasons. The quality of the photographs doesn't even approximate that of visual images formed by the eye and communicated by the nervous system to the brain. We can compare the human optical system to an analog computer. The eye analyzes and records momentary changes in the information arriving from perceived objects. In the case of Mars, this information is from more than 57 million kilometers away, through several hundred kilometers of atmosphere. However, the memory bank in the brain allows us to sort out similarities and singularities in the succession of images, from which we can select those with the clearest detail. Illustrations representative of the Martian surface thus become largely a matter of interpretation or delineation skill of the observer. On the other hand a camera system may be likened to a

digital computer. The information arriving at the focus of the camera lens in the form of light energy is stored at millions of discrete points in a chemical matrix, the photographic negative or plate. During the exposure interval required to obtain sufficient "strength" in the latent image, atmospheric disturbances cause random displacements of many of the image points. The developed negative consists of the *average* character of the image at each point; as a result the total picture is blurred or, more accurately, lacks resolution. Fortunately, space probes on fly-by missions are able to secure a variety of data, including high resolution television pictures from comparatively close proximity to selected planets. The results of the Mars fly-by missions have been nothing less than spectacular. We shall review the status of these missions presently.

Mars has always attracted the interest and whetted the imagination of mankind. From the dawn of recorded history and legend, Mars has symbolized war or battle, probably because of the association of its ruddy color with war's

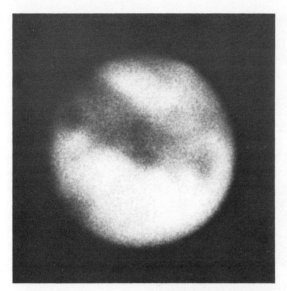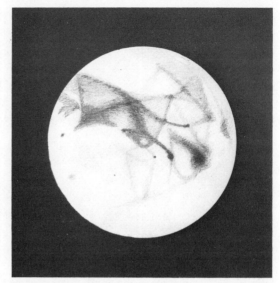

FIGURE 8-10 Photograph of Mars compared with a drawing of the surface made simultaneously during the opposition of 1926 (Courtesy Lick Observatory).

blood, carnage, and destruction. The ancient Sumerians so viewed Mars; the Chaldeans more than 3000 years ago named Mars *Nergal,* master of battles. To the ancient Greeks it was *Ares; Mars* is of Roman origin. The names are those of the god of war in the respective cultures. As we have already seen, Kepler used the data on Mars assembled by Tycho Brahe in the final development of his revolutionary solar system hypothesis. Up to the advent of the space probes, modern man has seen Mars only from his terrestrial vantage point; but he has identified many similarities between Mars and Earth. Mars and Earth have days of nearly the same length; Mars and Earth experience a similar progression of seasons. We have observed seasonal color changes and seasonal waxing and waning of the polar caps (Figure 8-11). Man has even gone so far as to develop weather maps that depict the circulation of the Martian atmosphere. Man has seen "canals" and "oases," and associated with these, presumably irrigated vegetation belts (Figure 8-12). Not unexpectedly man has speculated

about the existence of intelligent beings who are capable of manipulating their environment to suit their needs.

The light and dark areas on Mars' surface were thought to be islands and continents, and oceans or seas, respectively; the names assigned to the visible (and imagined) surface details reflect man's attempt to equate Mars' observed physical characteristics and environment with those of the earth (Figure 8-13). Let us review just what we did know about Mars before the space-probe explorations.

Physical characteristics. Mars' average distance from the sun is 228 million kilometers, or 1.52 AU. Its orbital eccentricity is high for a planet, 0.093; as a result Mars' distance from the sun varies by 40 million kilometers. Of more importance to us, the large eccentricity causes Earth-Mars distance at opposition to vary between 56 million kilometers and 101 million kilometers. The most favorable oppositions occur late in August every 15 to 17 years (Figure 8-14). At closest approach Mars is 3

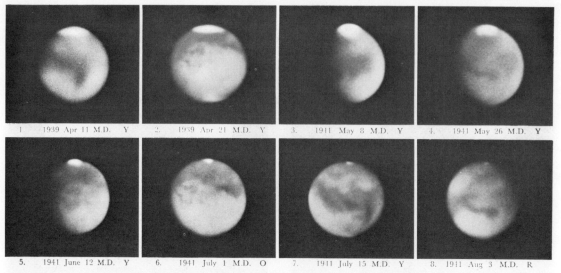

FIGURE 8-11 Changes in the south polar caps of Mars. South is at top as it would be seen in a telescope (Courtesy Lowell Observatory).

times brighter than Sirius. Only Venus among the planets can outshine Mars at this time. When Mars is at superior conjunction it appears as much as 7 times smaller and 49 times fainter than when it is at most favorable opposition.

FIGURE 8-12 Percival Lowell's globe of Mars, 1907 (Courtesy Lowell Observatory).

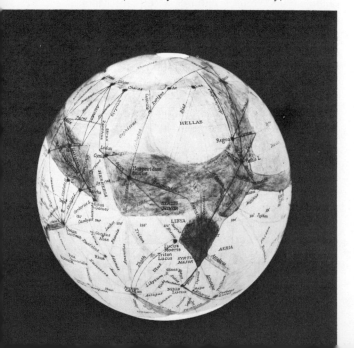

Mars' axis of rotation is inclined 24° from the perpendicular; the Martian day is just over 24^h37^m long, earth time. Because of the close similarity to the length of our day, 40 nights of observation are needed in order to see Mars make one apparent retrograde rotation on its axis relative to the earth (Figure 8-16).

Figure 8-17 shows Mars at both equinoxes and both solstices in an orbit that is exaggerated for clarity. Because Mars passes through aphelion between the spring equinox and summer solstice, the northern hemisphere spring is the longest season, 199 days. As the planet speeds up toward perihelion, summer shortens to 182 days. Perihelion passage occurs during autumn; this season is only 146 days long. Then as Mars slows down on its journey toward aphelion, winter is lengthened to 160 days. Table 8.3 shows the Martian seasons for both hemispheres.

Because the northern hemisphere has a longer spring and summer, we might at first suppose it to be hotter than the southern hemisphere during these seasons. Such is not the case because of Mars' greater distance from the sun at aphelion; at this time the planet receives

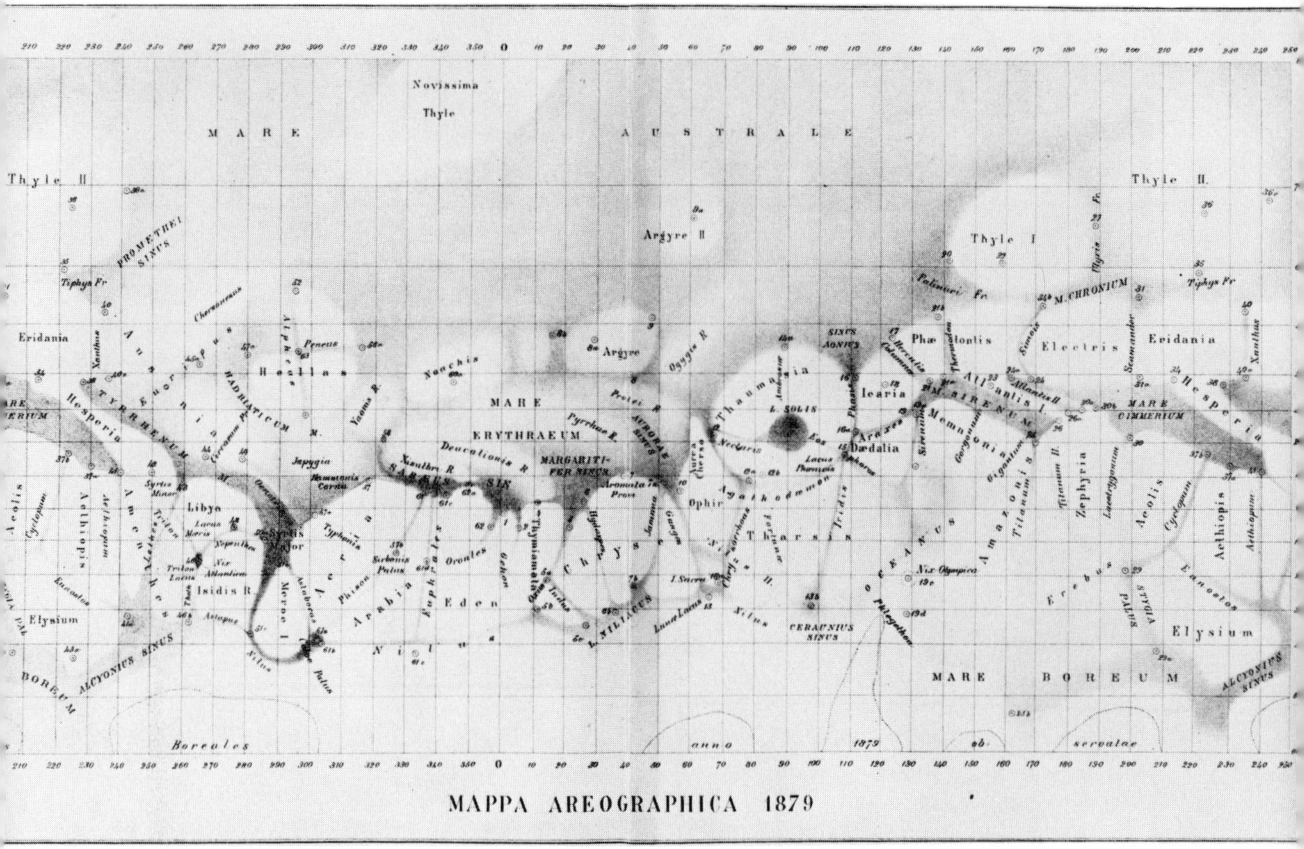
Novissima Thyle
MARE AUSTRALE
Thyle II
Thyle I
Argyre II
Argyre
Hellas
Eridania
Electris
Icaria
Dædalia
MARE ERYTHRAEUM
Libya
Ophir
Tharsis
Eden
Arabia
Elysium
MARE BOREUM
MAPPA AREOGRAPHICA 1879

FIGURE 8-13 Schiaparelli's map of Mars, with nomenclature standardized according to international agreement (Courtesy Lowell Observatory).

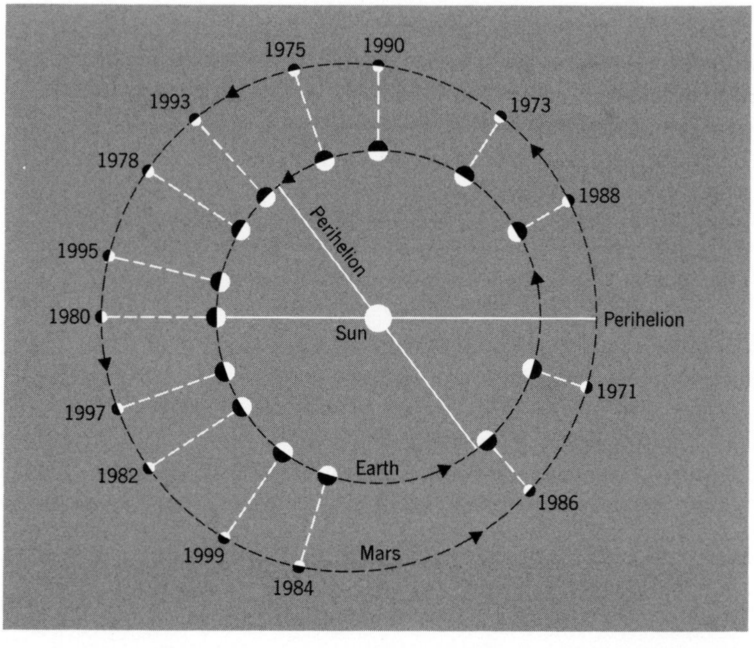
1975
1990
1993
1973
1978
1988
1995
Perihelion
1980
Sun
Perihelion
1997
1971
1982
Earth
1986
1999
Mars
1984

FIGURE 8-14 Relative positions of Mars and the earth at all oppositions from 1971 through 1999. The next favorable oppositions will occur in 1986 and 1988.

FIGURE 8-15 Relative sizes of Mars as seen at favorable and unfavorable oppositions (Courtesy Lowell Observatory).

44% less solar radiation than during perihelion passage. Even though the spring-summer period is relatively shorter in the southern hemisphere, the heating is more intense by the same 44%. The south polar cap frequently disappears during summer; the north polar cap has never been observed to do so. We can summarize Mars' seasonal phenomena thus:

Northern Hemisphere

Winter: shorter and warmer
Summer: longer and cooler

Southern Hemisphere

Winter: longer and colder
Summer: shorter and hotter

From the observed perturbations of Mars' two small satellites (Chapter 9), we can compute Mars' mass to be slightly less than 11% (0.1078) that of the earth. From the known distance and corresponding angular size (which reaches 25" of arc at most favorable opposition), Mars' diameter is estimated to be 6786

km with an uncertainty of 16 km. These data, size and mass, combine to yield a density of 4.1 times that of water. Mars' surface gravity is only 38% that of earth. The velocity of escape for atmospheric gases for a planet of Mars' size and mass is just over 4.8 km/sec. Thus even though the lighter constituents such as hydrogen and helium could escape, Mars is assured *some* atmosphere.

On the average Mars receives only 40% as much solar energy per unit area as the earth. Accordingly the Martian average temperature is 90–100°F colder. Noontime summer temperatures in the Martian ''tropics'' reach as high as 50°F, whereas the daytime winter temperatures at a Martian pole remain below −110°F. Even in summer the temperatures at polar noon are barely above freezing, 32°F. Because of the large orbital eccentricity, all of Mars' daytime temperatures can vary as much as 50°F between aphelion and perihelion. The daily range in temperature is in the order of 150°F.

Mars' equatorial diameter exceeds its polar diameter by 35.4 km; proportionately a greater difference than is the earth's case. Mars' oblateness is 1/192 compared to 1/298 for the earth. This larger value suggests that Mars' mass is much less concentrated at the center. Perhaps Mars has little or no heavy metallic core.

The presence of a Martian atmosphere is deduced from observed differences in Mars' apparent size and markings when photographed in different colors. With red light surface details are clearer and the disk appears smaller; in blue light the diameter appears larger and the surface appears obscured with a considerable haze (Figure 8-18). Mars also exhibits a sizable twilight zone, a phenomenon only possible in the presence of an atmosphere. That Mars' atmosphere is thin though deep is deduced from the speed with which light from occulted (eclipsed) stars is cut off, and also from the nature of the twilight diffusion effects. Mars' atmospheric pressure at the surface is estimated to be no more than 10–15% that of the earth. Spectrographic analysis of the light reflected from Mars surface and occasional cloud or haze cover discloses that carbon dioxide is the dominant atmospheric gas. Mars' atmosphere can have at most 0.15% as much water vapor as the earth's; the oxygen content must be less than 1/1000 as much.

What about surface features? All of Mars' surface can be seen at the limb at one time or another. No appreciable roughness or irregularity, such as the moon's limb exhibits, is ever seen. Mars' surface irregularities, if they exist, are of low profile, that is, there are low hills rather than mountains. But extensive elevated plateaus are not ruled out.

Most of the dark markings are concentrated near the equator. No "sunlight" from the dark areas is ever observed as would be the case if they were large bodies of water. In spite of the names assigned by early observers denoting seas, bays, swamps, lakes, and channels, Mars appears to be without appreciable water.

Traces of water vapor have, however, been definitely detected in the Martian atmosphere. Together, carbon dioxide and water account for approximately one-half the estimated atmospheric pressure. The identity of the remaining constituents, if any exist, is unknown, even though detected nitrogen and argon are possibilities for the other 50%. Neither exhibit identifiable spectral characteristics in reflected light, so their presence cannot be completely ruled out. On the other hand the magnitude of the atmospheric pressure at Mars' surface may be overestimated, and the quantity of carbon dioxide may be underestimated. It is distinctly possible that carbon dioxide comprises more than 50% of the total atmosphere. We shall presently compare the above values with the data provided by the Mariner space probes.

The recession and growth of the polar caps proceeds quite regularly from year to year. The polar cap growth is hardest to observe because a cloud cover that conceals the details of growth forms over the entire polar cap in early autumn. Recession of the caps is more easily observed. About the middle of spring the cloud cover dissipates, and we observe an irregular edge to the cap. Isolated areas become detached as though a snow pack were lingering only on the tops of high hills or mountains or in the depths of valleys (Figure 8-19). The growth of one cap occurs simultaneously with the disappearance of the other.

From earth the composition of the caps cannot be established for certain. By the middle 1960s earlier beliefs that the caps were solid carbon dioxide had largely given way to the conviction, since refuted by the Mariner-Mars Missions (p. 181), that the polar caps were mostly ice crystals or snow. In any event, the speed of growth and recession of the caps, nearly 40 km/day, together with the scarcity of water vapor in the Martian atmosphere, preclude any great depth in the snow cover. Best current estimates place it between 5 cm and 100 cm, although drifts could be considerably deeper.

Many hypotheses have been advanced in the attempt to account for the seasonal color changes, often referred to as the wave of

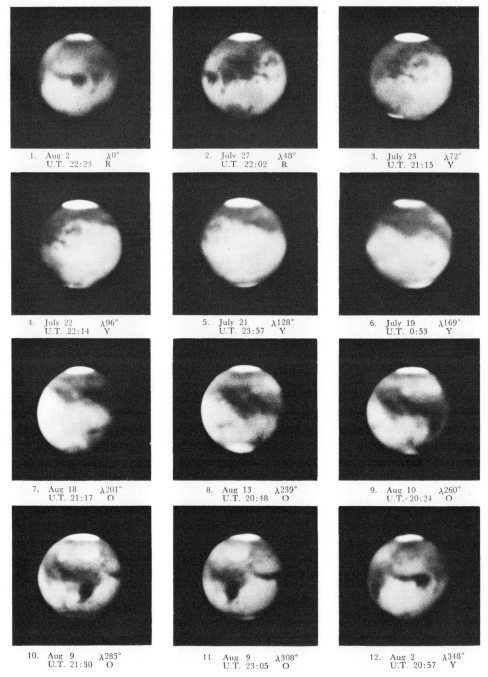

1. Aug 2 λ0° U.T. 22:23 R	2. July 27 λ48° U.T. 22:02 R	3. July 23 λ72° U.T. 21:15 Y
4. July 22 λ96° U.T. 22:14 Y	5. July 21 λ128° U.T. 23:57 Y	6. July 19 λ169° U.T. 0:53 Y
7. Aug 18 λ201° U.T. 21:17 O	8. Aug 13 λ239° U.T. 20:48 O	9. Aug 10 λ260° U.T. 20:24 O
10. Aug 9 λ285° U.T. 21:30 O	11. Aug 9 λ308° U.T. 23:05 O	12. Aug 2 λ348° U.T. 20:57 Y

FIGURE 8-16 Series of photographs of Mars showing complete rotation as seen from earth. Mars' rotation rate is so nearly the same as that of the earth that it takes nearly 40 days of observations to record the entire Martian surface (Courtesy Lowell Observatory).

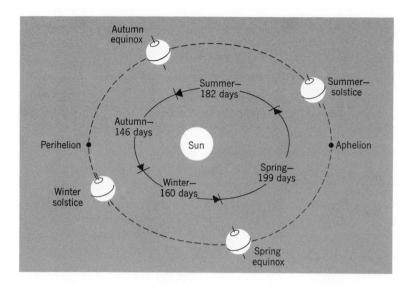

FIGURE 8-17 Seasonal positions of Mars relative to the sun and the elapsed time in earth-days between each position. The elapsed time clearly marks the length of each season.

darkening. This phenomenon is ushered in each spring near the edge of the polar cap. The darkening spreads toward the equator, reaching a maximum in the late summer. Indeed, the darkening reaches the equatorial zone from the spring hemisphere before the darkening fades in the fall hemisphere. One popular view holds that the seasonal flush of vegetative growth is responsible. But the dearth of surface water seems to be an overwhelming obstacle to any substantial plant growth, unless perhaps the vegetation is in the form of lichen and mosses.

Other hypotheses for the darkening include such events as the removal of a light-colored dust cover from a darker substrate by wind action or chemical changes in the minerals forming the planetary crust. No solution to this problem seems possible from earth-based observations alone.

The most controversial of the reported surface details concerns networks of fine, straight-line structures called "canals." Between 1877 and 1886 Giovanni Schiaparelli, director of the Brera Observatory in Milan, Italy, produced the most accurate maps of Mars' surface features yet recorded (Figure 8-13). He also initiated a revised system of nomenclature for the Martian surface features that is in general acceptance today. But Schiaparelli is best remembered for

TABLE 8.3 Martian Seasons for Both Hemispheres†

Location	Spring	Summer	Autumn	Winter
Northern Hemisphere	199 (aphelion passage)	182	146 (perihelion passage)	160
Southern Hemisphere	146 (perihelion passage)	160	199 (aphelion passage)	182

† Duration in earth-days.

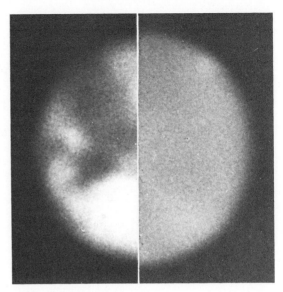
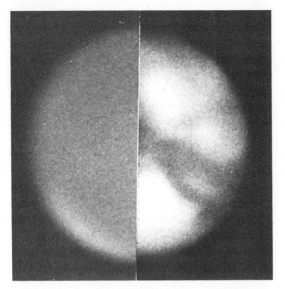

FIGURE 8-18 Apparent difference in the size of Mars when photographed in
red and blue light. Red light penetrates the atmosphere revealing the surface
below; thus the planet appears smaller. Blue light does not penetrate to the
surface; the larger apparent size is that of Mars *plus* its atmosphere (Courtesy
Lick Observatory).

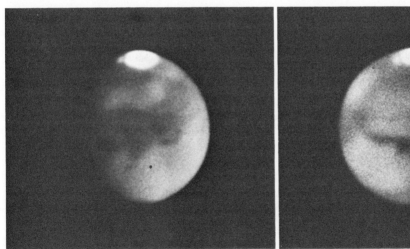
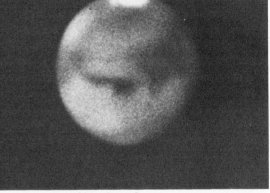

1. 1909 Aug 21   2. 1924 Sept 8

FIGURE 8-19 The *Mountains of Mitchel*. The retreat of the south polar cap of
Mars leaves a small detached patch that was first discovered by Mitchel at
Cincinnati in 1845. The patch is observed at practically the same identical
seasonal date at every opposition, lending strength to the hypothesis that the
snow patch is in truth lingering at the top of a mountain peak (Courtesy
Lowell Observatory).

his depiction of a number of straight, often doubled, features that he thought were natural watercourses. He called them *canali,* Italian for channels, though he named them after rivers on earth. While he is generally credited with both the discovery of the features and the introduction of the term *canali,* the papal astronomer, Fr. Pierre Secchi of Rome, is known to have used the term in connection with Mars as early as 1869. Schiaparelli himself wrote that the *canali* had been seen earlier by such excellent observers as Kaiser, Lockyer, and Green. Dawes and Procter also showed the canals on their early maps.†.

Unfortunately, the term *canali* was literally interpreted, in the United States especially, to mean canals. Thus the features became identified with artificial waterways constructed by intelligent beings in order to conserve the dreadfully scarce water. Percival Lowell constructed the famous Lowell Observatory at Flagstaff, Arizona and devoted a large part of his later life to the study of the planets, especially Mars (Figure 8-20). Lowell was particularly imaginative in his reports concerning the Martian canals. That these features would have to be nearly 20 km wide to be observable from earth did not seem to discourage any of the canal enthusiasts.

The Mariner-Mars Missions. In 1965, and again in 1969 and 1971, a total of four Mariner space probes scrutinized Mars from distances of as little as 1600 km. Mariner 4 in 1965, and Mariners 6 and 7 in 1969, were fly-by missions, that is, they each flew past Mars once, then went into orbit around the sun (Figure 8-21). Mariner 9 arrived at Mars on November 13, 1971, and achieved orbital insertion the next day during a planetary dust storm of unusual intensity. On October 27, 1972 after a useful lifetime of almost 516 days, Mariner 9 was silenced permanently on command from ground control. During the mission the space-

† For an excellent resource book, see Glasstone, S., *The Book of Mars,* NASA, Washington, D.C., 1968.

craft returned 7329 TV pictures of Mars and its satellites, and over 54 billion bits of science data. Only preliminary data from the latest Mariner are available to date, but the results provided by the Mariner missions have necessitated many changes in our ideas about Mars. Others of them have been confirmed; this latter success reinforced the confidence that space scientists and astronomers have in their earth-based techniques.

Mars' dimensions, mass, and rotation rate were all essentially confirmed. The difference between Mars' equatorial and polar diameters appears to be nearer 42 km than the 35.4 km obtained by optical measurements. Accordingly the oblateness is nearer 1/162 than 1/192; Mars is flattened more than was supposed. This fact, however, lends increasing support to the hypothesis that Mars' internal composition is more homogeneous than that of the earth.

Mars' atmosphere is different than was earlier believed. Carbon dioxide comprises more than 90% of the Martian atmosphere, gaseous nitrogen less than 1%. The existence of a small amount of oxygen in the upper atmosphere was verified, but the bulk of the haze is caused by particulate, or solid, carbon dioxide. Most of these crystals are confined to a layer some 16 km thick whose base is 16–24 km above the surface. On the other hand some of the obscuration in the vicinity of the polar caps does appear to be caused by ordinary solid water in the form of an ice fog. The composition of the polar caps appears to be either mostly solid carbon dioxide, with perhaps a small amount of snow, or else a thin fog or cloud of carbon dioxide crystals is shrouding the cap.

The Mariner data show that earlier estimates for the atmospheric pressure at Mars' surface were too large. The pressure is more nearly 1% of the earth's surface pressure, rather than 10–15%. Over one extensive region, the *Hellespontus,* situated near 58° S latitude, pressure determinations were substantially less

Lowell Observatory

MARS—1905.

FIGURE 8-20 Percival Lowell's map of Mars (Courtesy Lowell Observatory).

than others made nearby. The inference is that the Hellespontus region, an area of roughly 5800 km², is from 3 to 5 km higher than the surrounding surface. Thus, elevated plateaus such as this may be more significant as surface features of Mars than mountains or hills.

The most spectacular disclosures about Mars came in the form of high-resolution photographs televised back to earth by Mariner 7. Mariner 9 improved upon the data, and in addition showed other features missed by Mariner 7. Three distinct types of terrain occur on Mars: *crater fields*, *featureless* terrain, and *chaotic*, or jumbled terrain. All are substantially different from corresponding features found on the moon.

On September 22, 1971 earth-based observers detected a bright yellow dust cloud that had developed in the Martian southern midlatitudes. Within 2 weeks the storm had spread over the entire planet, obscuring even the prominent south polar cap. Even though the dust storm reached its peak about November 1, Mariner 9 found that the obscuration was still severe. Not until late December could the mapping of the surface proceed according to plan.

The preliminary results of the Mariner 9 mission are startling. Mars is a distinctly different kind of planet than was previously believed, even considering the results of the Mariner 7 mission. Mars is an active planet, not quiescent like the moon, in that geologic processes are still in progress. Dust storms are much more prevalent and are more severe than previously supposed. The dust has a high silica content, resembling the terrestrial crust in this respect, showing that differentiation of light and heavy matter is in progress. It now appears that most, if not all, of the color changes outside the polar

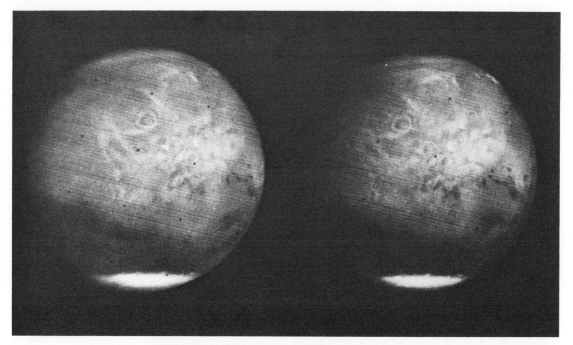

FIGURE 8-21 Two far-encounter views of Mars made by Mariner 7 at 10^h28^m and 11^h14^m, respectively, on August 4, 1969. The distances from Mars: 471, 750 km (right image) and 452,100 km (left image). The bright ring-shaped feature is *Nix Olympica,* a large crater or volcano (Courtesy NASA).

caps result from periodic covering and exposure of the subsurface by wind transported material.

Mariner 7 photographs show that most of the Martian craters are shallower with less steep walls than their lunar counterparts (Figure 8-22). But Mariner 9 showed that some craters are associated with young, though probably inactive, volcanoes in which the crater walls are still undergoing active erosion (Figure 8-23).

Mariner 9 discovered other erosional features not previously seen. Examples are a great chasm with branching canyons (Figure 8-24) and a braided channel resembling a dry terrestrial watercourse, familiar in the southwestern deserts, sweeping past a well-defined crater (Figure 8-25). Another erosional feature,

together with a profile of the elevations above the mean radius of Mars, is shown in Figure 8-26.

Erosion, perhaps by both wind and evaporation, is revealed in a layered structure of an oval tableland near the south pole (Figure 8-27). Rilles, so common on the moon, were thought nonexistent on Mars until the Mariner 9 mapping disclosed them (Figures 8-28 through 8-30). A high-resolution picture of the floor of a crater in the Hellespontus region disclosed a hitherto unsuspected field of sand dunes (Figure 8-31).

Another martian surface feature appropriately called *Featureless Terrain* appears to be unique, with the possible exception of the Great Plains region of the North American continent (Figure 8-32). The largest example so far

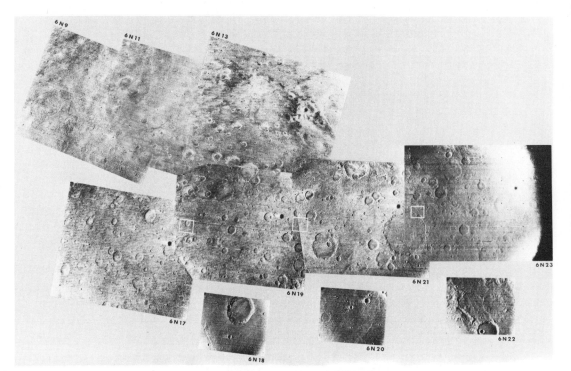

FIGURE 8-22 A mosaic of Mars' surface fashioned from views taken by Mariner 6 on July 30, 1969. The three individual frames at bottom are narrow-angle photographs of the regions outlined in white in the lower four wide-angle pictures. Craters as small as 300 m in diameter are visible in the high-resolution pictures. The four lower frames cover an area 625 km wide by 4000 km long parallel to and 15° south of the Martian equator (Courtesy Jet Propulsion Laboratories and NASA).

FIGURE 8-23 The Martian feature formerly called *Middle spot* is revealed by Mariner 9 to be (probably) a shield volcano. The summit crater is approximately 40 km across. The lower picture was taken with a telephoto lens and is of the area within the inscribed rectangle in the upper photograph. The smooth crater floor is probably a former lava lake. The crater walls are vertically scarred where loose material has slid downslope and has been removed by the Martian wind (Courtesy NASA).

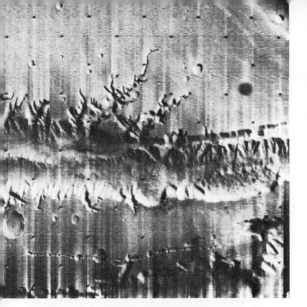

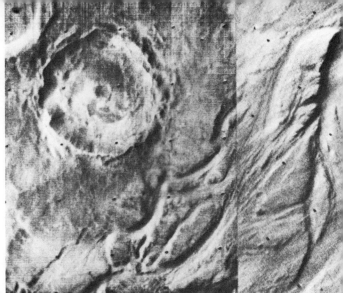

FIGURE 8-24 A vast Martian chasm with branching canyons eroding the adjacent plateau, photographed by Mariner 9 on January 12, 1972. At first the resemblance to watercourse erosion was held to be superficial. Discoveries in mid-1972 of more snow and more water vapor in the Martian atmosphere than was previously thought possible led to strong convictions that watercourse erosion is indeed responsible for such features (Courtesy NASA).

FIGURE 8-25 A braided channel sweeping past a Martian crater strongly resembles fluid-erosion dry washes in the Southwestern desert of the United States (Courtesy NASA).

identified is a bright, circular "desert" called *Hellas* centered about 40° S latitude, generally adjacent to the low-latitude portion of the Hellespontus. This region seems devoid of craters, at least down to the limit of resolution of about 300 m. Contrarily, the Hellespontus is heavily cratered; the craters disappear quite abruptly at the borders of the Hellas. Again, wind-blown dust or sand might well have filled in any and all craters in the latter.

The featureless terrain of the Hellas, a region plagued with frequent dust storms, is now known to be some 6000 m below its western rim and 4000 m below the mean Martian radius. In fact the Mariner 9 data show a vertical difference of over 12,000 m between the lowest and highest elevations. According to estimates made before the Mariner missions, and before radar determinations, the elevation difference was thought not to exceed 3 km.

A mosaic of several hundred TV frames was assembled to provide a panorama of about one-third of Mars' equatorial region (Figure 8-33). Prominent, and almost exactly on Mars' equator, is a magnificent chasm 4000 km long, 110 km wide, and more than 6000 m deep. This canyon is 6 times wider and 4 or 5 times deeper than the Grand Canyon of the Colorado River in Arizona.

Also newly discovered in the Mariner 9 photos is a complex of shield volcanoes with large summit craters in the vicinity of the *Nix Olympica,* a feature visible from earth (Figure 8-33, upper left).† Nix Olympica is the largest of the group, an enormous shield volcano 500 km wide at the base with a summit crater complex (probably a *cauldera* similar to those on earth) 65 km across (Figure 8-34). Elevations just part way up the slope determined by Mariner 9 are over 6000 m above the base.

† Shield volcanoes result from successive, rather quiescent, outpourings of lava that build up into mountains with gently sloping flanks and with a large central crater. With other types of volcanoes, explosive emission is the predominate mountain building process.

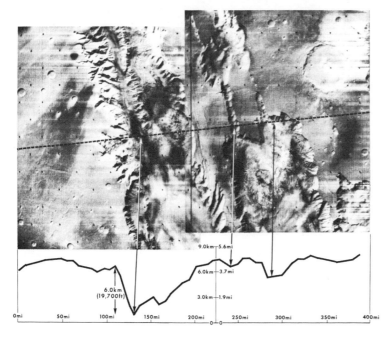

FIGURE 8-26 A mosaic of two photographs of the *Tethonius Lacus* region of Mars, together with a graphical profile of the vertical extent of the canyons. The white arrow on the left points to a canyon 6 km deep and 120 km wide. By comparison, the Grand Canyon of the Colorado River in Arizona is only 1.6 km deep and 21 km wide (Courtesy NASA).

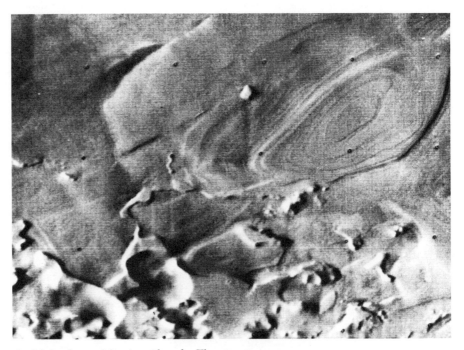

FIGURE 8-27 An oval tableland near Mars' south pole. The structure appears to consist of alternate layers of dust or volcanic ash and either or both carbon dioxide and water ice. Shadows in the pits indicate depths exceeding 150 m (Courtesy NASA).

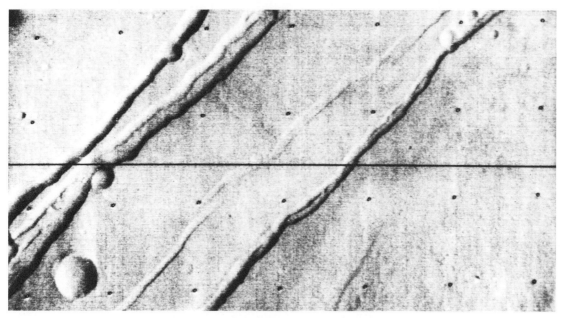

FIGURE 8-28 The first clear indication of the existence of rilles in the Martian crust, photographed on January 7, 1972 by Mariner 9. The widest rille (upper left) is nearly 2 km wide (Courtesy NASA).

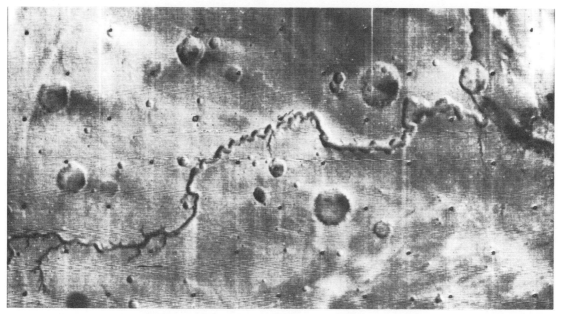

FIGURE 8-29 A sinuous rille resembling a terrestrial *arroyo*, or watercut gulley, fairly common in southwestern United States. The rille is 400 km long, and 5 to 6 km wide. No known lunar rille shows branching such as this (Courtesy NASA).

FIGURE 8-30 The *Rille of Rasena*, a sinuous, dendritic valley 700 km long (a Mariner 9 photograph). A rille may represent the collapse of a roof over subterranean lava flow, although the possibility that rilles are the result of water erosion in the early history of Mars cannot be ruled out (Courtesy NASA).

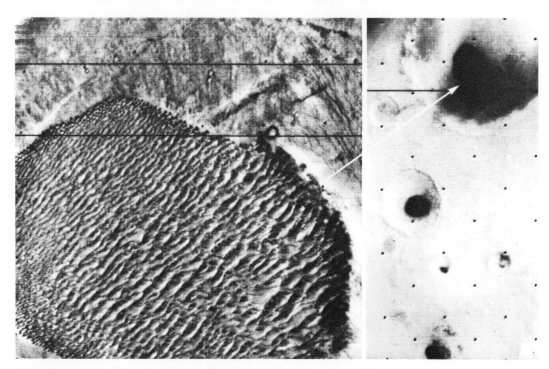

FIGURE 8-31 Sand dunes (right) in the floor of a Martian crater (left), a Mariner 9 photograph. The numerous long dunes are spaced about 1.5 km apart and were evidently formed by a strong prevailing wind. The crater, in the Hellespontus region, is 150 km wide (Courtesy NASA).

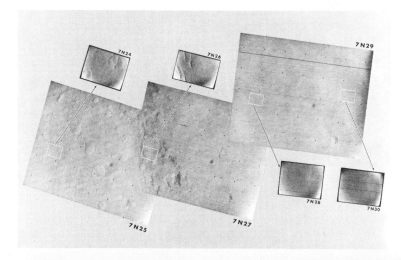

FIGURE 8-32 A mosaic of three Mariner 7 photographs (1969) of the bright circular desert, *Hellas* (Courtesy NASA).

FIGURE 8-33 A magnificent panoramic mosaic of several hundred individual Mariner 9 TV frames, covering one-third of the circumference of Mars. The Martian equator horizontally bisects the panorama. In the center of the mosaic is an enormous canyon 4000 km long, 120 km wide, and nearly 6500 m deep. On earth the feature would stretch from Los Angeles to New York; San Diego and Los Angeles would be on opposite rims (Courtesy NASA).

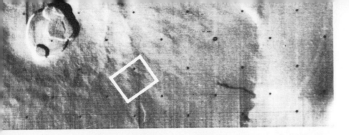

FIGURE 8-34 A portion of the gigantic shield volcano, *Nix Olympica,* showing the complex of summit craters. The high resolution detail of the area outlined by the white rectangle reveals a characteristic downslope flow pattern. Observe the large crack running along the ridge. Nix Olympica is plainly visible as a circular pattern in the Mariner 7 views (Figure 8-21), and in the extreme upper left of the equatorial mosaic (Figure 8-33) (Courtesy NASA).

The largest shield volcano on earth forms the Hawaiian Islands complex; this volcano is only 225 km across the base, and rises some 9000 m from the ocean floor to the summit of Mauna Loa. Nix Olympica is more than twice this diameter; it may well rise as high, or even higher.

A third kind of surface feature, the *chaotic terrain* (Figure 8-36), is likewise not found on the moon, although it may have its counterparts in landslide areas on earth. The chaotic terrain consists of irregularly jumbled short

ridges and depressions, essentially uncratered, and much different in pattern from ejecta sheets or the ropy pattern of lava flows on earth. The terrain appears to have collapsed as though from the withdrawal of substrata, possibly permafrost, from the underlying layers.

The Mariner missions failed completely to discover any signs of life-forms, vegetative or otherwise, and the missions also failed to disclose any evidence for the canals or oases. They did not show any evidence of the seasonal or areal darkening. The nature of these phenomena, if they exist, must be determined by further exploration, possibly by the soft-landing Viking missions planned for 1975.

The quest for evidence of life-forms was neither encouraged nor discouraged by the Mariner missions. All that can be inferred from

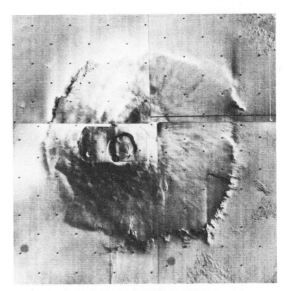

FIGURE 8-35 A mosaic of Mariner 9 photographs showing *Nix Olympica* in its entirety. The volcano is 500 km across at the base, compared to 225 km for the volcano whose highest summits form the entire Hawaiian Islands chain. Steep cliffs at the base appear as though they fell away into a water depth at one time. The height of Nix Olympica rivals that of *Mauna Loa* (Hawaii), which rises 9000 m above the ocean floor (Courtesy NASA).

FIGURE 8-36 Chaotic, or slump terrain, apparently caused by a landslide detached from the smooth higher plateau. Parallel grooves in the slide area resemble terrestrial avalanche tracks. This Mariner 9 photograph covers an area of 450 km by 350 km (Courtesy NASA).

the mounting evidence is that higher life-forms as we know them are impossible in the Mars environment. For one thing ultraviolet radiation from the sun probably reaches Mars' surface in lethal quantities. The absence of appreciable water and oxygen is a serious, if not fatal, shortcoming. But we cannot as yet rule out the possibility of simple life forms such as bacteria and lichens. Hopefully the Viking missions with their avowed intention of concentrating on the search for life will resolve this intriguing issue.

Mars is easily the most interesting target for continued intensive exploration. It is the most accessible and most hospitable of all of the planets; thus Mars favors direct manned exploration. Mars is an essential link in our quest for

knowledge about the origin, and possible fate, of the solar system.

Jupiter. Jupiter, the fifth planet outward from the sun, is aptly named after the lord of Olympus. Jupiter is the Roman equivalent for the Greek *Zeus,* chief among the legendary gods, god of moral law and order, protector of suppliants and punisher of guilt. Except for the sun, Jupiter dominates the solar system in size, in mass, in volume, and in ability to gravitationally perturb the rest of the family. It is 318 times more massive than the earth, indeed, 2½ times more massive than all of the other planets put together; only 1000 times less massive than the sun. Jupiter's equatorial diameter is 11 times that of the earth, only 10 times smaller than the sun, and its volume could hold 1300 tightly packed tiny earths. Jupiter's density is low compared to the terrestrial planets, only 1.33 times that of water. A person weighing 150 lb on earth would tip the scales at 347 lb on Jupiter. Thus Jupiter's surface gravity is 2.31 times that of the earth.

Jupiter rotates about an axis inclined only 3° from the vertical (Jupiter's orbit is the horizon-

FIGURE 8-37 Jupiter, photographed with the 36-in. refractor at Mount Hamilton, California (Courtesy Lick Observatory).

Physical Characteristics **191**

tal reference) with a period of 9ʰ55ᵐ. Jupiter rotates faster than any other planet. A point on its equator travels more than 48,000 km/hr relative to the celestial sphere, over 30 times faster than a corresponding point on earth. As a result, Jupiter is very oblate. Its equatorial diameter is 143,900 km, while its polar diameter is 134,130 km. The flattening of more than 9700 km represents an oblateness of 1/15; it is easily discernable in a small telescope.

Jupiter's orbit is nearly circular with an eccentricity of 0.048. Thus Jupiter's distance from the sun varies by 76 million kilometers; the mean distance is 780 million kilometers or 5.2 AU. Nevertheless, climate variations because of seasonal effects are nonexistent on Jupiter.

Jupiter's sidereal period is 11.86 earth-years, in good agreement with the predictions of Kepler's laws. The synodic period is only 1 year and 34 days, small compared to the length of the sidereal period. This is because Jupiter travels only about one-twelfth of its orbital path between successive oppositions.

Jupiter is not easy to observe in detail; it is too remote from the earth (607 million kilometers even at the most favorable oppositions. Even so the surface displays a striking alternation of light zones and dark bands that are parallel to the equator (Figure 8-37). The dark bands are more or less uniformly gray, although some lighter-colored spots are occasionally seen. The lighter bright zones vary somewhat in color both in position and in time. They may appear light yellow, or again light red, and even at times light bluish in cast. Moreover each band and zone rotate about Jupiter's axis at a different rate; the variation in periods ranges from 9ʰ59ᵐ in high northern latitudes, to 9ʰ49ᵐ at the equator. The rotation is asymmetrical, that is, speeds in the southern hemisphere are somewhat higher than at corresponding northern hemisphere latitudes. The broad equatorial zone, some 16,000 to 24,000 km wide, travels some 320 km/hr faster than the adjacent regions; it gains one

full lap on its neighboring bands in 103 Jovian days!

Quite obviously we do not observe a solid surface; only a cloud cover or a totally liquid or gaseous body could rotate at differential rates. Experienced observers are convinced that the banding is associated with zonal winds traveling parallel to the equator. One interpretation is that the light zones are regions of rising, warmer currents that are heated from below. The dark bands are cold sinking currents, or perhaps clearing regions through which the surface of Jupiter may be visible.

Jupiter's atmosphere consists predominantly of hydrogen, with substantial amounts of methane, more familiarly known as ordinary household natural gas, and ammonia. This is the kind of atmosphere the earth is thought to have originally had, the kind from which amino acids and long-chain protein molecules were formed in the presence of electrical discharges. These compounds are the forerunners, or building blocks, of living structures. This atmosphere has long since vanished from the earth. In Jupiter's case, however, the average low temperature (−150°C) and large mass combine to produce an escape velocity of over 58 km/sec, greater than the average speed of any gaseous molecule at this temperature. Presumably Jupiter's atmosphere is primordial, that is, it has existed since the birth of Jupiter from the nebula that mothered the entire solar system.

Jupiter is a very different kind of object than either the sun or the terrestrial planets. Theoretical studies indicate that the total atmospheric thickness cannot exceed 165 km. At this distance below the cloud tops the hydrogen begins to liquify under the high pressures existing there. The percentage of liquid material gradually increases like a thickening gruel until finally it becomes a liquid ocean of hydrogen. But this isn't all. Within 245 km below the ocean surface, the pressure becomes so enormous that the liquid becomes slushy, then solid. Whether or not the solid hydrogen encloses a rocky core cannot be hypothesized.

Jupiter gives off significantly more energy than it receives from the sun, some of it in the form of "bursts" of radio noise. Jupiter also has a magnetic field similar to, but about 10 times more intense than, the earth's. Indeed the most reliable data concerning Jupiter's rotational rate are obtained by radar tracking of the bursts of radio noise that seem to be fixed with respect to unseen surface features. Both the radio bursts and the magnetic field are characteristic of a high-temperature, electrically conducting core. Hydrogen in the liquid or solid state is metallic in character and thus is a good electrical conductor. Gravitational pressures associated with a planet of Jupiter's size could produce internal heating effects whereby temperatures could be as high as 11,000°C. Because Jupiter's core is not overlaid with an insulating rocky mantle, some of the internal heat may escape. Perhaps this is the explanation for Jupiter's mysterious energy source. In any event as seen from the earth Jupiter is the most powerful radio source in the sky save the sun.†

The most intriguing visible feature of the Jovian atmosphere is the Great Red Spot, first reported in 1831. It may have been seen by Cassini in 1660 and used by him in determining Jupiter's rate of rotation. But the first authenticated report dates only to the first third of the 19th century at which time the Great Red Spot appeared to be a great hollow in the clouds (Figure 8-38). By 1878 the spot had become deep red, 48,000 km long by 11,000 km wide. The spot was large enough to hold the earth and its moon, and the planets Mercury, Venus, Mars, and Pluto with room to spare.

The Great Red Spot has long since faded but it is still plainly discernible. No one knows just what causes the phenomenon; one suggestion

† Other radio sources are known to emit vastly more energy than Jupiter. But their strength is attenuated across the enormous distances that separate them from the earth, thus these sources are perceived only very weakly.

FIGURE 8-38 The Great Red Spot on Jupiter, photographed in blue light with the 200-in. reflector at Mount Palomar (Courtesy Hale Observatories).

is that it is the top of a huge column of heated atmospheric gases that is rising up over some immense surface obstruction. The spot wanders somewhat from a mean position. One calculation shows that the spot has drifted slowly around the planet, completing at least three revolutions since it was discovered. If this is so, we would find it difficult to explain how the spot could be associated with any elevated surface feature.

Pioneer missions. Much of what we know about Jupiter may be either corroborated or refuted in the very near future. As this book goes to press, NASA's Pioneer 10 and Pioneer 11 spacecraft are winging their way toward a near encounter with Jupiter in December 1973 and January 1975, respectively. Pioneer 10 was launched March 2, 1972 on a trajectory that will carry it within 130,000 km of Jupiter on December 3, 1973. During a week in the immediate vicinity of the giant planet, Jupiter's magnetic field will be measured and its radiation belts mapped. If all systems remain in

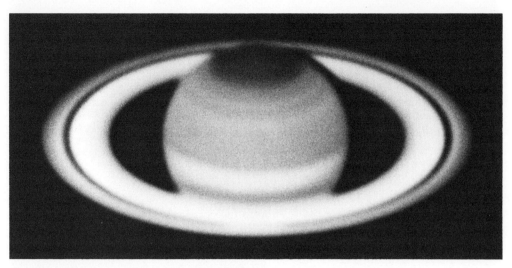

FIGURE 8-39   Saturn, showing the planet and ring system, photographed in blue light with the 100-in. reflector at Mount Wilson (Courtesy Hale Observatories).

working order, Pioneer 10 will take Jupiter's temperature and will also take about 20 pictures of the banded cloud cover. Thereafter, Jupiter's massive gravitational field will act like a slingshot in propelling the spacecraft out of the solar system toward the constellation Orion. Perhaps it will eventually encounter another unknown "solar" system, and thus be the first manmade vehicle to signal to other unsuspecting civilizations that there is intelligent life elsewhere in their universe.

Between July 15, 1972 and February 15, 1973, Pioneer 10 successfully negotiated the 328 million km wide asteroid belt (Chapter 9) without incident. The region proved to be potentially far less dangerous than scientists feared; Pioneer 11 is accordingly expected to traverse the region without significant danger from collision with celestial missiles in the asteroid zone. Pioneer 11 is a back-up mission for Pioneer 10. Depending upon the success of the latter, Pioneer 11 may be guided toward a near or far pass over Jupiter's equator, or over the south pole which some scientists believe may be cloud free. If all goes well, Pioneer 11 may then swing by Saturn for a close look

before it departs the solar system headed for its ultimate destiny.

Saturn.   Saturn, named after the god of seed-sowing and the mythological father of Jupiter, is the most remote of the planets that are readily seen with the unaided eye. Indeed, only one more planet, Uranus, has even the potential of being seen without the aid of a telescope. Saturn is the last planet outward from the earth in the ancient heirarchy. Because of its remarkable ring structure (Figure 8-39), Saturn is, except perhaps for the moon, easily the most dramatic, if not the most beautiful, telescopic object in the sky.

Saturn completes one trip around its orbit every 29½ years (its sidereal period); its synodic period is only 378 earth-days. Direct measurements confirm the predictions of Kepler's laws that Saturn's mean distance from the sun is 9.5 AU or 1426 million kilometers. This places Saturn a prodigious 648 million kilometers out beyond Jupiter. It is so remote from the sun that a Saturnian observer would see a solar disk only one-tenth the size we observe. At this distance the earth, which never is closer to Saturn than 1240 million kilometers, will always appear within 6° of the sun. Indeed, Saturn's orbit is so large compared to that of the earth that Saturn's apparent disk varies in diameter between only 14 and 20″ of arc at superior conjunction and opposition, respectively. Thus Saturn is always almost as bright as the brightest stars. Saturn is so remote from the sun that the intensity of solar radiation at the planet's average distance is only 1/95 that arriving at the earth. It takes light 1 hour and 20 minutes to arrive at Saturn from the sun.

Saturn's orbit is only slightly more eccentric (eccentricity = 0.056) than the orbit of Jupiter. But even this relatively small eccentricity for such a large orbit means that Saturn's distance from the sun varies by 160 million kilometers between perihelion and aphelion passage. Saturn's orbit is inclined $2\frac{1}{2}°$ to the ecliptic, and its axis of rotation is tipped $26\frac{3}{4}°$ from the perpendicular to its orbit. Thus Saturn will experience a cycle of seasons, if a distinction can be made between very cold and colder still.

Saturn's diameter is just under 121,000 km, second only to Jupiter. But its mass, determined by perturbations of its satellites is only 95 times that of the earth. Thus Saturn is a lightweight; the combination of large diameter and small mass gives Saturn a density only 0.7 that of water. This means that given an ocean large enough to hold it Saturn would float!

Saturn's rotational period cannot be determined as accurately as can that of Jupiter. Clouds that are less conspicuously marked than Jupiter's conceal Saturn's surface. Doppler-shift measurements made on the light reflected from the cloud cover, first from one limb and then the other, yield a fairly reliable period of $10\frac{1}{2}$ hours. Saturn's high rate of spin, together with its small density, produces the largest oblateness known in the solar system, nearly one-tenth. Thus, a point at either pole is 6000 km closer to Saturn's center than any point on the equator.

Saturn's surface gravity is only 17% greater than the earth's. The large oblateness would have a peculiar effect on a Saturnian traveler; his weight would vary by 30% depending merely on whether he is at one of the poles or on the equator. At Saturn's average surface temperature of −230°F, the velocity of escape is 35 km/sec. Thus Saturn, like Jupiter, has retained its original atmosphere; it resembles that of Jupiter except that ammonia is absent. Presumably it has crystallized out as snow because of the lower temperature. Cloud bands are visible on Saturn, but they lack the distinctive markings and the color contrasts seen in Jupiter's bands.

Saturn's temperature at the top of the clouds is nearly 40°F warmer than it should be at this distance from the sun. Saturn probably has an internal structure and an internal heat source similar to those of Jupiter.

Saturn's Rings. Saturn's striking appearance, one that never fails to create a feeling of awe among observers seeing Saturn through a telescope for the first time, is due to the system of rings surrounding Saturn almost precisely in the plane of its equator. Three concentric rings are observed spanning a distance of 12,900 km to 217,000 km from Saturn's surface. Because Saturn's axis of rotation is fixed in space, terrestrial observers alternately see the rings 28° "open" (Figure 8-40) from below the plane of the rings, then edge-on in which they disappear from view, and then finally 28° open from above the plane of the rings. The complete cycle coincides with Saturn's sidereal period.

The three rings, separated by narrow gaps, vary in brightness from the broad *bright* central ring to a fainter, narrower *outer* ring, and an extremely faint, gossamer-like inner *crape* ring. The ring system is very thin, perhaps only 16 km thick. Indeed, theoretical studies show that the rings may be only a few meters in thickness. Doppler-shift studies of the reflected light demonstrate that the inner ring rotates around Saturn's axis with the shortest period; the periods of the other rings increase outward in agreement with Kepler's laws (Figure 8-41).

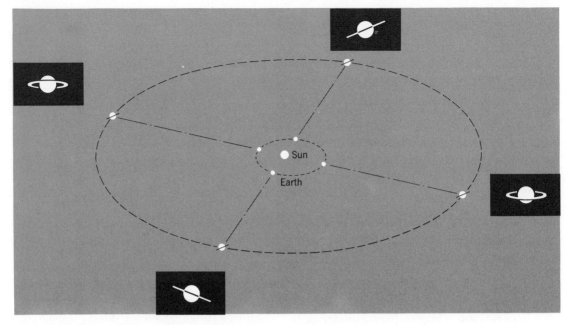

FIGURE 8-40 Four different aspects of Saturn's ring system as seen from earth at approximately 7½ year intervals.

Thus the rings are independent of each other. Spectral analysis of the light reveals that they are composed of billions of solid particles, ices perhaps, or ice-coated grains of solid rock varying from microscopic up to marble size. In spite of the apparent brightness of the rings, they are quite transparent. Variations in the brightness show that the solid material makes up only 6% of the volume of the rings. Occasionally stars can be seen shining through the rings. On the other hand the rings are dense enough to cast a shadow on Saturn's cloud cover as well as to show a shadow cast by Saturn on them.

The rings are very likely the remains of a satellite that strayed into the "forbidden zone" (Chapter 6) where gravitational tidal disruptions exceed the cohesive forces holding the body together.

Uranus. As our inquiries take us farther and farther outward through the solar system, we

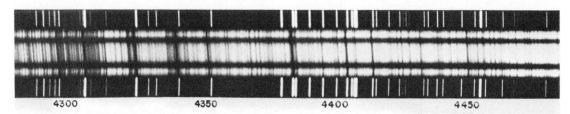

FIGURE 8-41 Spectrum of Saturn and the ring system showing differential Doppler displacement of the spectral lines resulting from the relative differences in the rotational rates and directions (Courtesy Lowell Observatory).

become increasingly aware of the limitations of our earth-based instruments. Optical images in the telescopes become more diffuse; errors in the precision of angular measurements become relatively more serious, and we must resort to, and rely more and more on, secondary measurement techniques such as Doppler-shift phenomena. As a result our knowledge about celestial objects, including the more distant members of the sun's family, becomes couched more and more in general terms; the details of their nature and behavior become inaccessible to us.

So it is with Uranus, seventh planet from the sun, the first addition to the classical solar system (or earth system from the Ptolemaic reference) family. Uranus is a faint object just barely within the limit of visibility of the unaided eye (Figure 8-42). At least 5000 stars appear as bright or brighter than it does. The famous English astronomer, Sir William Herschel (German by birth and English by adoption) discovered Uranus on March 13, 1781 while he was making a routine survey of the sky in the constellation of Gemini. Once he had spotted the planet he found it easily distinguishable because Uranus displays a small but visible disk; stars do not.†

Herschel thought he had discovered a comet, but after watching it for several weeks it became evident that the object was indeed a new planet. When past records—the object had been seen and recorded (as a star) for more than 100 years—were analyzed along with Herschel's new data, a nearly circular orbit beyond Saturn, inclined only 46' to the ecliptic emerged. In keeping with tradition the planet was named after one of the gods in Greek mythology; Uranus was the father of the Titans, and the grandfather of Jupiter.

Uranus' distance from the sun is measured in billions, not millions, of kilometers. At 19.2 AU

† The images of all stars formed by a telescope appear subtly diffuse rather than as points. This is an aberration in the telescope to which we shall pay more attention in Chapter 11.

FIGURE 8-42 Uranus and its satellites photographed with the 120-in. reflector (Courtesy Lick Observatory).

or 2 billion 868 million kilometers, Uranus is more than twice as far from the sun as Saturn. If we could voyage to Uranus while maintaining a constant speed, say, equivalent to the 29,000 km/hr velocity of low-orbiting earth satellites, the trip would take 10 years! A radio message traveling at the speed of light from earth to listeners on Uranus would take 2^h42^m to traverse the distance; we wouldn't hear the response for at least 5^h24^m after the message was sent.

Uranus' orbit is so huge that 84 years must elapse for the completion of a single sidereal period. On the other hand the synodic period is only 369½ earth-days. The earth is always seen within 3° of the sun from Uranus; sunlight intensity here is only 1/370 as intense as it is at the earth's distance. Thus the temperature is always below −300°F, bewilderingly cold.

Uranus' diameter is somewhat uncertain; however, it is in the neighborhood of 47,400 km. Perturbations of its satellites lead to an estimate that its mass is 14.5 times that of the earth. Thus Uranus' density is about 1.56 times that of water, a value characteristic of other of the major planets. Uranus' surface shows no distinct markings; its rotational speed must be found by spectrographic methods (Doppler-shift determinations). This value, again uncertain, leads to a period of 10^h45^m, again characteristic of the major planets. Its oblateness at 0.06 is greater than Jupiter's but less than that of Saturn. Uranus' polar diameter is between 3500 and 4500 km less than its equatorial diameter.

Uranus appears distinctly greenish in the telescope; the green hue is caused by a substantial fraction of methane in an otherwise hydrogen atmosphere. No ammonia can be detected; presumably as on Saturn it has crystallized out of the atmosphere and is deposited in the surface. The structure of Uranus is unknown, but the available evidence points to great similarities with Jupiter and Saturn.

Uranus' rotational spin is retrograde, opposite to its orbital motion which is, like all the other planets, direct. Moreover, the axis of rotation is inclined 82° from the vertical so that the planet is in effect lying on its side with the axis pointing in a fixed direction in space. As a result Uranus' seasons are very different from our own. At the solstices, the sun describes an apparent path around the respective pole only 8° from the polar zenith. The sun arrives at the equator 21 earth-years later, and both poles are in near darkness. The sun finally sets at one pole as it is about to rise at the other. The period between midsummer and midwinter is 42 earth-years.

How it is that Uranus is spinning in a retrograde sense around an axis that is almost horizontal arouses some interesting speculation. Let us consider the situation shown in Figure 8-43. In (a) we see a planet with axis of spin perpendicular to its orbit; the spin direction is direct. In (b) we see the configuration that would result if the planet were tipped 45°. In (c) the axis is horizontal as if the tipping were continued in the same direction through another 45°. The rotational motion is now perpendicular to the orbit and both of the terms, retrograde and direct, become ambiguous in this situation. But in (d) we see the *former* north pole depressed below the plane of the orbit; with respect to this pole the direction of spin is unchanged. By definition, however, *the former south pole is now the north pole;* with respect to *this* pole the spin is retrograde. We wonder: *Has* Uranus somehow tipped over from an original position such that its spin was direct? If so, what kind of force could have produced the upset? How long did it take?

By 1821 Uranus had appreciably departed from the theoretical positions calculated by the French astronomer, Alexis Bouvard. Bouvard had derived his tables from the combined data of more than 20 sightings dating back 131 years, including those of Herschel. He had also carefully included the expected perturbations

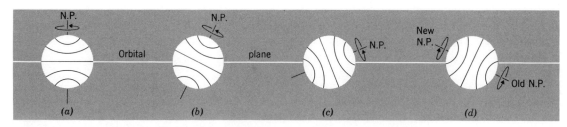

FIGURE 8-43 Rotation of a planet relative to the plane of its orbit. (*a*) Initial position; (*b*), (*c*) sense of rotation is unchanged even though planet's axis is displaced from the vertical; (*d*) planet continues to spin in the same direction, but relative to the orbital plane the opposite pole is now the north pole. Thus the sense of rotation is reversed in (*d*).

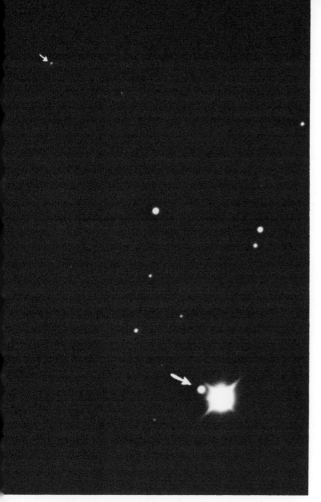

FIGURE 8-44 Neptune and its satellites (Courtesy Lick Observatory).

from Saturn's gravitational tugging. But by no amount of "forcing" could Uranus' observed positions be made to fit the calculated Keplerian ellipse. In desperation Bouvard rejected all of the data previous to Herschel's in the belief that it was in error. He then recast the position tables, "leaving to the future," he wrote, "the task of finding out . . . whether (there is) some strange and undetected influence acting on the planet."

In spite of these corrections Uranus deviated from the revised predictions even more severely during the next 20 years. The serious question arose among astronomers, whether to reject the hitherto infallible universal law of gravitation as being inapplicable at such great distances from the controlling body (the sun), or to search for a possible perturbing body. Almost no one dared tackle such an extraordinarily difficult problem, except for a 23 year-old undergraduate student at Cambridge. After 2 years of prodigiously laborious calculations (a task that could now be completed on a modern high-speed computer in 20 minutes), John Couch Adams proved the perturbations in Uranus' motions could be explained by a trans-Uranian planet. Moreover he had computed its expected mass and orbit. But Adams' work was not published and, inexplicably, no one in England who knew of it bothered to look for the planet.

Meanwhile in France a lecturer in astronomy at the Paris *École Polytechnique,* U. J. Le Verrier, unaware of Adams' work, independently set about the task. In mid-1846 he published his convictions that a new planet was responsible for the perturbations. He communicated the predicted coordinates to the Berlin astronomer Galle on September 23, 1846. That same night Galle found *Neptune* in the constellation of Aquarius. Both Adams and Le Verrier have since been jointly credited with the discovery.

Neptune. Neptune plods along in its orbital path, 30 AU from the sun, at only 5 km/sec, one-sixth the orbital speed of the earth. It will take Neptune 165 years to complete 1 sidereal period since its discovery, a task that will not be fulfilled until the year 2011.

The solar system from a Neptunian vantage point presents a bleak and strange vista. Light takes 4 hours to cover the 4 billion 495 million kilometer trip from the sun to Neptune. At this unimaginably great distance the sun appears no larger than Venus does to us, but it is about 500 times brighter than the full moon as we see it. Mercury, Mars, and Uranus could not be seen from Neptune without the aid of a telescope. (They are too faint.) Jupiter and Saturn would appear as inconspicuous "stars," never

Physical Characteristics 199

more than 10° and 17°, respectively, from the sun. The earth and Venus could be seen very faintly, but only when the sun was eclipsed by one of Neptune's moons. Earth and Venus are always within 2° and 1½°, respectively, from the sun and are normally lost in its brilliance. From a vantage point on earth, if Neptune were 5 times brighter, the planet would be just at the limit of naked-eye visibility.

Neptune's constants are determined with difficulty and the margins of error are understandably large (Figure 8-44). Neptune's diameter is approximately 45,000 km; its mass is 17.2 times the earth's mass, and it has a mean density of 2.45 that of water. Neptune's rotational period is 15^{h}48^m, and at its largest its disk is only 4″ of arc in diameter. The disk appears bright green due to the presence of methane in its atmosphere. In every visible respect Neptune is sufficiently like Uranus as to warrant the appellation, "Uranus' twin."

Of passing interest is the fact that Neptune is the first of the planets that does not "fit" the Bode-Titius scheme within acceptable limits.

Pluto. As if it were too shy and wary of venturing into the light, Pluto circles endlessly around the rest of the known solar system in a region of nearly perpetual darkness. Its path is 39.5 AU or 5 billion 900 million kilometers from the sun (on the average); this is nearly 1.5 billion kilometers out beyond Neptune. Pluto's orbit displays a mysterious, intriguing anomaly, however. Its eccentricity is 0.25, larger than that of any other planet except Mercury. Furthermore the orbit is inclined a phenomenal 17° to the ecliptic. This is 2½ times the inclination of the orbit of Mercury, 5 times the inclination of the orbit of Venus, and more than 8 times the inclination of the orbit of any of the other planets except Saturn (Figure 8-45).

The large eccentricity results in Pluto being more than 50 AU from the sun when at aphelion, yet only 29.7 AU distant when at perihelion. Pluto will not make its first perihelion pas-

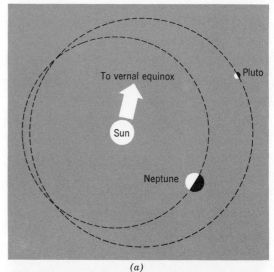

(a)

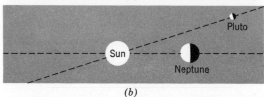

(b)

FIGURE 8-45 (a) Plane view of Pluto's orbit relative to that of Neptune; (b) elevation view showing the relative inclination of Pluto's orbit to that of Neptune.

sage since its discovery until the year 2010. At that time Pluto will be 56 million kilometers *inside* Neptune's orbit! Yet there can never be a collision; because of the orbital inclination Pluto never comes closer to Neptune than 385 million kilometers.

Pluto is 4000 times too faint to be visible with the naked eye. The planet is so small no existing telescope can assure an accurate determination of its angular size. Pluto's diameter is estimated from the way it occults faint stars; statistical treatment of widely separated observations yields an upper limit of 6800 km with a probable value of 6200 km. This latter

 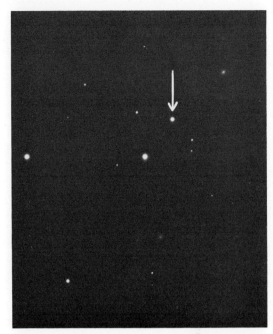

FIGURE 8-46 Pluto (Courtesy Hale Observatories).

estimate is in good agreement with the most reliable measurements obtained with the 200-in. Hale telescope at Mount Palomar in California.

A recent determination of Pluto's mass based upon the way it perturbs Neptune is 0.6–0.8 times that of the earth. Assuming the validity of such an uncertain quantity, together with acceptance of the 6200 km diameter, Pluto's density is 1.4 that of the earth, a figure that seems inordinately high. Yet any lower density would require a corresponding decrease in the mass. It would not take much of a decrease such that Pluto could no longer produce a measurable perturbation on Neptune. This problem awaits solution.

Pluto's sidereal period is 248 earth-years; its synodic period exceeds 1 year by only 1½ days (Figure 8-46). Pluto's brightness appears to fluctuate with a regular period as though the albedo is different for different parts of the

planet. This period, 6.39 earth-days, is assumed to be that of its rotation. No atmosphere can be detected. Presumably those gases that are not frozen in the perpetual cold of −400°F have escaped; Pluto's low mass could not have held them in gravitational captivity. The intensity of sunlight as seen from Pluto is only 1/1550 as intense as it is on earth.

Considerable speculation exists concerning Pluto's role as a planet. Its small size and its highly eccentric and greatly inclined orbit suggest that Pluto *may* have been one of Neptune's satellites that somehow was perturbed into an escape orbit around the sun. But all efforts toward verifying this possibility through dynamical analysis have thus far proven futile.

Pluto was discovered, as it were, "by mistake." Rather, the discovery was made legitimately but the search was based on questionable data. With the discovery of Neptune, all of the discrepancies in Uranus' motion save about 1/60 could be accounted for. This small residual, plus unaccountable perturbations in Neptune's orbit, suggested the existence of still another planet farther out. Percival Lowell deduced the position coordinates for the unseen planet from an analyses of these perturbations, but he died in 1916 before he could institute a search for it. The planet, unrecognized as such, actually appeared near the predicted position on four plates taken at Mount Wilson Observatory in 1919.

A more perservering search was eventually instituted at Lowell Observatory. In March, 1930 Clyde Tombaugh announced to the world that the suspected planet had been located on plates taken the preceding January. The name Pluto was adopted, partly because this mythological god was ruler of darkness, but mostly because the first two letters, PL, were the initials of Percival Lowell. Thus joint tribute could be made with this most appropriate selection.

We now recognize that the residuals in the perturbations of Uranus that prompted Lowell's search were *less* than the instrumental errors in the observations. Indeed, Lowell predicted a mass for the intruder that was 6 times that of the earth; any lower value could not have produced the perturbations. As we have seen the evidence supports a value some 5 times smaller. That the planet was located close to where Lowell predicted it would be found must stand as one of the most remarkable coincidences in the history of celestial mechanics.

Planet "X". An intriguing possibility of the existence of a trans-Plutonian planet was reported in April, 1972 by a group of computer scientists at the University of California Lawrence Laboratories at Livermore. The group is headed by J. L. Brady, a mathematical astronomer.

In a paper appearing in the *Journal of the Astronomical Society of the Pacific,* Brady and his co-workers hypothesized that previously unexplained discrepancies in the prediction for the return of several periodic comets (p. 221), notably Halley's comet, were due to perturbations by a tenth planet at a mean orbital distance of 64.5 AU, some 24.5 AU out beyond Pluto. Computations showed that the mass of the hypothetical planet is 300 times that of the earth, and that its orbital period is 512 years.

More importantly, the "new" planet appears to be traveling in a retrograde direction in an orbit that is inclined some 60° to the plane of the ecliptic. If the planet exists it should be at present in the constellation Cassiopeia, near the plane of the Milky Way. Because the planet will be considerably fainter than Pluto, and because of possible obscuration by the Milky Way, the physical discovery of such a planet may elude searchers even longer than did the discovery of Pluto.

In October, 1972, the above hypothesis for the existence of a tenth planet out beyond Pluto was strongly refuted by three prominent astronomers. Among their objections was the belief that a planet having the mass, mean orbital distance, and orbital inclination as developed by the Brady group, would have been bright enough to have surely been visible on the plates of a photographic survey of the northern sky made at Lowell Observatory during the years 1929–1945. The plates reveal stars at least two magnitudes fainter than the apparent brightness of the planet. Considering the formidable difficulties involved in examining the plates in search of a single tiny speck (recall the search for Pluto) that would be the planet, we should consider the case open. The dispute should spur further search for *indisputable* proof, either for, or against, the planet's existence. Controversies such as this are the very stuff of which the whole cloth of astronomy is woven.

SUMMARY

Examination of the motions of the planets known to antiquity, those out to and including

Saturn, shows that they all closely obey Kepler's laws of celestial mechanics. A peculiar coincidence between the mean planetary distances measured with respect to the sun and a certain geometric progression of numbers (p. 156) was publicized by Bode of the Berlin Observatory in 1772. No real scientific significance has been discovered for the relationship, known as Bode's law. When Neptune and Pluto were subsequently discovered, their distances did not fit the Bode's law pattern, thus effectively disproving its validity as a law.

Planetary positions as seen from earth are described in terms that depend in part on whether the planetary orbit is inside of or exterior to the earth's orbit. All planets can achieve *superior conjunction* in which the planet is in alignment with but on the opposite side of the sun as seen from earth. Inferior (interior) planets can be seen at *inferior conjunction* wherein the planet is in alignment with but on the near side of the sun. Superior (exterior) planets reach *opposition* in place of inferior conjunction, a position that places them diametrically opposite in direction from the sun as seen from earth.

Maximum displacements east or west of the earth-sun line are termed *maximum elongations* and these only apply to inferior planets. For superior planets the position is *quadrature*, since the maximum displacement occurs when the planet is at a right angle relative to the earth-sun line. The adjectives *eastern* or

western indicate on which side of the line the planet is to be found.

Five *elements* uniquely fix the position and orientation of a planet's orbital plane relative to the ecliptic and the orientation of the orbit in the plane. The elements are: a, semimajor axis; e, *eccentricity;* and i, inclination of the orbit in question; (these describe the size and shape of the orbit, and the relative position of the orbital plane); Ω, the heliocentric (sun-centered) longitude of the ascending node, Ω and ω, longitude of the perihelion point, (these describe the orientation of the orbit within its orbital plane).

The planets Mercury, Venus, Mars, and Pluto are termed *terrestrial planets* because of their close resemblance in size to the earth. The much larger planetary relatives, Jupiter, Saturn, Uranus, and Neptune are known as the *major planets*.

Each of the planets displays some familiar terrestrial characteristics, in size, in composition, or in motion. Yet each planet has more numerous characteristics that are startlingly different. With the possible exception of Mars, no other planet appears capable of supporting life in the form or chemistry as we know it. Mars alone appears amenable to manned exploration by earth's inhabitatnts. Unquestionably man will utlimately explore the Martian surface in person. Such exploration will unquestionably yield valuable insights into the nature and history of the evolution of the solar system.

Questions for Review

1. Can Venus or Mercury ever be seen at opposition from the earth?

2. What planets can be seen at maximum eastern or western elongation from earth?

3. What point in the solar system do all planetary orbits have in common?

4. How is the mass of a planet such as Mercury or Venus determined?

5. Why is the surface of Venus much hotter than that of Mercury, even though the latter is very much nearer the sun?

6. What phases does Mars exhibit as seen from the earth? From Jupiter?

7. Why would a person on Jupiter weigh only a little over twice his weight on earth, even though Jupiter is over 300 times more massive than the earth?

8. Exluding the sun, where is most of the rest of the mass of the solar system concentrated?

9. In what major respects do Jupiter and Saturn differ from either the sun or the terrestrial planets?

10. What is the current explanation for Saturn's rings?

Prologue

Even though the Bode-Titius scheme appears to be an implausible explanation for the apparent geometric spacing of the planets, the gap between Mars and Jupiter is puzzling. Could some kind of phenomenal event have created it? Is the debris that on the average occupies an otherwise empty orbit the result of paroxysm of nature that annihilated one of our member plants? Or do we instead see the kind of rubble from which the earth itself took form?

Take the planetary moons, the natural satellites: not all planets have them, and some planets control groups of very small moons together with groups of quite large ones. Why? Even if we concede that the small moons represent captures of planetoids, this still doesn't explain where the larger ones came from. Also consider the comets, those mysterious, inscrutable long-haired-appearing objects that consist of largely empty space; objects that terrified people in the past and still create anxieties among some of us today. What are they, and why and how did they "get" that way? These are some of the questions that intrigue us more than the answers to them could edify us. Unfortunately we do not have the answers yet to these kinds of questions and the search for the answers is bedeviled with frustration, astronomically speaking.

The Lesser Bodies
of the Solar System

9

The sun's family is complex indeed. The largest and most obvious members are the fewest in number. Several of these have orbiting bodies of their own. Except for the earth's moon, or companion planet depending on one's point of view, the only objects other than planets that are visible with the unaided eye are occasional comets and meteors. Even the meteors are not themselves seen; we see only incandescent trails that mark their path through the upper atmosphere. In fact, without extremely careful scrutiny of their motions, we could not be sure that any of these transient objects, meteors and comets, *are* members of the solar system.

A relatively modest telescope reveals two additional classes of objects other than comets and meteors: planetary satellites and the minor planets, or asteroids (like a star) as they are usually (erroneously) called. A better appellation would be *planetoid* (resembling a planet). The planetoids easily escape detection unless one knows when and how to look for them or, more importantly, how to recognize them. Let us begin with the satellites and examine the lesser bodies in as much detail as we can. Only then will our view of the organization of the solar system be complete.

Planetary Satellites

Satellites are not common to every planet. So far as is presently known 32 natural bodies are orbiting some of the planets under gravitational

bonds that keep them captive. Their distribution is thus: from our terrestrial frame of reference, the earth has 1 satellite. Mars has 2 satellites; Jupiter has 12; and Saturn has 10. Then the numbers decrease to 5 for Uranus and only 2 for Neptune. Evidently no satellites (of detectable size at least) are associated with Mercury, Venus, and Pluto. At least none have been discovered and no gravitational perturbations of the motions of these planets that would reveal the presence of satellites have been detected. As to the number, 32 may not be final. Saturn's tenth satellite was discovered as recently as 1967. Before that, one of Uranus' 5 satellites was first seen in 1948, and the second of Neptune's companions was discovered in 1949. There may be more of these bodies awaiting the searching eye of some future discoverer.

Satellites of Mars. In 1610 Johannes Kepler predicted that Mars has 2 satellites. His prediction, of course, was correct, but its basis was anything but scientific. The anomalous gap between Mars and Jupiter led Kepler to speculate that another planet might be orbiting in the void. If so, the planet ought to have 3 satellites. According to Kepler this would yield a simple progression from 1 terrestrial satellite to 2 (postulated) for Mars, 3 (also postulated) for a hypothetical planet and 4 satellites for Jupiter. (The Galilean moons of Jupiter had already been discovered.) What Kepler's reaction would have been to the existence of 12 Jovian satellites is difficult to imagine.

In any event, Jonathan Swift in his famous satire, *Gulliver's Travels* (1726) placed Kepler's prediction for Mars' satellites in literary perspective. He wrote of the inhabitants of the flying island of Laputa who had "discovered lesser stars or satellites which revolve about

Mars in such a way that the squares of their periods of revolution were equal to cubes of their respective distances from Mars." Swift obviously had known about Kepler's harmonic law. Voltaire mentioned two Martian satellites in his *Micromegas,* published in 1750. Apparently Voltaire had read *Gulliver's Travels.*

Among the searchers for the elusive moons was Sir William Herschel who, having failed to discover them in 1783, became convinced that Mars lacked satellites. An even more careful search by d'Arrest at Copenhagen Observatory in 1864 was unsuccessful. But during the close opposition of 1877 the search finally paid off for Asoph Hall at the United States Naval Observatory in Washington, D.C. Hall discovered the outer satellite which he named *Deimos* on the night of August 11, 1877. While watching the object on August 17, he was rewarded by the discovery of the other moon. Hall named it *Phobos.* Phobos and Deimos are usually taken to be the names of the steeds that pulled Mars' chariot. But Hall noted that Bryant's translation of the fifteenth book of Homer's *Iliad* identified Phobos and Deimos as attendants or sons of Mars whom he summoned to yoke his steeds. Phobos means fear, Deimos is translated as flight or panic.

The Martian satellites are so small that neither displays a visible disk. Estimates of their size are made on the bases of their apparent brightness and the further assumption that their composition is not too different from the Martian planetary material. This latter assumption is necessary in assigning values for their albedos. From these considerations we obtain an estimate (highly uncertain) of 16 km for the diameter of Phobos, and 8 km for the diameter of Deimos.

A single TV picture taken of Phobos by Mariner 6 (p. 70) revealed that the satellite was not spherical. Rather, it resembled a fat sausage, or a potato, although no other detail was discernible. In order to examine the satellite in more detail, Mariner 9 was programmed to photograph Phobos from much closer range.

The beautifully detailed image of Phobos shown facing the opening of Part Three was captured by the Mariner 9 TV cameras on the 34th orbit of Mars. The TV picture was computer-enhanced to bring out the surface detail. Not only was Phobos' elongated shape verified, but also the surface was shown to be profusely cratered. The irregular edge at top left in the photograph may represent a loss of material, perhaps by a meteoroid impact (p. 226). The cratered surface suggests that Phobos is very old, perhaps as old as Mars itself, and that the satellite possesses considerable structural strength to have withstood so many impacts.

Phobos' average distance from the surface of Mars is 5600 km; its orbit is very nearly circular. Its distance from Mars' center is only about 2.8 times the radius of Mars. We see therefore that Phobos is very nearly at the Roche limit. (Chapter 6) of 2.44 times the planetary radius. According to the evidence Phobos appears to be slowly spiraling in toward Mars. If such proves to be the case, it is probable that tidal forces will eventually dominate over the gravitational adhesion holding the satellite together; Phobos would then disintegrate. The debris would eventually form a ring around Mars similar to one of Saturn's rings.

Phobos revolves around Mars in the same direction as Mars' diurnal rotation with a sidereal period of 7^h39^m, about one-third of Mars' rotational period. To a Martian observer Phobos would rise in the *west* and set in the *east* twice each Martian day! Because Phobos is so close to the surface of Mars, and because its orbital plane is essentially in the plane of Mars' equator, the satellite would forever be invisible at surface positions poleward of 70° N or S latitude (Figure 9-1a).

The radius of Deimos' orbit is approximately 23,500 km. Thus the satellite circles Mars at a distance from the surface of roughly 20,000 km. Deimos' sidereal period is slightly over 30¼ hours. Because this is so similar to Mars'

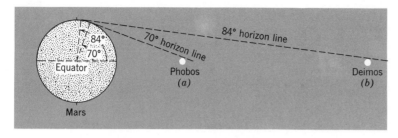

FIGURE 9-1 Respective horizon lines for hypothetical Martian observers at 70°N latitude and 84°N latitude. (*a*) Phobos is invisible for all observers north of 70°N latitude; (*b*) both Phobos and Deimos are invisible for all observers north of 84°N latitude.

sidereal period, Deimos has a synodic period of 5½ days. To a Martian observer Deimos would rise in the east and not set in the west until 2½ days later. Then 3 more days would elapse until the satellite rose again.

Deimos can be seen from the surface only up to 84° N and S latitude (Figure 9-1*b*). Thus an observer on Mars who had never ventured equatorward below 84° latitude would be unaware of the existence of the moons of Mars unless a more knowledgeable individual told him about them.

From our point of view here on earth, the principal astronomical value of Mars' satellites is two fold. First, determination of their orbital perturbations allows an accurate determination of Mars' mass. The secondary value is considerably more exotic. There are two plausible explanations concerning the origin of the satellites. Either they were cast in the original mold along with Mars at the same time that the planet was formed, or else they are captured asteroids. For the first eventuality, any internal heating the asteroids might have experienced, such as occurred in the earth and probably occurred in Mars, would have been rapidly dissipated away. The satellites are too small to have retained any appreciable heat. Thus the bodies would represent perpetual repositories of the primeval material out of which the solar system emerged.

For the second possibility, captured asteroids would represent readily accessible examples of otherwise inscrutable solar system material. In either event direct manned exploration of these objects is within range of our existing technical capability. Their exploration should provide an exciting and immeasurably valuable chapter in the story of man's conquest of space, and in his increasing understanding of where we came from and where we are going.

Jupiter's Satellites. On January 7, 1610, Galileo pointed a telescope at Jupiter for the first time in man's experience. He saw four bright pinpoint objects near Jupiter that he at first mistook for stars. But as he watched on successive nights, Galileo saw to his astonishment that the four objects crossed Jupiter from right to left, first on one side of the planet, then on the other. (The field of view of Galileo's telescope was inverted, hence the apparent right-to-left motion.) Moreover the periods of motion for each body differed from those of the others. Galileo realized that these were satellites of Jupiter, a miniature Copernican system as it were. As we have seen earlier, the discovery constituted a devastating blow to the immutability of the heavens.

In order of increasing distance from Jupiter, the Galilean satellites (so-called because of Galileo's discovery) carry two designations.

Galileo identified them as I, II, III, and IV. Subsequently the names Io, Europa, Gannymede, and Callisto, taken from Greek mythology, were assigned. Their respective distances from Jupiter range from 420,000 km to 1,880,000 km; their periods vary from 1¾ to 16⅔ earth-days. All show perceptible disks in large telescopes, and all display discernible (barely) surface markings. The diameters of the Galilean satellites vary from 3200 km to 5150 km. Only one is smaller than the moon, two are larger than Mercury. The latter two would be eligible for classification as planets were they not orbiting Jupiter. From the mutual perturbations of their motions, the masses of the four satellites are estimated to range from 60% that of the moon to twice the moon's mass. Callisto is the most unusual; its density is only 0.6 that of water. Presumably this satellite's composition is similar to that postulated for the upper layers of Jupiter's surface.

Satellite V escaped detection until 1892. It is only 110,000 km from Jupiter's surface. Jupiter's overwhelming brightness coupled with the satellite's small size (less than 160 km in diameter) effectively concealed it from ready discovery. The seven satellites subsequently discovered fall readily into two distinct groups. Three are orbiting Jupiter at a distance of approximately 12 million kilometers, the other four are roughly 24 million kilometers distant. All are estimated to be less than 160 km in diameter.

The four outer satellites revolve about Jupiter in the retrograde sense. It is a distinct possibility that the bodies are captured asteroids (Figure 9-2). Yet, the retrograde motion is not necessarily proof a priori. Celestial mechanics shows that a gravitationally powerful sun can rob a planet of its very distant satellites more easily if their motion is direct rather than retrograde. Perhaps the outer satellites are the survivors of a larger family whose direct-motion members have long since been kidnapped by the sun. The possibility also exists that Jupiter may in time capture other asteroids; under

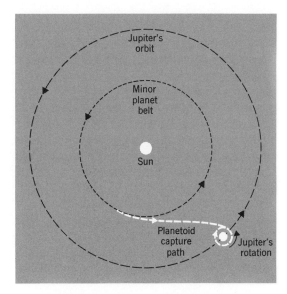

FIGURE 9-2 Capture of a planetoid by Jupiter. A trajectory similar to the one shown can account for the existence of retrograde satellites.

these conditions the outer satellite family could be described as having a "floating" population consisting of occasional permanent departures and even rarer new arrivals.

Three of the Galilean satellites are very nearly in Jupiter's orbital plane. Even the orbit of the fourth one is not appreciably inclined. As a result Jupiter occults, eclipses, or is eclipsed by one or more of these satellites almost daily. Only one of them can pass from one side of Jupiter to the other above or below the planet's disk as we see it, and even then such passage does not occur in every cycle. An occultation occurs when a given satellite disappears behind Jupiter's disk; a transit occurs when it crosses in front of Jupiter. A satellite is eclipsed when it passes through Jupiter's shadow. Depending upon the relative orbital positions of the earth and Jupiter, a satellite can be first occulted and then eclipsed, or first eclipsed and then occulted (Figure 9-3). An observer on Jupiter within the shadow

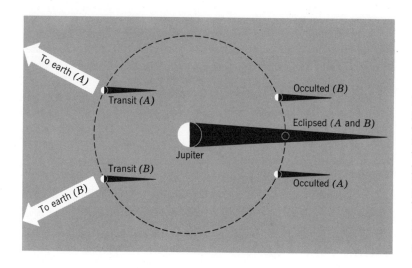

FIGURE 9-3 The geometry of Jovian eclipses. When the earth is in direction *A* a satellite will be first occulted as it passes behind Jupiter, then eclipsed as it passes into Jupiter's shadow. When the earth is in direction *B* a satellite will be first eclipsed then occulted.

cast by a satellite in transit would experience an eclipse of the sun.

The eclipses of Jupiter's Galilean satellites played a significant role in the advancement of astronomy. As early as the 5th century B.C. Empedocles was satisfied that light travels with a finite speed. Galileo attempted to measure the speed of light by timing successive flashes of light from lanterns held by himself at the summit of a hill and by an assistant on a hill 13 km distant. Initially Galileo flashed his lantern and started timing. At the instant his companion saw the flash he opened his lantern; Galileo then timed the instant of arrival of the returning light beam. The experiment of course failed; it is not possible to measure a significant time interval with ordinary timekeeping devices for the travel of light across such a short distance. Galileo also failed to take into consideration the human reaction time, an interval immensely longer than the travel time of light between the two points.

In 1676 Olaus Roemer at the Paris Observatory demonstrated the "progressive propagation" of light by timing a succession of eclipses of the nearest of the Galilean satellites. He found that when the earth was approaching Jupiter succeeding onsets of the eclipses were

too early; when the earth was receding the eclipses occurred too late. By carefully summing the discrepancies Roemer concluded the maximum cumulative variation was $16\frac{1}{2}$ minutes and he ascribed the phenomenon to the time it takes light to travel across the earth's orbit (Figure 9-4). Roemer's determination was within 5% of the modern value of 3×10^{10} cm/sec. This was a surprisingly good result considering the observational difficulties and the uncertainty in the diameter of the earth's orbit.

Saturn's Satellites. Saturn has ten satellites, unless one includes the countless numbers of miniscule particles of which the rings are composed. The first was discovered in 1655, the latest in 1967. *Janus,* the most recent addition to the family, was discovered by A. Dolfuss during the interval when the aspect of Saturn was such that the ring system appeared to be a line as viewed from earth. The satellite is extremely difficult to see because it is barely outside the outer edge of the ring system. The brilliance of the rings overwhelms the apparent brightness of Janus.

Phoebe, the outermost satellite is 12,900,000 km from Saturn and its motion is

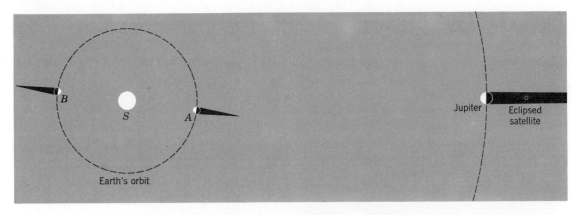

FIGURE 9-4 The geometry of Roemer's determination of the speed of light. Eclipses of a given Jovian satellite differed by 16½ minutes from the predicted times when the earth was at *B* compared to *A*. Roemer attributed (rightly) that the error was really the travel time of light across the earth's orbit.

retrograde. Phoebe was the first solar system satellite discovered whose motion was opposite to the rotation of its parent planet; as such it was quite a curiosity. Of all of Saturn's satellites, *Titan,* seventh outward in order, is the only one larger than earth's moon. It is also the only one known to have an atmosphere; spectrographic studies show that the atmosphere is not unlike that of Saturn in composition. Methane is the dominant constituent.

Uranus' Satellites. Five known satellites attend Uranus; all have orbits essentially in the plane of Uranus' equator. All, like Uranus, display retrograde motion. All of the satellites are reasonably close to Uranus in contrast to some of the satellites of Jupiter and Saturn; their distances from the parent planet vary from 106,000 km to 585,000 km. In diameter they range from 645 km to 1600 km. Again, these are estimates based largely on their albedos. As they are seen from earth, however, all of the satellites are extremely faint because of their great distance from the sun.

Because Uranus' equatorial plane is nearly perpendicular to the plane of its orbit (p. 198) the satellites alternately display a linear configuration and an open aspect (Figure 9-5). The interval between each configuration is approximately 21 years, one-fourth of Uranus' sidereal period.

Neptune's Satellites. The behavior of Neptune's satellites with reference to those of other planets is as different as night and day. *Triton,* the nearest and best behaved, is only 357,000 km from Neptune, almost the same as the earth-moon distance. Nevertheless Triton revolves about Neptune in a retrograde sense contrasted to the planet's direct rotation. Its diameter, estimated from its apparent brightness, is on the order of 4800 km.

The second of Neptune's satellites, *Nereid,* is too faint to be seen by direct telescopic observation; it is purely a photographic object. Nereid's orbit is highly inclined to Neptune's equatorial plane (160°); its orbit is highly eccentric ($e = 0.76$). Thus its distance from Neptune varies between 1,400,000 km and 96 million kilometers. Its orbital motion, however, is direct. The peculiarities in the orbital motions of Neptune's satellites, wherein the innermost is retrograde and the outermost is direct (cf. with the outer satellites of Jupiter and Saturn), raise interesting questions with regard to a common origin with the parent planet. If

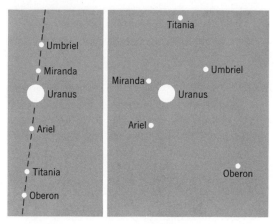

FIGURE 9-5 Uranus' satellites are alternately seen approximately in line when their orbits around Uranus are seen edge-on, and open when the orbits are seen in plane view from above one of Uranus' poles.

Nereid and Triton were indeed evolved at the same time as Neptune, we would expect a family resemblance which does not exist as far as motion and orientation are concerned.

The Minor Planets: Planetoids

When William Herschel discovered Uranus in 1781 at almost exactly the distance from the sun predicted by the Bode-Titius relation, interest in the gap between Mars and Jupiter was rekindled. Bode, at the Berlin Observatory, organized a loose-knit association of 24 astronomers throughout Europe. He asked each one to scrutinize one of the hours of right ascension in the immediate vicinity of the ecliptic for the missing planet. The systematic search had not yet achieved any results when word was received in Berlin that Piazzi in Palermo, one of the 24 searchers, had discovered what appeared to be a new comet. The discovery was made in one of the designated search regions of the celestial sphere, in the constellation Taurus. January 1, 1801, was the date.

Illness forced Piazzi to temporarily suspend his observations of the object on February 11 after only 41 days. By the time his letter concerning the discovery reached Bode in March, the brilliance of the sun had made the region inaccessible. For a while it appeared that the matter was permanently closed; the span of observations was too short for the application of existing techniques of orbital computation. But a brilliant student of mathematics came to the rescue. Karl Frederick Gauss, later to become one of the most celebrated mathematical physicists of all time, invented a method whereby three separate observations were all that were necessary in order to calculate an orbit. The method is now a classic. By November, 1801 the premier application of Gauss' new method yielded for the body a period 1680 days, an orbital eccentricity of 0.08, and a mean distance from the sun of 2.77AU. On New Year's Eve, 1801 the planet was relocated in the constellation Virgo, almost precisely where Gauss said it should be.

The orbital elements, however, were not like those of any known comet; they were very reminiscent of planetary orbits. As a result, the astronomical world applauded the work of Piazzi and Gauss; the missing planet had been found! Not only that, it was almost exactly at the distance predicted by Titius. Piazzi named the new planet *Ceres* after the mythological protectress of Sicily.

But this was not the end of the matter. Not quite 3 months later, on March 28, 1802, Olbers in Bremen discovered another planet in the constellation Virgo. The new planet, named *Pallas*, was at the same distance as Ceres, but its orbit was highly eccentric (0.23), and it was inclined 35° to the ecliptic. The two new planets were in no danger of collision. The discovery of Pallas in the gap was considerably less than satisfying. A new tenant (Ceres) had already moved in. The "no vacancy" sign was up. But when on September 1, 1804, right on the heels of Pallas' discovery, Harding at the Lillienthal Observatory near Bremen discovered a third planet, *Juno,* in the gap at 2.67 AU the evidence could no longer be ignored. The orbit was occupied by a family of planets, fragments perhaps, of a disintegrated larger body. Or possibly it is primeval material that never consolidated into a single planet.

Vesta, the brightest of the minor planets was discovered in 1807 at 2.36 AU from the sun. Then 38 years passed before the next planetoid, *Astrea,* was discovered. Since then the new finds have been frequent. By 1868 the number of known planetoids had reached 100; by 1879 it was 200; by 1890 as many as 300 had been identified and cataloged. Then in 1891 a discovery technique utilizing photographic plates (Figure 9-6) was perfected and the discoveries accelerated. To date the number of known planetoids hovers around the 1600 mark. Probably another 55,000 are in range of the 100-in. Mount Wilson telescope. Most of the known planetoids travel entirely between the orbits of Jupiter and Mars; 88% of them are at a mean distance from the sun of 2.77 AU. The total mass of all of the planetoids is estimated to be less than 0.003 that of the earth.

Some notable exceptions to the average orbital scheme are *Icarus, Hidalgo, Eros, Adonis, Apollo,* and *Hermes.* There are many others. Some of these planetoids have large orbital eccentricities. As a result they penetrate well into the central region of the solar system. Conversely some wander far out past Jupiter. Icarus, for example, travels in an orbit with an eccentricity of 0.83, perilously close to a path that would free it from the sun's gravitational entrapment. This little flying rock pile (it is too small for its central gravity to mold it into a spherical shape) penetrates to within 27.5 million kilometers of the sun, half again as close as the planet Mercury. The planetoid is well named. Icarus, in Greek mythology, built wings so that he could soar like the birds. Tragically he maneuvered too near the sun; his wings which he had ill-advisedly made of wax, melted. Icarus' end was ignominious.

FIGURE 9-6 Planetoid trails on a photographic plate. Trails such as these lead to the discovery of these faint, visually insignificant bodies (Yerkes Observatory Photograph).

On the other hand Hidalgo, with an orbital inclination of 43° and an eccentricity of 0.65, comes in almost to Mars' orbit. At the other extreme of its orbital journey it ventures out nearly as far as Saturn. Hidalgo requires 13.7 years to complete one round trip; most of the asteroids complete their orbits in 4.7 years on the average.

Eros, discovered in 1898, can come within 22.5 million kilometers of the earth; Apollo approaches to within 3.2 million kilometers. Adonis, in the year of its discovery (1936), approached within 1.6 million kilometers; in 1937 Hermes passed by the earth only a scant 960,000 km distant.

The astronomical value of the planetoids, aside from the fact that they may represent primeval material and thus should be investigated in manned exploration, rests in these close approaches to the earth. Parallax determinations are highly precise; thus their distances in kilometers are accurately known. In the computations of their orbits, the length of their semimajor axes in astronomical units are also accurately determined. A knowledge of their distances in both sets of units establishes the exact relationship between them, that is, the number of kilometers per astronomical unit. This method has yielded very accurate determinations of the length of the astronomical unit. To be sure, in recent years radar echoes from the moon and nearby planets have yielded precise distance measurements.† A comparison between the two methods is reassuring. The older method yields values that are in excellent agreement with the electronics methods.

That some of the planetoids are irregular in shape is attested to by the periodic irregularities in their apparent brightness. Presumably a long dimension and then a shorter one are alternately presented toward the sun. The quantity of reflected light is proportional to the area of the illuminated surface. At least one planetoid seems to be essentially cigar-shaped: 8 km across by 24 km long. It is tumbling end over end. One wonders how this splinter survives in this shape. Calculations show that collisions among the asteroids are relatively frequent. How, then, has such a strange-shaped object escaped being pulverized in the celestial grinding machine that is the asteroid belt? This is indeed a puzzle.

The asteroid belt. Statistical treatment of the orbital data of the known minor planets yields a doughnut-shaped region occupying roughly the gap between Mars and Jupiter in the Bode-Titius scheme. Best estimates of the dimensions of the region, commonly referred to as the *asteroid belt,* place the width in the range of 280 million kilometers and the thickness about 80 million kilometers.

† The speed of travel times the duration of the journey is just the length of the trip. For radar, the speed is that of light, c. The duration is the time interval Δt between the transmitted signal and the received echo. Because the time for a round trip, the distance is just one half of the product $c\Delta t$.

In addition to the planetoids, the material in the belt is thought to consist of billions of particles ranging in size all the way down to fine dust. Most of the particles are believed to range between 1.0 and 0.1 mm in diameter. In spite of the prodigious number of particles, the volume of space occupied by them is so vast that a casual passage through the belt would make it appear nearly empty.

Such was the situation revealed by Pioneer 10 (Chapter 8) as it passed through the asteroid belt on its way toward Jupiter. The spacecraft is capable of observing celestial debris of all sizes depending upon its distance from the vehicle. For example, Pioneer 10 is capable of detecting particles 4 mm in diameter (pea-sized) as far away as 100 m, and particles the size of ping-pong balls at a distance of 1 km.

Pioneer 10 crossed the asteroid belt at an angle; thus its flight path through the region was about 330 million kilometers long and it took 7 months to reach the comparative safety of "empty" space beyond. During the traverse not a single encounter occurred with particulate matter larger than 0.1 mm in diameter. Moreover, encounters with smaller particles showed no increase within the asteroid belt over those elsewhere along the flight path. Initial interpretation of these data suggests that the smaller sized matter is cometary debris (p. 220) and is not asteroidal in nature.

Pioneer 10 made over 200 sightings of particles larger than 1.0 mm; some of them were house-sized and larger. This count is somewhat larger than was expected, but still far below a dangerous level. The matter in the asteroid belt is orbiting the sun at an average speed of 72,000 km/hr or 20 km/sec. It had a speed relative to the spacecraft of 54,000 km/hr or 15 km/sec. Particles with relative speeds this great are in effect extremely high-energy missiles; particles as small as 2 to 4 mm in diameter can penetrate 1 cm of aluminum. If Pioneer 10 had sustained even one direct hit from one of the larger particles, significant damage to the craft would probably have been sustained. But as this book goes to press, six months after Pioneer 10 left the asteroid belt, all is well; all systems aboard are working as expected, and the scientific world is awaiting with subdued excitement the first close look at Jupiter on December 3, 1973.

Comets

The most singular, at times the most awesome, and historically the most terrifying of the celestial objects are the comets (Gr., *komētēs*, long haired, from the Gr., *Komēs*, hairy). Most of them appear without advance publicity; they remain in the vicinity for a brief period, and then they steal away as mysteriously as they come, never to be heard from again.

From as far back as we can trace in antiquity up to surprisingly recent times, comets were regarded as omens of evil, foretelling such dire happenings as the demise of kings, the onset of pestilence and of devastating wars. As late as 1910 charlatans did a brisk business selling "comet pills" to the ignorant and unwary, to protect them against the demonic influence of these "hairy" wanderers.

<u>Physical Appearance of Comets.</u> A comet may, but does not always, consist of three distinct parts (Figure 9-7). There is always a fuzzy-appearing, nebulous *coma* (head), less often a bright compact *nucleus* within the coma. When the comet nears perihelion it occasionally develops a long, luminous "tail" that may stretch millions of kilometers out into space.

Comets are large. The coma is always larger than the earth; some comets have comas that exceed 250,000 km in diameter, nearly twice the diameter of Jupiter. Yet their masses are cosmically miniscule, averaging less than $\frac{1}{10,000,000}$ that of the earth. We shall see presently how one arrives at such an estimate. Because this relatively small mass occupies such a large volume, the average density of comets is very small. The matter remaining after the production of a very good laboratory vacuum is more dense than the comets, by far.

FIGURE 9-7 Morehouse's comet (1908 III). This photograph, made on November 15, 1908, shows many streamers in the tail structure. The designation 1908 III indicates the third comet discovery of that year (Yerkes Observatory Photograph).

Thus there is an element of truth in the description of a comet: a great big bag full of nothing.

<u>Composition of Comets.</u> Spectral analysis of the light from comets reveals the presence of both solid particulate matter and certain gases and their dissociation products. We are able to deduce these facts from several clues. One clue is the presence of an absorption spectrum that is almost identical to that of sunlight. A second clue arises from the appearance of an emission spectrum when the comets swing close to the sun. The emission spectrum is produced by the actual luminescence of gaseous material. A third clue is afforded by the occasional splitting of a cometary nucleus into several discrete fragments together with certain residual meteor showers (p. 219) following a

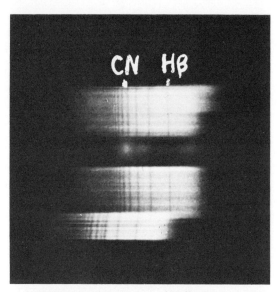

FIGURE 9-8 Spectrum of Halley's comet (Figure 9-14) shown with that of a comparison star (λ *Tauri*). Symbol *CN* indicates an emission line superimposed on the absorption spectrum; H_β is the hydrogen *beta* line (see text). (Yerkes Observatory Photograph).

comet's disappearance. Only particulate matter will behave in this manner.

When a comet is more than 3.0 AU from the sun it displays only an absorption spectrum. The essentially one-to-one correspondence of this spectrum with the solar absorption spectrum (Figure 9-8) is strong evidence that the comet shines by reflected light, as does the moon. This evidence is reinforced by the joint variation of the comet's apparent brightness with distance from both the earth and sun according to the inverse square law (Chapter 10). When a comet is much beyond Jupiter's average orbital distance from the sun, it is in general visible only through photographic techniques. If the comet were self luminous, its distance from the sun would not necessarily influence its apparent brightness.

When a comet approaches within 3.0 AU of the sun an emission spectrum superimposed on the absorption spectrum becomes visible. An emission spectrum is caused by the interaction of the high-energy (short wavelength) fraction of sunlight, principally ultraviolet light (Chapter 10), with gaseous molecules (or atoms). Apparently the intensity of such radiation is insufficient at distances exceeding 3.0 AU to produce an emission spectrum and the corresponding cometary luminescence.

Detailed studies of the emission spectra of many comets show that among the gaseous constituents are methane, nitrogen, carbon dioxide, ammonia, and water vapor. These are the familiar substances in the atmospheres of the major planets.

The existence of both "ices" of the constituent gases and mineral matter in the particulate fraction of a comet's composition can be argued analytically as follows. Gases selectively scatter light in inverse proportion to the fourth power of the wavelength of the radiation. We first discussed this in Chapter 6. The scattering effectiveness does not depend on the density of the gases producing scattering. It depends only on the number of molecules along the line of sight. Thus, the rarefied gases in the coma of a comet are capable of scattering light to the degree that we observe because the line of sight through the coma is so long. But since the light initially comes from the sun, we should only see the blue scattering we observe in our own atmosphere. This is not the case; a cometary spectrum shows all the visible solar light wavelengths in approximately the same concentration as in direct sunlight. Only solid particulate matter can produce such an effect. Since the ices of the gases in a coma can be expected to vaporize when a comet is near the sun, the blue portion of the spectrum ought to be intensified and the remaining wavelengths ought to be diminished in intensity if light scattering is the predominant feature. But the observed effect is much smaller than it ought to be. Hence we conclude that most of the visible light from a comet is sunlight reflected from particulate matter. Empirical support for this conclusion is postulated from the phenomenon of comet splitting and from the evidence of meteor showers which we shall presently review.

Now when the solid ices vaporize as a comet nears perihelion, radiation excitation dissociates some of the gaseous molecules just as it does in the earth's atmosphere. When further energized by high-frequency (short wavelength) radiation, these dissociation products, chemically known as *ions,* give off visible light just as the gases in the familiar fluorescent tube become luminous when electrically excited. The vapors and ionized particles stream out from the coma in the form of a tail under both radiation pressure and the influence of corpuscular (particle) ejection from the sun known as the *solar wind* (Chapter 10). Hence, a comet's tail always points *away* from the sun (Figure 9-9).

Comets invariably occult a number of stars in pursuing their paths around the sun. In all such observed cases no diminishing of a star's apparent brightness, either by a comet's central portion or by its luminous tail, has ever been detected. Occasionally a comet will transit across the sun's disk. In no case has a comet's nucleus (or coma) remained visible as a black spot against the background of the solar disk. Both of the above observational facts testify to the extreme low density of the comet material. The phenomenon of solar transits additionally places upper limits of about 20 km on the particle size. Theoretically a larger-sized fragment would be visible as it eclipsed a portion of the solar disk.

Splitting of a comet's nucleus into two or more fragments is generally conceded to mean that some of the cometary particles are large, perhaps on the order of several hundred meters. A striking example of comet splitting, especially in view of the subsequent events, is afforded by Biela's comet, named after the Austrian officer who discovered it on February 27, 1826. Gambart at Marseilles computed the orbital elements and recognized that the comet was identical with one discovered by Mon-

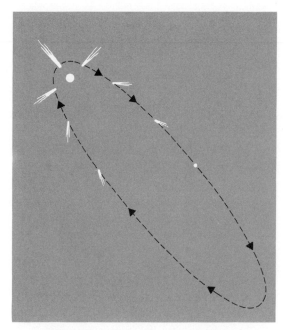

FIGURE 9-9 A comet's tail points away from the sun as a result of radiation pressure and the effects of the solar wind (Chapter 10).

tagne at Limoges in 1872 and by Pons at Marseilles in 1805. No apparitions (appearances) of the comet were observed in either 1812 or 1819, calculated to be the years of the two returns immediately preceding the apparition of 1826.

Biela's comet arrived at perihelion in 1832 within 12 hours of the predicted time. In so doing it passed within 32,000 km of the earth's orbit. No danger to the earth had been anticipated, however, since the earth was not due to arrive in the vicinity until late November, a month later. The comet would clear the earth by more than 80 million kilometers. Notwithstanding there was considerable public apprehension at the comet's imminence. Contrary to public expectation no adverse events occurred.

Biela's comet was undetected in 1839, the next apparition, because the sun was in the way. It reappeared on schedule in 1846, and

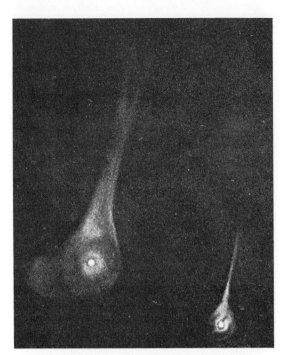

FIGURE 9-10 Splitting of the nucleus of Comet Biela after a drawing by Struve in 1846, 1 month after the observed splitting. The two portions were already more than 240,000 km apart.

while under close scrutiny, the comet broke in two! The smaller fragment drifted slowly away from the main nucleus and grew considerably in brightness (Figure 9-10). The "twins" returned in 1852, by this time separated by nearly 2.5 million kilometers. They were not observed in 1859, again because of interference by the sun's brilliance. Thus the return in 1865 was anticipated with considerable interest. But in spite of intensive search by several major observatories, and by extremely accurate recalculations of the orbital elements, Biela's comet(s) went undetected. It— or they— have never been seen since. One wonders what kind of fatal accident must have befallen not one, but both parts of the comet.

This is not, however, the end of the tale. In November, 1872, the earth again reached the point where the orbit of the lost Biela's comet crosses the earth's orbit. On the night of the 27th a veritable rain of meteors emblazoned the sky for 6 hours between 7 P.M. and 1 A.M. London time. Shower after shower of "shooting stars," accompanied by blinding fireballs and noiseless explosions descended almost vertically from a point in the sky near the bright star γ Andromedae. At the Collegio Romano 13,900 meteors were counted; at Moncalieri 33,400 were tallied. One lone observer in England reported 10,600 meteors in the 6 hour span. Estimates place the total count at near 160,000. Had Biela's comet still been in existence, it would have passed the point of the meteor shower some 12 weeks earlier. That the shower represented disintegration fragments following the comet's splitting received substantiation when, in November, 1885 another spectacular meteoric shower was encountered. The display was visible all over Europe. In backtracking over previous meteoric showers that appeared to radiate from the same place in the heavens on the same date in November, apparitions were found to have been recorded in 1798, 1830, and 1838. Calculations show that the original Biela's comet was attended by a swarm of particles reaching 320 million kilometers behind it along the orbit and extending 480 million kilometers in front. Evidently the disintegration process was of long standing.

Motion of Comets. A given comet has no distinguishing physical characteristics whereby it can be positively identified on successive appearances. A comet's only badge of identification is the orbital path it pursues around the sun. The orbits of many comets are so eccentric that it is impossible to tell from the small, observable portion whether the orbit is elliptical and hence typical of a captured object, or parabolic or hyperbolic, both of which are solar-escape paths. Mathematical analysis leads to the suggestion that comets with parabolic paths (eccentricity = 1.0) have previously been members of the solar system. Such comets have been disturbed out of their orbital paths, usually by Jupiter's gravitational influence, and they presumably will never appear again. Estimates of their original distances place them on the order of 100,000 AU from the sun. Comets with eccentricities greater than 1.0 are following hyperbolic paths; presumably such comets are not members of the solar system. We must not overlook the possibility, however, that their orbits may have been elongated ellipses, and that accelerations induced by the major planets may have given them hyperbolic speeds. In any event, having once achieved a hyperbolic path, a comet is sure to leave the system forever. The origin of these comets is a deep mystery. In both the parabolic and hyperbolic cases the revolution of the comets with respect to the sun appears to be evenly divided between direct and retrograde motion.

Comets with elliptical orbits are more certainly close members of the solar system since most have periods less than a few hundred years. Most of the periodic comets have orbits that are less highly inclined to the ecliptic, and most of their motions are direct. The majority of the periodic comets are associated with the planet Jupiter. That is, the aphelion distances of these comets are roughly at Jupiter's distance from the sun, and their periods are on the order of 10 years. Others of the short period comets have periods of from 10 to 20 years and are

thus associated with Saturn; still others are associated with Neptune and Uranus. The grouping of periods in such definite families shows almost conclusively that the short period comets are gravitationally influenced by the major planets. Table 9.1 lists some of the periodic comets whose returns have been observed.

Comet Schwassmann-Wachmann I is a comet of more than passing interest since its entire orbit lies between Jupiter and Saturn. The orbital eccentricity brings the comet at perihelion nearly to Jupiter's path. Sooner or later the comet will arrive at perihelion about the same time that Jupiter is in closest opposition to it. We may expect large perturbations in the comet's orbit which conceivably could eject it from the solar system.

Most comets are invisible to the unaided eye. Indeed, even small telescopes cannot gather enough light from these nebulous objects so that they become visible. Thus the general public and even casual astronomical observers remain ignorant of the apparitions of most comets. On the other hand some comets occasionally make spectacular appearances, becoming readily visible and remaining so for extended periods up to several weeks. Not infrequently their tails extend several tens of degrees across the sky. Recent notable examples include the Arend-Roland comet of April, 1957 (Figure 9-11), comet Mrkos of August, 1957

TABLE 9.1 Some Recently Observed Periodic Comets.

Comet	First seen	Last seen	Period (years)	Closest approach to sun (astronomical units)
Encke	1786		.3	0.34
Pons-Brooks	1812	1953	70.9	0.77
Pons-Winnecke	1819	1970	6.3	1.23
d'Arrest	1851	1967	6.7	1.37
Giacobini-Zinner	1900	1965	6.4	0.93
Neujmin 1	1913	1966	17.9	1.54
Honda-Mrkos	1948	1969	5.2	0.56

FIGURE 9-11 Comet Arend-Roland, April 27, 1957. The leading "spike" is an illusion of perspective; in reality it is the extremity of the fan-shaped tail farthest from the viewer (Courtesy Lick Observatory).

FIGURE 9-12 Comet Mrkos, August 22, 1957 (Courtesy Hale Observatories).

(Figure 9-12), and comet Ikeya-Seki of October, 1965 (Figure 9-13).

The most famous of all of the bright comets, Halley's comet, is also periodic. It last appeared in 1910 and is due again in 1986 or 1987 (Figure 9-14). This comet is named after Edmund Halley, an English astronomer, who shrewdly surmised that the bright comets of 1531, 1607, and 1682 were one and the same. Halley based his supposition on computations which showed that all of these comets were following almost identical orbital paths.

Halley predicted that the comet would return in 1758. Sceptics noted that Halley would almost certainly be dead by then (he was born

in 1656), and therefore he would not have to suffer the ridicule of his peers when the predicted comet failed to appear. The sceptics were wrong. The comet appeared on schedule, as it did also in 1835. Subsequent backtracking along the historical appearances of other bright comets shows with certainty that Halley's comet was first observed in 239 B.C. Possibly the record extends back to 466 B.C. for a total of 31 perihelion passages.

No bright comet has appeared from 1965 to the time this book goes to press. But on March 7, 1973 a new comet, one that promises to be

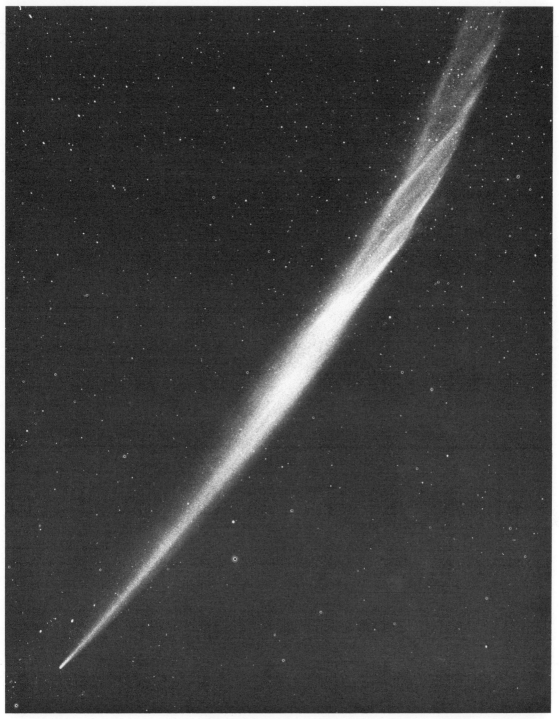

FIGURE 9-13 Comet Ikeya-Seki, October 29, 1965 (Courtesy Lick Observatory).

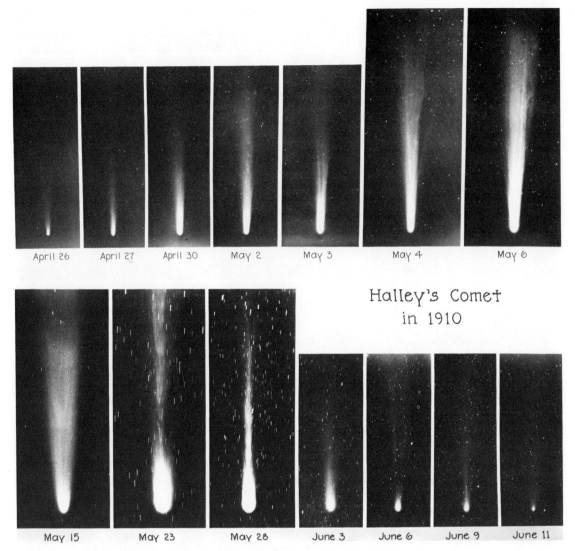

April 26 April 27 April 30 May 2 May 3 May 4 May 6

Halley's Comet
in 1910

May 15 May 23 May 28 June 3 June 6 June 9 June 11

FIGURE 9-14 Fourteen views of Halley's comet taken between April 26 and June 11, 1910 (Courtesy Hale Observatories).

a veritable giant, was discovered by Lubos Kohoutek of the Hamburg Observatory, West Germany. Preliminary measurements indicate that comet Kohoutek should blossom into a celestial spectacle of a lifetime, ten times brighter than any of its recorded predecessors. At perihelion passage the comet is expected to extended across 1/6 of the southern sky.

When first discovered, the comet was 730 million kilometers from the sun. By comparison, Halley's comet is never visible at distances from the sun exceeding 275 million kilometers. Moreover, when discovered, comet Kohoutek was far more distinct than Halley's comet ever is on initial appearance at each successive apparition. Recent tracking by observatories in the United States, Japan, and Great Britain confirm that Kohoutek is a long period comet that should figuratively eclipse Halley's comet in brilliance and duration. Indeed, it could rival the full moon in brightness.

Comet Kohoutek will make its closest rendezvous with the sun on December 28, 1973, and it is expected to provide a spectacular celestial display all through late November, 1973 to early February, 1974. While at this writing in July, 1973 the comet has not yet developed a tail, comet experts expect it to do so at any time.

Comet Theories. Many hypotheses, some dating back to the time of Tycho Brahe, have been proposed in an effort to account for cometary sources and behavior. Among those most plausible are the theories of J. Oort of Leiden, and F. L. Whipple of Harvard and of the Smithsonian Astrophysical Observatory. In 1950, after a lengthy study of comet orbits, Oort proposed that the local supply of comets is replenished from a vast reservoir that reaches out as far as 100,000 AU. According to Oort the sun is able to hold vast numbers of these distant comets for periods up to 5 billion years. During such a period many stars will pass through the swarm of comets. Some will be perturbed away from the sun and thus become lost before they are ever known. Others will be accelerated into eccentric orbits which swing the comets for the first time in close perihelion passages. These are the new comets. If the perihelion distances are less than 1.0 AU, the life expectancy of these comets is short; gravitational disintegration occurs.

A very large fraction of the new comets is lost when the orbits are further perturbed, as they are when they come within gravitational range of the larger planets. Still others suffer perturbations by these planets that sometimes lengthen, sometimes shorten their periods.

One of the major objections to the "gravel pile" theory of a comet's constitution—the theory that we have tacitly accepted thus far in our discussion—concerns the gaseous portion that accompanies the solid material. How could such a miniscule quantity of mass hold gaseous molecules (and atoms) gravitationally captive, even at the extremely cold temperatures of free space? We are forced to conclude that the gases are frozen into solid ice or snow, or at least that the molecules are absorbed onto the surface of the mineral particles. Even this last compromise is not without fault. When the comet approaches near the sun, solar heating will vaporize the gases. As we have seen, the gaseous material along with some of the mineral matter is diffused into the tail from whence it subsequently is lost to space forever. How can a comet repeat the phenomenon over and over again? It would seem that all of the gases would be lost on the first, or at most on the second or third perihelion passage.

Whipple's theory appears to resolve these difficulties. According to his hypothesis a comet nucleus resembles a "dirty snowball," that is, ices of the various gases are intermingled with earthy minerals which are present in the form of dust and gravel. When a comet nears the sun, the gases in the outer layers of the nucleus vaporize forming the coma and leaving a more or less impervious shell of mineral material surrounding the inner mixture. This shell insulates most of the frozen gases from direct solar heating, although in time heat is conducted inward and the gases are slowly released. If the nucleus is on the order of kilometers in diameter, a comet could survive hundreds, even thousands, of perihelion passages before the gases are exhausted.

Whipple's theory also explains the anomalous motion of comets such as that of comet Encke. This comet has consistently arrived at perihelion from 1 to 2½ hours ahead of schedule. No external forces can be postulated which account for this speeding up. According to Whipple, the comet nucleus is presumably rotating like all known planets and other celestial objects. The vaporized gases from the interior of the nucleus are finally discharged through a weak place in the rocky shell, forming a jet of escaping gas. This rocket-like jet will either accelerate or decelerate the comet, depending upon the direction of the jet. In the case of Encke's comet a loss of as little as 0.2% of the mass per perihelion passage could produce the observed change in period.

Finally, Whipple's theory is consistent with the observed splitting of cometary nuclei into several discrete parts. Only a compact, essentially solid nucleus would experience a sudden gravitational tidal disruption resulting in two or three fragments. A loose agglomerate would simply disintegrate and vanish unobtrusively.

Meteors — Meteoroids — Meteorites

Almost assuredly everyone who has attained the age of cognizance and who possesses the faculty of sight has seen at least one shooting star or *meteor* (Gr., *meteoron*, p. *meteora*, things of the air). The term meteor encompasses such diverse phenomena as clouds, rain, and hail as well as rainbows, halos, and whirlwinds. Customary usage, however, reserves its use for those transient celestial objects that enter the earth's atmosphere and almost immediately vanish in a blaze of glory (Figure 9-15).

This striking astronomical phenomenon has always fascinated mankind. Some primitive societies (and others not so primitive) thought that a shooting star marked the intervention of diety in man's troubled affairs. It represented the escape of another spirit to the happier and more secure hereafter. Other peoples thought that the wish which a young maiden made at the appearance of a shooting star was sure to come true.

The popular name shooting star is not astronomically appropriate because in no way, other than during the brief moment of its visibility, does a meteor resemble a star. In strictest accuracy the term meteor refers to the visible phenomenon. We now know that a solid particle of cosmic matter, very often no bigger

FIGURE 9-15 Meteor trail with a portion of the constellation Orion in the field of view (Yerkes Observatory Photograph).

than a grain of sand, is heated to incandescence through friction with the earth's atmosphere. The outer layers vaporize in part and melt in part; friction sweeps the material away in the form of a long, brilliant streamer, or tail.

The object itself is a *meteoroid* until it becomes visible. If the particle is sufficiently large that a portion of it survives the fiery passage through the atmosphere and strikes the earth, it becomes known as a *meteorite*. On the other hand millions of meteoroids reach the surface of the earth without ever becoming visible; these are the *micrometeorites*. The name is appropriate; the particles are so small that their surfaces are large compared to their masses. Such particles can radiate the frictional heat away fast enough as they pass through the atmosphere so that the surface remains relatively cool, certainly below the temperature where light is emitted. Occasionally, extremely brilliant meteors are seen; these are called *fireballs*. If, in addition, loud sharp reports resembling the thunder accompanying a nearby lightning bolt are heard, the meteor is called a *bolide*. The noise comes from the explosive expansion of the frictionally heated air along the meteor's path, and it is testimony that the meteor is very low in the atmosphere.

Meteorites are classed according to their composition as stones, irons, or stony-irons. Stones are similar in composition to the rocky portion of the earth's crust, although some minerals are occasionally identified that are not found in terrestrial rocks. Stones are difficult to identify because of this similarity. The presence of *chondrules*, spherical globules embedded in the crystalline matrix, almost certainly identifies a stone as having an extraterrestrial origin; apparently rocks of the earth's crust do not possess them.

The iron meteorites are an alloy of approximately 90% iron and 9% nickel. Presumably these meteorites are similar in composition to the earth's core. Stone-irons are rare. They consist of a mixture of stony minerals and iron fragments. Even though some of the minerals in meteorites are not found on earth, there are no unknown chemical elements. The range of meteoric composition from stones, through stony-irons, to irons raises an intriguing question: Do meteorites, along with the planetoids, represent debris from the destruction of a larger planetary body? Certainly the sequence of composition is similar to what geologists and geophysicists believe is the internal structure of the earth. Again, as suggested earlier in this chapter, a direct manned examination of some of the larger planetoids should shed considerable light on the mystery.

A casual observer far from city lights on a dark moonless night will see on the average up to ten meteors per hour. How can we turn this statistic into an estimate of the daily meteor fall? First, we can multiply the rate per hour by 24 and secure the average number of visible meteors per day within the field of view of the observer. We next multiply this quantity by the ratio of the area of the whole celestial sphere to the area of the fractional field of view. This technique leads to an estimate of upwards of 100 million visible meteors per day over the whole earth. Finally, this result is compared with samples of meteoric dust and debris collected from various test surfaces such as rainwater, snow, and greased plates. The grand total of all meteoric material is variously estimated at between 100 and 1000 metric tons per day. Even so, the earth is so large that this rate of fall in the aggregate would at best have added only a centimeter or two to the earth's radius during the whole of its life history.

Meteor Heights, Frequencies, and Speeds. The heights at which meteors appear and vanish are obtained by taking simultaneous photographs of the several occurrences from two stations a known distance apart. The measured parallaxes (a form of triangulation) yield a range of heights, for the 500 or so meteors measured to date, from 90 to 115 km for the appearance, and from 70 to 100 km for extinction. The upper limits are associated with the faint meteors, the lower values with the bright ones. By employing an ingenious shutter rotating at a known speed, the trail of a given meteor is "chopped" into segments during the exposure of the photograph (Figure 9-16). The triangulation technique which yields the meteor heights also allows calculation of the length of the meteor trail segments. From the combined data meteor speeds are determined; the range is from a low of 11 km/sec to a high of 72 km/sec.

If we statistically analyze the distribution of meteor speeds during the night hours a curious fact appears. Meteors are traveling *faster, on the average, after midnight than before midnight*. In addition the statistical treatment shows that the frequency of meteors is greater after midnight than before midnight. Let us see if we can qualitatively account for these statis-

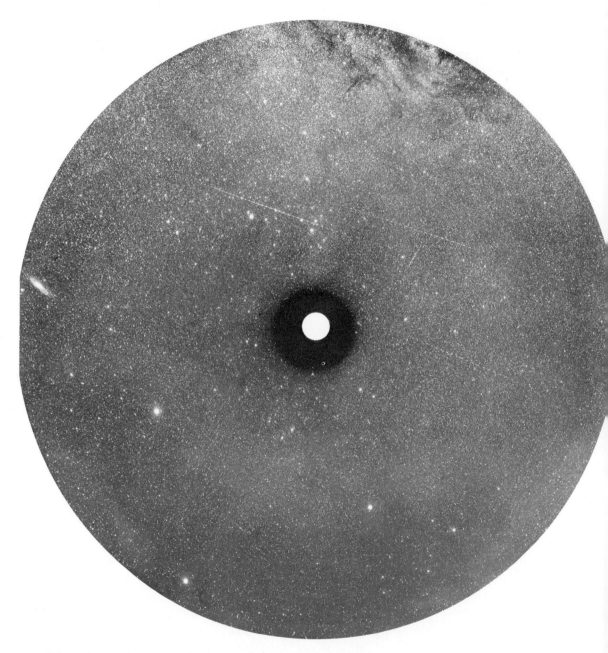

FIGURE 9-16 Three Perseid meteors whose trails have been "chopped" into segments 60 times per second by a rotating shutter mounted at the center of the camera. A portion of the Milky Way is to the north (at top), and the Andromeda Galaxy is to the east (at left) in the photograph (Courtesy Smithsonian Astrophysical Observatory).

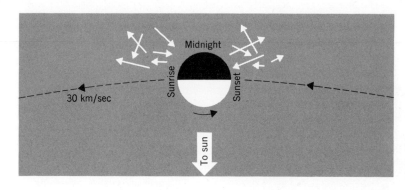

FIGURE 9-17 Frequency of meteors is less before midnight than after (see text).

tical results. In Figure 9-17 we see the random distribution of meteor paths in relation to the earth's orbital motion. Between sunset and midnight meteors moving away from the earth (opposite the earth's motion) will never be seen; indeed only those meteors that are traveling faster than the earth's orbital speed of 30 km/sec can overtake the earth and become visible.

But from midnight to dawn not only will the earth overtake slower meteors, it will sweep up all the meteors coming toward it on a collision course. Thus the higher count after midnight is accounted for. As for the speeds, unless a meteor is following a parabolic or hyperbolic path (never yet observed for certain), it must be traveling at less than the velocity of escape at the earth's distance from the sun, 42 km/sec. Thus from midnight to dawn the maximum observed velocity is on the order of 12 km/sec, the difference between the earth's speed and the maximum speed for any overtaking meteor. After midnight the speeds add; the relative meteor speeds can be as high as 72km/sec. Since the post midnight count is greater, we can readily see that the statistical average of the speeds is also greater after midnight.

Meteor Swarms and Showers. More than 75% of the meteors appear at random times, traveling in random directions. Because their periods of visibility are so short it is impossible to determine anything about their orbital char-

acteristics. The remaining 25% appear in groups on definite dates, almost at definite hours. The meteors of a given group all appear to originate from a common point in the sky. This point is called the *radiant* because the perspective effect makes the parallel meteor paths appear to radiate from a point. This is identical to the effect we observe in the apparent convergence of streets, sidewalks, lines of trees, and poles to a vanishing point. The periodic appearances from various radiants are collectively referred to as meteor showers.

Almost certainly many meteor showers constitute the remnants of comets that have been disrupted and thus ceased to exist as comets. Biela's comet mentioned earlier is an example. On the other hand some meteor showers appear to be related to meteoric material that is distributed along the orbits of still existing comets.

Meteor showers occur when the earth encounters a meteor swarm or meteor stream. Figure 9-18 shows the geometry associated with both types of showers. Clearly a meteor shower will not be visible unless a swarm of meteors and the earth both arrive at the common intersection of their orbits at the same time. Such a meteor shower will be periodic but not necessarily an annual occurrence. When the meteroids of a particular group are dispersed all along their mutual orbit, we refer to it as a meteor stream. This distribution leads to annual displays of meteors. The annual

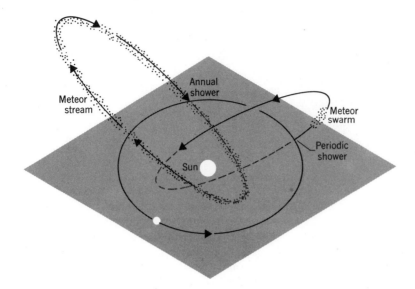

FIGURE 9-18 Meteor swarm and meteor stream. A stream produces an annual shower if it intersects the earth's orbit; some of the meteoroids will be at the intersection when the earth arrives. A swarm of meteors produces sporadic displays. Both the swarm *and* the earth must arrive at the intersection of the orbits simultaneously in order to produce a shower.

showers vary in intensity according to whether or not the particles are irregularly distributed in density along the orbital path.

We are able to gain some insight into the vast dimensions of a meteor stream and at the same time determine the probable concentration of the meteoric particles that make it up. The direction of the flight of a group of shower meteors from their radiant point, with respect to the direction of the earth in its path, is first determined. Analysis shows whether or not the earth crosses the meteor swarm or stream at a sharp angle. (In most cases it does.) After determining the angle of intersection it is a simple matter of geometry to calculate the width of the stream from the time it takes the earth to cross it. The elapsed time is marked by the beginning and end of the meteor shower. Thus, the Perseid stream has an estimated diameter of nearly 80 million kilometers; the Draconids stream is only 650,000 km wide.

The average separation of the particles in a given swarm or stream is estimated from the speed of the earth, the speed of the meteors, and the dimensions of the stream or swarm. The average separation of the meteoric particles in the Perseid shower is on the order of 120 km

while that of the Leonids at one of the maximum displays (in 1833, for example) is only 15–20 km. Names of the showers are for the most part taken from the constellation containing the radiant. Other meteor showers are named for vanished comets whose orbits the stream or swarm is following.

Notable Meteorites and Meteoric Craters. While most of the meteors are estimated to be no larger than a grain of coarse sand, some rather extraordinary meteoric finds that run to sizable masses have been made. The earth bears scars of some comparatively large-scale impacts. These must be of recent origin since the natural processes of erosion remove all traces in less than 100,000 years.

The largest meteorite found thus far is the *Hoba West*, discovered near Grootfontein, South West Africa. Its mass is estimated to be in excess of 50,000 kg (50 metric tons). Admiral Peary brought back to the United States a large meteorite that his expedition discovered in Greenland in 1897. It is now on display in the American Museum of Natural History in New York City. At 34 metric tons, it is the largest meteorite on display in this country.

FIGURE 9-19 Meteorite brought back by Admiral Peary from Cape York, Greenland. The meteorite is now in the Hayden Planetarium, American Museum of Natural History, New York (Courtesy, American Museum of Natural History).

millions of small fragments have been found in the surrounding area. There is some evidence that a large iron mass rests some 230 m below the floor of the crater, under the southeast rim. The impact raised a rim 50 m above the surrounding plain. The crater itself is 1400 m across and 200 m deep.

A large crater in Quebec now holds a lake nearly 3 km across. Fromerly called the Chubb Crater, it is now known as the New Quebec Crater (Figure 9-21). It is in solid granite, a circumstance that almost certainly rules out volcanic activity as the cause. Granite is not a volcanic rock. It appears likely, even though no meteoric fragments have been found, that this crater was formed by one of these extraterrestrial intruders.

Throughout the world there are more than 1000 cataloged finds (Figure 9-19).

The most recent evidence of large scale meteorite impacts includes a fall in Central Siberia in 1908 and fall near Vladivostok in Siberia in 1947. The tremendous explosion and shock wave accompanying the first fall felled trees radially outward for more than 30 km in every direction. A nearby herd of 1500 reindeer was annihilated; the shock from the impact was detected as far away as Europe. No pieces of the meteorite have ever been discovered, but the original mass is estimated to have exceeded 100,000 metric tons. The second fall produced more than 100 craters up to 30 m in diameter, scattered over an area of about 5 km². Thus far only small fragments aggregating 5 tons have been found. One, and possibly both of these events was the result of a cometary collision.

The Barringer Crater near Winslow, Arizona (Figure 9-20) was the first one discovered that was known to be the result of a meteorite impact. Various estimates place the time of fall between 5000 and 75,000 years ago. The main mass has not yet been discovered, but literally

SUMMARY

In addition to the nine planets, the sun's family includes 32 natural satellites, uncounted numbers of planetoids ranging in size downward from 780 km in diameter; it includes perhaps billions of comets in orbits ranging in the hundreds of thousands of astronomical units distant. The sun's family also counts myriads of meteoroids among its membership, many of which are traveling in organized groups as if they are the remnants from the formation or the final destruction of comets.

The planetary satellites are useful in providing earth-bound observers with the means of determining the masses of their parent bodies. The most recently discovered was Janus (1967), the tenth member of Saturn's family. It is also the closest satellite to Saturn. Satellite sizes range from 8 km for the diameter of Deimos, to two Jovian satellites that are larger than Mercury. Titan, Saturn's largest, is the only satellite known to have an atmosphere.

The four largest Jovian moons were discovered by Galileo. This discovery in effect initiated the final collapse of the geocentric uni-

FIGURE 9-20 The *Barringer* meteor crater near Winslow, Arizona (Yerkes Observatory Photograph).

verse concept in favor of the Copernican heliocentric system. Later, in 1676, Olaus Roemer in Paris used the anomalies in the observed versus the predicted times of eclipses for one of the Galilean satellites in his demonstration that the velocity of light was finite. His value for the speed of light was within 5% of that presently accepted, a remarkable achievement for the time.

The four outer Jovian satellites, Saturn's outer satellite, and Neptune's inner moon all revolve around their respective planets in the retrograde direction. With the possible exception of the Neptunian satellite there is a strong possibility that these retrograde moons are captured asteroids.

The principal utility of the minor planets (planetoids) rests in the fact that we can determine or refine the number kilometers per astronomical unit. Several of the planetoids come well within range of Mercury and Venus. These planets disturb the planetoid orbits in such a way that estimates of the masses of the two planets can be made.

Collectively, the known planetoids are at an average orbital distance that fills the void in the Bode-Titius scheme between Mars and Jupiter. We do not know if the minor planets represent the debris from the cataclysmic destruction of an earlier planet, or if they represent material that never consolidated into one.

FIGURE 9-21 The *New Quebec* crater, now a lake (Courtesy Dominion Observatory).

Comets are the most exotic members of the sun's family, consisting as they do of almost nothing but space. Most of the solid matter is probably granular in size, resembling coarse sand, although widely separated particles up to kilometers in diameter cannot be ruled out. At distances exceeding about 3.0 AU from the sun the comets shine by reflected light. On closer approach to the sun radiation pressure from the latter, together with the solar wind (Chapter 10), pushes the gaseous constituents as they vaporize into a luminescent, tenuous tail that always points away from the sun.

Except for random gas molecules and specks of dust, the meteoroids are the smallest members of the solar-system family. Most do not exceed a grain of coarse sand, on even gravel, in size. When a meteoroid is heated to incandescence by friction on entering the atmosphere it becomes known as a meteor. Any fragment that survives the frictional ablation and succeeds in reaching the earth is termed a meteorite. Several craters of obvious meteoric origin have been discovered. The most notable example is the Barringer Crater near Winslow, Arizona. Such craters testify to the large masses of some of the meteorites.

Questions for Review

1. From our terrestrial perspective what is the chief astronomical value of planetary satellites other than our own moon?

2. Outline Roemer's method for determining the speed of light using one of Jupiter's satellites.

3. What is the principal astronomical utility of the planetoids?

4. Discuss why more meteors per hour are visible, on the average, after midnight than before?

5. Why is the average speed of meteors greater after midnight than before?

6. What are the essential physical conditions for an annual periodic meteoric display?

Prologue

The sun is the most fascinating of all the solar system members because it is so totally different from all the rest. Because of its extreme surface temperature, over 5770°K, it is incapable of supporting life, or anything else for that matter, except violently heaving gases and plasmas. Yet in the absence of the sun, the planets—all of them—would be equally impotent as life-support systems.

The sun is a smallish, undistinguished star, 109 times the diameter of the earth, and 330,000 times more voluminous. Yet its density is only one-fourth that of the earth. Each second the sun emits an incredible amount of energy, and it has been doing so without perceptible change for 5 to 10 billion years.

In this chapter we will look at the sun from two different perspectives: as the gravitationally dominant, tyrannical yet benevolent benefactor of its family, and as a star, the nearest one to us and the only one accessible to our detailed scrutiny. We are particularly interested in how this gaseous ball of matter, too hot for the existence of solids or liquids, can emit energy at a rate, mass for mass, more than a million times greater than any known chemical conflagration.

The Sun

10

The sun is vastly different from the rest of the members of its personal family. It gives off heat and light at an enormous rate, whereas all the rest of the solar-system objects, save the comets, shine by reflected sunlight. The sun is much larger and more massive than the other solar-system members. We conclude its larger size from the sun's apparent equality in diameter with the full moon even though it is at least 370 times further away. The sun's gravitational dominance over the rest of the system attests to its larger mass. The geology of the earth offers irrefutable proof that the sun has been shining with undiminished brilliance for several billion years. Thus its energy source must be far different from those we commonly experience or use.

That the sun is an object worthy of our detailed attention goes far beyond these obvious reasons. From time immemorial the sun has been worshipped as the giver of life. Even primitive societies sensed the sun's indisputable importance. Today, we know that the sun is essentially the earth's sole source of energy. True, energy is released from the earth in its geological writhings (earthquakes) and by volcanoes and hot springs. Man has induced nuclear energy release, both controlled and uncontrolled. But these energy sources are miniscule when compared to the sun.

Perhaps 75% of the earth's utilizable energy is derived from the combustion of fossil fuels. But these fuels merely represent stored solar energy; they were formed from living organisms that required the sun's light and heat in their cycle of birth, death, and decay. Hydroelectric power accounts for almost all of the remaining 25%. It is solar energy, however, that evaporates water and drives the atmospheric engine which deposits the water on mountain slopes in the form of precipitation. The release of this energy under the gravitational descent of water down the slopes is utilized in driving turbines that generate electricity.

That the fixed stars are visible at all when our three outer planets cannot be seen without optical aid almost assuredly places them in the same category as the sun. Thus, an apparent inconsistency in placing the sun at the end of the list of solar-system objects, even though we examined the rest in descending order of importance, is not so inconsistent after all. This arrangement provides for a smooth transition in our inquiries from the inactive solar captives, to the sun, to the stars, and ultimately to the wider universe of stellar galaxies. For the sun is indeed a star, the only one near enough so that its surface is accessible to our detailed scrutiny.

Physical Properties

What can we learn about the sun from our rather remote position? The answers to this question should surely encompass such quantities and properties as the sun's distance, its size, and its mass. We might also inquire, does the sun rotate on an axis? What is the meaning of its surface markings? What is the behavior of its atmosphere, if it has one? Most important, we want to now the nature of the sun's energy source and the mode of energy release. In addition, the sun's temperature, its density, its surface gravity, its composition, its age, and its destiny are all matters of interest and importance. Information about all of these things will help us in our quest for knowledge about stars in general.

The Sun's Distance. The method used to determine the earth-moon distance (Chapter 7) is applicable to the determination of the earth-sun distance which averages 149.6 million kilometers. In practice, however, the results are less than satisfactory when this method is used. The sun is on the order of 400 times more distant than the moon. Consequently the solar parallax (p. 129) is very small, 8.8 of arc by best direct measurement. But by using Equation 7.4 the reader can see for himself that an error in measuring the parallax as small as 0".1 of arc can lead to an error of 6.5 million kilometers in estimating the earth-sun distance. The apparent angular difference between the diameters of a dime and a penny when they are viewed from a distance of 4.1 km is 0".1 of arc!

Techniques far superior to the one above for determining the earth-sun distance are available to us. Kepler's laws define 1.0 AU as the mean earth-sun distance. But we must know precisely the number of kilometers per astronomical unit if the latter is to be useful. Fortunately the formulas already developed in Chapter 7 are adequate to the task *provided* they are applied to bodies whose parallax is sufficiently large to overcome instrumental or reading errors in their determination. Venus, when near inferior conjunction, and Mars, when at opposition, have been thus routinely employed in the past. Even better are the close approaching minor planets. From these, the *length of an astronomical unit, 149.6 million kilometers,* is obtained.

In recent years radar beams have been bounced off the surfaces of Venus and Mars. The product of the delay time between the transmitted signal and the returning echo, and the speed of propagation of both the signal and echo (speed of light) yield a very high precision

for the distance of the planet at any given instant. At the instant the radar measurement is made, the planet's relative distance in terms of a fraction of an astronomical unit is known. Hence the number of kilometers per astronomical unit can be precisely determined. This method further substantiates the value of 149.6 million kilometers for the mean earth-sun distance (1 AU).

The Sun's Diameter. On the average the sun's disk has an apparent angular diameter of about 0.5 (30") of arc. We must compute an average value because the variation in the earth-sun distance as the earth travels its elliptical orbit makes the sun alternately appear slightly larger, then smaller. Calculations employing Equation 7.5 yield a solar diameter of 1.39 million kilometers. This is 109 times the earth's diameter and 10 times the diameter of Jupiter. By solar-system standards, the sun is large indeed. The moon's orbit, if centered at the center of the sun, would everywhere lie more than 150,000 km below the sun's surface!

Volume, Mass, and Density. With the sun's diameter established we can now easily obtain its volume in terms of earth units. From geometry we know the formula for the volume of a sphere: $V = D^3/6$ where D is the sphere's diameter. Next we form the ratios

$$\frac{V_\odot}{V_\oplus} = \frac{kD_\odot^3}{kD_\oplus} \qquad (10.1)$$

Here, k is the constant quantity $\pi/6$, and the subscripts $\odot$ and $\oplus$ identify the quantities with the sun and earth, respectively. Observe that these ratios, which are equal since they are constructed from equalities, allow us to compare the diameters of the sun and earth in terms of their measurable diameters. In evaluating Equation 10.1 we first observe that $k/k = 1$, and according to the rules of arithmetic we can rewrite $D_\odot^3/D_\oplus^3$ as $(D_\odot/D_\oplus)^3$. But in terms of earth diameters and volumes taken as unit quantities, the term in the above parentheses is just 109. Hence the sun's volume is just (109)³ or 1.3 million times the volume of the earth. Similarly, it is 1000 times that of Jupiter. Indeed, with a little straightforward computational effort, the inquisitive reader can verify that the volume of the sun exceeds the aggregate volume of all of the planets and their satellites by more than 700 times.

Determining the mass of the sun is considerably more complicated. It is accomplished with the help of Newton's laws applied to the observed acceleration of the earth in its orbit. Even though, for all practical purposes, the earth's speed in its orbit is constant, the earth is being accelerated because the direction of its motion is changing. A review of Chapter 4 may be helpful here. Recall that Newton's laws in this situation assert that the accelerating force $F = ma$ is directed toward the center of the orbital path. We use identifying subscripts as a reminder that this force acts on the earth's mass and produces a terrestrial acceleration thus: $F = m_\odot a_\oplus$.

Now the source of this force is the mutual gravitational attraction between the earth and sun. Newton's law of universal attraction applies.

$$m_\oplus a_\oplus = G \frac{m_\oplus M_\odot}{d^2} \qquad (10.2)$$

The perceptive reader will recognize that this equation combines Newton's second law and his gravitational law. This is permissible because the gravitational quantity on the right *is* an accelerating force. In this situation it is acting on the earth.

At this stage two quantities are undetermined: the mass of the sun *and* the acceleration that the earth is experiencing. We must now find the acceleration of the earth toward the sun independently of the above relationships. When this is accomplished the mass of the sun can be calculated by use of Equation 10.2. One way of doing it is with reference to the motion of the earth toward the center of its orbit. Of course it never reaches the center; the

earth's forward speed always keeps the earth falling *around* the center of the orbit. Recall that the earth's forward speed averages 30 km/sec (30×10^5 cm/sec). During each second the *direction* of motion of the earth changes by an angle ω measured at the center of the orbit. The *angular speed,* the rate at which the angle changes, is just $360° \div 32 \times 10^6$ sec. This follows from the 360° change in direction the earth experiences with each orbital revolution; this takes place in 1 year of 32 million seconds. Therefore, in 1 *second* the angular change in direction is $\frac{1}{32} \times 10^6$ of 360°. Figure 10-1a shows the relationships.

We shall find it convenient to show the changes in direction, together with the angular speed, with the aid of vectors. In Figure 10-1b vectors **A** and **B** show the scale speed and direction of the earth during 1 second. The magnitude of the change and its direction is represented by vector **C**. In vector mathematics the difference between two vectors can be

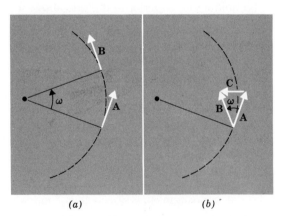

(a)   (b)

FIGURE 10-1   Vector representation of the angular change in the earth's orbital velocity. For practical purposes the speed remains constant, so the vectors are the same length. But the direction changes during a given time interval as shown by the different directions of the speed vectors. (*a*) Angular change ω in second (angle greatly exaggerated for clarity); (*b*) vector **C** is the vector *difference* in direction between **A** and **B**.

represented by a third vector joining the free ends of the two original vectors. In the figure **C** is this difference vector; we note that it points toward the center of revolution, that is, toward the sun. Because **A** and **B** are very large compared to **C** we can represent the latter as a small segment of the arc of a complete circle. This follows from the fact that a *chord* connecting very closely adjacent points on an arc is negligibly different in length from the arc. Therefore, we form the ratio of the length of the arc segment to the angle between vectors **A** and **B**, and 360°. In words, the small arc segment **C** generated by angle ω in 1 second is to the whole circumference as the angle ω is to 360°.

$$\frac{\mathbf{C}}{2\pi\mathbf{A}} = \frac{\omega}{360°} \qquad (10\text{-}3)$$

Equation 10.3 is a valid argument because 2π times the radius of any circle is the length of its circumference. In the figure **A** is the required radius; $2\pi\mathbf{A}$ is the total path length in Figure 10-1b. Upon evaluation we find **C** = 0.59 cm/sec during each second of travel. Hence the acceleration of the earth toward the center of its orbit is 0.59 cm/sec². Of course we have approximated the orbit as circular; but with this approximation *the sun will be occupying the center.*

Upon insertion of this value for a in Equation 10.2 we find the mass of the sun is 332,000 times that of the earth. Incidentally, each planet has a different value for its acceleration toward the sun because each is at a different distance from the sun. If each planet were to stop orbiting the sun and to begin to fall directly in toward it, the time that it would take a given planet to reach the sun depends upon its initial acceleration and its distance from the sun. Table 10.1 lists the times.

Because we are presently involved with central forces and accelerations, it is logical to digress slightly and examine by synthesis one of Newton's arguments concerning gravitation. With his universal law of gravitation

TABLE 10.1 Times of Fall to the Sun of the Various Planets

Mercury	15 days	Saturn	5 years
Venus	40 days	Uranus	15 years
Earth	65 days	Neptune	29 years
Mars	121 days	Pluto	44 years
Jupiter	766 days		

Newton was able to show that if a planet, small compared to the sun, stays in orbit rather than flying away in a straight line, it must be acted upon by a centrally directed force; $F = mv^2/d$. Here, m is the mass of the planet and v^2/d is the magnitude of the acceleration toward the center that produces the angular change in direction as shown in the preceding discussion.† As before, d is the distance of the planet from the sun's center, and v is the speed of the planet in its orbit. Now this force is produced by the mutual gravitational attraction between the sun and the planet. Thus we can write

$$G\,\frac{M_\odot m_\oplus}{d^2} = \frac{m_\oplus v^2}{d}. \qquad (10\text{-}4)$$

The right side of the equation *is* the gravitational force; the left side describes the earth's motion under the action of this force.

Equation 10.4 can be routinely solved for v except that the distance d must be precisely known for the moment that the value of v is desired.

In practice, Equation 10.4 is not of much use, even though it summarizes concisely the theoretical relationships needed to determine velocity at a given instant. We are usually more interested in the average velocity of, say, a planet. It is much easier to compute an average planetary speed from knowledge of the

† When we are interested primarily in magnitudes only, as is generally the case with orbital motion, we can dispense with vector notation. Quantities having magnitudes only are called *scalars*. For example, an object moving with a velocity **v** has a speed v.

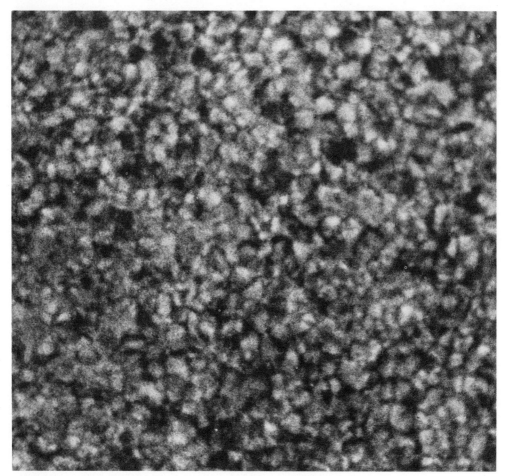

FIGURE 10-2 Solar granules. Granules are the tops of discrete convective cells of rising material in the photosphere. Each column is about 1000 km in diameter (Courtesy Hale Observatories).

total path length and the time it takes to complete the trip. Let the path length be the length of the orbit, which is just $2\pi a$ where a is the radius of the orbit. Let p represent the *orbital period*; this is the time required to complete the orbit. Then the average speed v is just $v = 2\pi a/p$, and v^2 is $(2\pi a/p)^2 = 4\pi^2 a^2/p^2$. We substitute this expression for v^2 in Equation 10.4.

$$G\frac{M_\odot m_\oplus}{d^2} = \frac{m_\oplus v^2}{d} = \frac{m4\pi^2 a^2}{ap^2} \quad (10.5)$$

Here the general distance d is now designated as a radius a. After routine mathematical rearrangement, which involves the substitution of the symbol k for all of the combined constant quantities, we have another solution for the sun's mass.

$$M_\odot = \frac{ka^3}{p^2} \quad (10.6)$$

Equation 10.6 is just precisely Kepler's harmonic law in slightly different form. Moreover, Newton's verification of Kepler's laws shows that it is general and universal in application to any central-force system. The mass of the sun did not appear in Kepler's original notation

242   The Sun

because it was taken as a unit quantity. This treatment worked for Kepler because the masses of the planets are negligible compared to the sun's mass. When both masses are appreciable Kepler's harmonic law is more precisely written,

$$(m_1 + m_2)p^2 = ka^3 \qquad (10.7)$$

We will find this form of Kepler's harmonic law of vital importance later on in our assault upon the mysteries of stellar masses in general.

Now that we have the sun's volume and the sun's mass in earth units we can at once write the relative density (ρ) of the sun, $\rho = m/v$. Its value is 332/1300 or 0.255 that of the earth. Since the earth is 5.5 times as dense as water it follows that the sun is 0.255 × 5.5 or 1.4 times as dense as water. In this sense the sun is a relative lightweight even though its mass is the unimaginable quantity

950,000,000,000,000,000,000,000,000

(9.5×10^{26}) metric tons!

One final quantity that we can calculate from the accumulated data, together with the application of Newton's universal law of gravitation, is the acceleration of gravity at the sun's surface. The computational exercise is of no importance here; we shall simply accept the value of 28 times the earth's surface gravity. This means that a given object would weigh 28 times more on the sun than it does on earth. Thus, if a petite young lady of 110 pounds on earth could arrive on the sun without vaporizing in the intense heat, imagine her consternation to find her weight has increased to a ton and a half!

The Sun's Appearance

Contrary to what we might expect, the sun's surface is extraordinarily difficult to observe. The turbulence of the earth's atmosphere blurs the details just as it does for lunar and planetary observations. Moreover, the sun is so brilliant that it must be viewed through intensity reducing filters or optical wedges. Both

contribute to additional loss of detail. In spite of these difficulties, on days when atmospheric turbulence is minimal, several characteristics of the solar surface stand out. First, the sun has a granulated appearance (Figure 10-2) as though the surface is boiling in violent convective upheavals from below. Second, around the sun's limb, or visible edge, the surface brilliance is substantially reduced. The phenomenon is termed *limb darkening*. The darkening gives a three-dimensional, spherical appearance to the sun's disk. Third, transient semipermanent markings, both dark and bright, are observed from time to time. We shall consider these features in detail just as soon as we can develop an adequate theory that explains their significance.

The semipermanent markings (Figure 10-3) do have one utility that is independent of their physical nature. They show that the sun turns

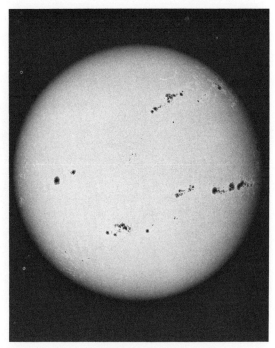

FIGURE 10-3 Sunspots, the most easily seen of all transient surface markings on the sun (Courtesy Hale Observatories).

on an axis that is inclined about 7° to the plane of the earth's orbit. Said another way, the earth's orbit is tilted 7° out of the plane of the sun's equator. The direction of the sun's rotation is west to east, in the same direction as the revolution of all of the planets. The surface features make their first appearance as they come around the west limb. They travel across the face of the sun in paths parallel to and near the sun's equator, finally disappearing around the east limb. The markings can often be tracked for several months. Strangely enough, and reminiscent of the behavior of the surface features of Jupiter and Saturn, the markings on the sun indicate rotational periods of 25 days for the sun's equatorial zone, 28 days at latitudes 45°, and 34 days near the poles. Obviously the observed surface is not solid. So far no measurable changes in the rotation rates have ever been observed, and the rate is so slow that no measurable equatorial bulge exists.

The Inverse Square Law

Let us now obtain a preliminary estimate of the rate at which energy, heat energy at least, is emitted by the sun. We must experimentally determine the rate of arrival of the energy at the earth's distance, and then extrapolate the quantity back to the sun. This will require a determination of the extent to which the energy per unit area is diminished or diluted during its 150 million kilometer journey to the earth.

First, consider a point source of energy emission from which the radiation travels outward uniformly in all directions. Clearly the energy rays must spread apart evenly; the quantity of energy passing through a small surface area near the sun is spead over a vastly larger area at a great distance from it. Suppose we examine a discreet bundle of energy rays emanating from the point as in Figure 10-4. For convenience the boundaries of the bundle will be in the shape of a regular pyramid, each side of which is an isosceles triangle (a triangle with two equal sides). The energy rays contained within these boundaries pass through selected cross sections of the pyramid, eventually departing through the base. We choose two such cross sections such that the distances from the source to the first, and from the first to the sec-

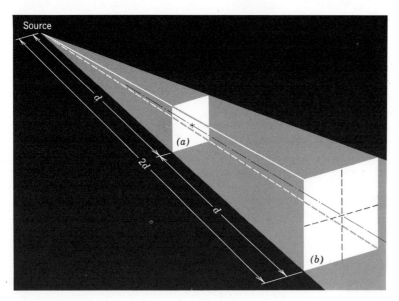

FIGURE 10-4 The inverse square law. Intercept area at distance d from source is just one-fourth that at distance $2d$; energy passing through area a is spread over any area b that is 4 times as large (from Cole, F. W., *Introduction to Meteorology*, John Wiley and Sons, New York, 1970).

ond, are the same. Let the distance be d in the figure.

Now an easily proven theorem in plane geometry asserts that if the midpoints of any two sides of a triangle are connected by a straight line, such line is parallel to the third side and is just one-half its length. Reexamination of Figure 10-4 shows that this relationship exists for each side of the pyramid; the connected construction lines enclose a square "surface." Let us further join the opposite midpoints of the auxiliary lines that girdle the pyramid forming its base. From the above theorem, each of the squares now formed is exactly the same size as the square constructed on the midpoints of the sides of the pyramid. If energy is conserved, no more can pass through the base of the pyramid than passes through the first square. Moreover if the energy is spread out uniformly, each base square passes only one-fourth of the total energy passing through the first square. But the base squares are twice the distance from the source as the first square. Hence we have this result: *the apparent intensity of the energy emitted from a point source varies inversely as the square of the distance from the source.* Symbolically, $E \propto 1/d^2$. The symbol $\propto$ means *proportional to.* More succinctly, the apparent intensity of radiation measured at some distance d from a source will be only one-fourth as great at 2 times the distance, one-ninth as great at 3 times the distance, and so on. Now let us use this information in our attack on the problem of the rate of energy release from the sun's surface.

The Solar Constant

By definition the solar constant is the amount of solar energy, both heat and light, per minute falling on or passing through a 1 cm² surface perpendicular to the sun's direction at the mean earth-sun distance. It is a derived quantity whose units are calories per square centimeter per minute. A calorie is the amount of thermal energy that is exactly sufficient to raise the temperature of 1 g of water from 14.5°C to 15.5°C. It would take about 80,000 calories to heat 1 liter (slightly over 1 qt) of water from an average room temperature of 20°C to the boiling point. Nothing in the unit itself tells us how long the task would take.

Clearly the calorie is a small quantity of thermal energy. Yet according to the results of exhaustive research extending over many years, the solar constant is determined to be no more than 2.0 cal/cm²-min. Until recently this estimate was subject to uncertainties in the amounts of atmospheric scattering, reflection, absorption, and reemission of the incoming solar energy. Modern orbiting geophysical satellites have confirmed the above estimate, which is based upon 71% atmospheric transmission. In truth the earth-based determinations turned out to be in surprisingly good agreement with the more reliable satellite data.

Even though the solar constant is a small quantity, the cumulative effect of this radiation rate over a large area for an extended period of time is surprising. Rough calculations show that the quantity of solar energy arriving at the earth each day, if totally utilized as heat, would be sufficient to raise the temperature of sufficient water to cover the 50 United States to a depth of 4 m from the freezing point to the boiling point!

With the aid of the inverse square relationship just developed, together with the solar constant, we can extrapolate the energy radiation rate from the sun's surface even though it is 150 million kilometers distant. One extra step is involved since the sun is not a point source of energy. Careful scrutiny of the inverse square relationship, especially as shown in Figure 10-4, discloses that the dilution of sunlight intensity with distance is proportional to the ratio of the *areas* of the successive surfaces through which it passes. The computed radiation rate from 1 cm² of the sun's surface is thus 91,000 calories per min. In more familiar terms, this emission rate is equivalent to 72,000 hp per yd², or 53,700 kilowatts per m².

The total energy emission per minute from the sun is a prodigious 31.5×10^{24} hp (31.5 followed by 24 zeros), or 23.5×10^{24} kW. This quantity beggars the imagination. Even so it is plain that the sun's energy is not released by an ordinary combustion process. Suppose for example that the sun's mass consisted only of high-grade coal, and that ample oxygen was available to effect combustion at the observed furious rate. This quantity of coal would be totally consumed in roughly 2700 years. But geological evidence on earth indicates that the sun has been radiating essentially unchanged for more than two million times 2700 years. The 19th century astronomers were well aware that an entirely unknown energy conversion process was at work in the sun. Let us trace the development of scientific thought that finally unveiled the truth.

The Sun's Temperature

Direct determination of the sun's surface temperature is hopelessly beyond our capabilities. Even the crudest calculations indicate a temperature so high that any imaginable sensing instrument would be instantaneously vaporized on contact with the sun. Indeed, its destruction would occur long before an instrument package would reach even close proximity to the surface. The solar constant is the only datum in the form of direct information. It is a measure of the sun's radiant intensity or brightness, but the question we must resolve is whether or not there is direct correlation between the sun's brightness, the color of emitted light, *and the temperature of the emitting surface.* If so, here is a potential way out of the dilemma concerning the sun's temperature because two of the three quantities, brightness and color, are known.

Radiation Characteristics of the Sun

It is beyond our scope or need to pursue the details of the development of scientific thought that provides the theoretical method for determining the sun's temperature. A reasonably expanded discussion of radiation theory appears in Appendix 5; it can profitably be read at this point by those who are so inclined. For others, the following summary should suffice.

We know from experience that the more intensely an object is heated the higher its temperature becomes, and the more intensely it radiates its acquired energy away. If the heating continues long enough both radiant heat and light are given off. The color of the light progresses from darkest red through all of the colors of the spectrum to bluish white, even blue and violet, corresponding to the temperature increase, providing, of course, that the object is not annihilated in the process. Figure 10-5 shows qualitatively the representative relationships between energy emission rates versus wavelengths at which the emission occurs for several temperatures.

In general, different materials radiate differently for a given temperature; usually the color of the light given off amply demonstrates this behavior. However, when viewed through small holes connecting them to the exterior, cavities within the different materials *all radiate exactly the same.* The name *cavity radiator* is applied to the phenomenon; such cavities radiate essentially as ideal radiators, better known as *blackbody radiators* (Figure 10-6). Blackbody radiation theory predicts the intensity of the radiation emitted by an ideal radiator for each wavelength, over the entire array of wavelengths involved. It makes these predictions for any and all possible temperatures. Figure 10-7 shows qualitatively the relation between radiation intensity and wavelength for several selected temperatures. The temperature line can be thought of as a graph connecting discretely measured radiation intensities at each of a large number of wavelengths. Clearly the graph shows that the over-all intensity, represented by the area of the region between the temperature curve and the base of the graph, increases with increasing temperature. Also,

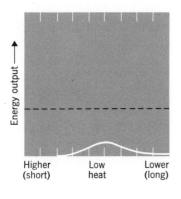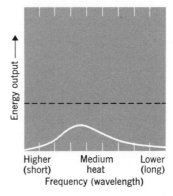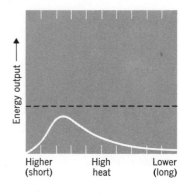

FIGURE 10-5 Representative emission curves for an object heated to successively higher temperatures (from Cole, F. W., *Introduction to Meteorology*, John Wiley and Sons, New York, 1970).

the magnitude of the intensity maximum, represented by the highest point of the temperature curve, increases with increasing temperature. Finally, the graph illustrates that the wavelength at which the maximum intensity of radiation occurs shifts toward shorter wavelengths as the temperature increases. That is, the higher the temperature, the more blue and the less red is the color of the emitted light.

Three famous physical laws are associated with these three blackbody radiation characteristics. *Planck's law,* the mathematical form of which is of no concern to us here, predicts the *shape* of the temperature curves. *Wien's law* relates the temperature of the radiating object to the wavelength at which the radiation intensity is a maximum, $T = k\lambda_m$, where λ_m represents the maximum wavelength and k is an arbitrarily determined constant. *Stefan's law,* also

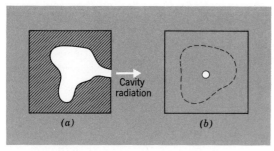

FIGURE 10-6 Cavity radiator: (a) cross-section view and (b) front view. Cavity radiation emerges from the tiny hole leading to the interior (from Cole, F. W., *Introduction to Meteorology*, John Wiley and Sons, New York, 1970).

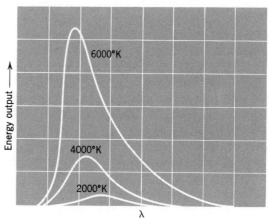

FIGURE 10-7 Characteristic radiation curves for a blackbody at different temperatures (from Cole, F. W., *Introduction to Meteorology*, John Wiley and Sons, New York, 1970).

known as the Stefan-Boltzmann law, shows that the energy emission rate is proportional to the fourth power of the absolute temperature, $E = k'T^4$. In this expression k' has a different value from the k in Wien's law. For example, if the absolute temperature of a blackbody radiator is doubled, the rate of energy emission is increased sixteenfold; a threefold temperature increase produces 81 times the original rate of energy emission.

Comparison of the sun's radiation curve to the appropriate blackbody radiation temperature curve shows a good fit, that is, the sun is nearly a blackbody radiator in its energy emission behavior. The curve for best fit yields a solar surface temperature of 5770°K. Application of Wien's law yields a solar temperature of 6120°K. The correct temperature is somewhere between these two values; for our purposes we can approximate it as 6000°K.

The Sun as a Blackbody Radiator. Only one critical problem remains: Is the sun an ideal radiator? The answer is no. Exhaustive analyses of the solar spectrum radiation characteristics show significant departures from the theoretical models in the ultraviolet and the near infrared (these wavelengths will be defined presently). But the departures are minor in importance (Figure 10-8). We are safe in approximating the sun's surface temperature at 6000°K in round figures; high altitude research, especially with the aid of geophysical research satellites outside the earth's atmosphere, is slowly resolving most of the questions with regard to the wavelength region of the solar spectrum that is absorbed in the atmosphere. Best modern information indicates that the sun is within 5% of being an ideal radiator. Presumably, since the sun appears to be a typical star, we are now equipped to extend blackbody radiation theory to stellar astronomy in hopes of unlocking further mysteries of the universe. But, as we shall soon see, formidable roadblocks still obstruct the path.

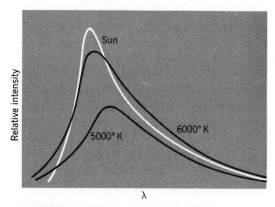

FIGURE 10-8 Departure of the sun's radiation curve from the theoretical blackbody model for 5000°K and 6000°K. The departure, exaggerated by this relative sketch, is less than 5% from the curve for a 5770°K object.

Electromagnetic Spectrum

We usually speak of electromagnetic radiation phenomena with much more confidence than we do about the way in which they are generated and propagated. Familiar terms such as radio waves, microwaves, x rays, radiant heat, and light are actually descriptive of the ways in which electromagnetic radiation interacts with particular "sorting" or sensing devices.

Careful examination of the operation of these detectors reveals that no sharp cut-off in the electromagnetic radiation distinguishes one kind from another. We can tune radio and television sets over quite a range; our eyes are sensitive to a range of colors; our nervous system detects differences in radiant heat. What, we are inclined to ask, differentiates one form of radiant energy from another? Let us see.

From the physicist, we accept evidence that all frequencies of electromagnetic radiation are propagated at the same speed in a vacuum and at very nearly the same speed in a material substance. Thus, from the relation established in Chapter 5 ($c = \nu\lambda$), since c is constant the difference in behavior must involve *both* wavelength *and* frequency. This is indeed the case.

Let us borrow again from physics. We shall construct a chart or diagram which gives us some insight into the array of electromagnetic wavelengths (Figure 10-9). These we can relate to familiar manifestations. We must always keep in mind, however, that our subdivisions are arbitrary; the array is continuous. We shall call the array the *electromagnetic spectrum.*

An examination of Figure 10-9 discloses several important relationships. Longer wavelengths and lower frequencies are associated with power and radio waves. These appear at the right end of the figure. As we move toward the left, wavelengths become shorter and frequencies become higher. Radio phenomena give way to heat phenomena. Next is visible light which in turn gives way to ultraviolet radiation, x rays, and forms of nuclear radiation.

We observe that the heat phenomenon is associated with frequencies below the threshold of visible red light. The name *infrared* (below the red) is assigned to this form of radiation. Similarly, the array of frequencies immediately above the threshold of violet light is termed *ultraviolet* radiation.

Our principal concern is with a small segment of the total electromagnetic spectrum.

the sun emits radiant energy principally in the ultraviolet, visible light, and infrared frequencies. These are the frequencies of concern to us here. As we have seen, such energy is emitted in small, discrete bundles or packets called *photons*. But the oscillatory nature of photons emphasizes the wave characteristics of electromagnetic radiation. Thus we can think of sensing devices as being susceptible to particular photons and insensible to most of the others.

Absorption and Emission Spectra. In Chapter 5 we learned that certain frequencies of the emitted light from a star sometimes appear to be missing, or are at least reduced intensity. We now need to turn this behavior to our further advantage. It will not be necessary to pursue the subject of atomic or molecular emission or absorption in detail; some generalizations will suffice.

Gustav Kirchhoff over a century ago gave a qualitative description of spectra in general. Three major types exist: a *continuous* spectrum wherein all radiant frequencies are present, an *emission spectrum* of certain discrete frequencies, and an *absorption spectrum* of similar frequency characteristics. Modern physics describes and explains these spectra to the everlasting advantage of the astronomer.

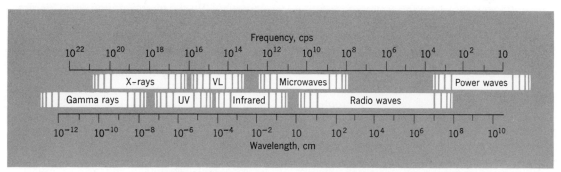

FIGURE 10-9 Schematic representation of the ranges of various electromagnetic radiation manifestations. Note that a distinct overlap with adjacent categories is present (from Cole, F. W., *Introduction to Meteorology,* John Wiley and Sons, New York, 1970).

Energy emission, or absorption by atomic or molecular oscillators, involves transitions of certain orbiting electrons from one energy state to another. The precise transition depends upon the location of the electron orbit or orbits in the molecule or atom, and upon the energy of the photons, or quanta, absorbed or emitted by them. Each atom or molecule is capable of emitting or absorbing energy only at certain discrete frequencies although the emission often consists of a cascade, or stepwise energy release. The precise combination of frequencies involved constitutes an identifiable signature of the particular oscillator. Once we know the temperature at which the energy transitions occur, we can immediately identify the substance involved. Conversely, if the chemical composition of the oscillator is known, the nature of the absorption or emission spectrum yields valuable information about the existing temperature.

There are, however, major obstacles that torment our investigations. For example, individual atoms or molecules absorb or emit in their characteristic fashion *only* if the proximity of other oscillators does not interfere. In the case of incandescent solids, liquids, or high-pressure gases, the atoms or molecules are so close to each other that interference is very severe. The result is a *continuous emission spectrum* covering the entire range of permissible frequencies as determined by the temperature. All substances show the same type of continuous spectrum *under the above conditions;* the identifying discrete absorptions and emissions are effectively concealed.

On the other hand, spectrographic analysis of the light emitted from a stimulated or excited low pressure gas reveals an *emission spectrum,* a series of bright line images of the spectrograph slit; all other frequencies are absent. Moreover, the discrete series of emissions is uniquely characteristic of the particular substance (Figure 10-10a–c). If instead a beam of white light (all visible frequencies present) is passed *through* a sufficient quantity of a cool or cold (nonemitting) gas, the spectrograph reveals an *absorption spectrum* wherein certain frequencies, again characteristic of the particular substance, are missing, or are at least in very low brightness contrast (Figure 10-10d). Most important, the missing frequencies are precisely those the gas would emit as a bright line spectrum if it were excited to emission. Thus, both bright line and absorption spectra are unique and identical for a given substance, provided emission takes place at reasonably low temperatures. Even under extreme temperature conditions the emission spectrum is still identifiable with the substance.

Painstaking and exhaustive research employing laboratory samples has provided spectroscopists with a catalog of spectra, each associated with a particular substance. They are thus able to identify any known naturally occurring element and countless other molecular compounds simply by spectral comparisons with the laboratory standards.

The qualitative explanation for the reduced intensity associated with the absorption spectra lines is not difficult to understand. In Figure 10-11 a representative quanta, or photon, of radiant energy is intercepted and absorbed by a particular oscillator. In a few hundred-millionths of a second the energy is reemitted, either at the same or at a different frequency, or at combinations of frequencies. Whichever the emission mode, the oscillator, invariably an electron, emits the energy uniformly *in all directions,* in contrast to the specific direction of the absorbed photon. Since energy is conserved, the quantity continuing along the original direction is markedly reduced by the scattering. But since we look *toward* the source, the energy depletion is obvious. Actually we observe the cumulative effects of countless millions of oscillators all engaged in an identical performance, although not necessarily in unison.

Here, then, is a powerful tool at our disposal. Not only does it give us the means for identifying the chemical constitution of remote systems, it also provides a way of

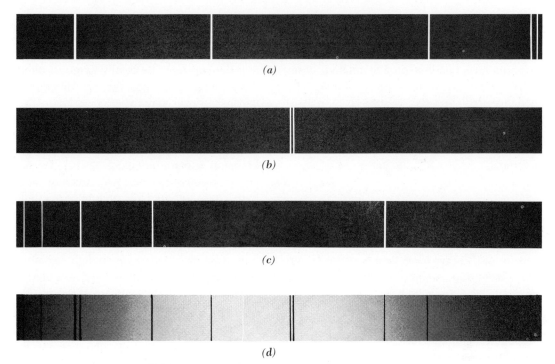

(a)

(b)

(c)

(d)

FIGURE 10-10 Several simple emission spectra (*a–c*), and a corresponding (composite) absorption spectrum. Represented are the well defined doublet of sodium (*b*), the 5 lines of hydrogen in the visible light range (*c*), and the H and K lines of calcium, extreme left in the absorption spectrum (not shown in the emission spectra). See text for discussion.

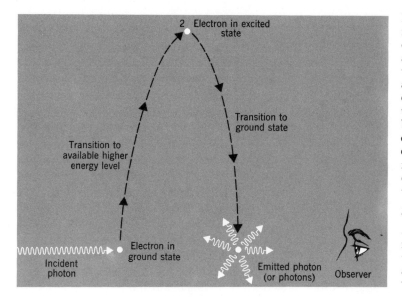

FIGURE 10-11 Diagram illustrating the principle of radiation depletion along the line of sight by the excitation of an atomic electron. Although only the transition from the excitation energy level back to the ground (unexcited) state is shown, the electron may "cascade" from several intermediate energy levels. Each transition releases photons with frequency characteristics. The energy is radiated in all directions, effectively reducing the intensity of the background spectrum at each particular wavelength (see text).

determining the relative abundances of the constituent substances. The intensity of the emission or absorption spectra is known to be partially dependent on the relative abundances of the substances. More dramatically, even, we shall find that statistical analyses of the spectra of the stars can give us great insight into the composition of the universe. No other method of doing this has yet been discovered.

Composition and Structure of the Sun

The obvious starting point for our intellectual attack on the larger cosmos is right here with the sun. Already we have seen how the continuous spectrum of sunlight can be interpreted in terms of the sun's effective surface temperature.† It tells us even more than this. Because of its continuity over a precise range of frequencies, a range that is temperature dependent, the solar spectrum informs us that the sun, or at least the light emitting portion, is one or another of an incandescent solid, liquid, or high-pressure gas. We must now decide between these three alternatives, either singly or in combination. Then we must bring all the evidence to bear on the problem of the sun's composition and structure.

Two experimental conclusions rule against the possibility that the sun is solid or liquid. We know from chemical thermodynamics that the solid, liquid, and vapor phases of a given substance can coexist up to a temperature called the *triple point*. Above this temperature, the solid phase is impossible although the liquid and vapor phases can coexist. But there exists a *critical temperature* above which even the liquid phase is impossible. The critical temperature is unique for each substance and is independent of pressure, no matter

---

† If the sun were an ideal blackbody radiator, the spectrum analysis would yield a *true* surface temperature. Since it is not, all we can claim for the sun is knowledge about an effective temperature, that is, the temperature that an ideal blackbody with the same surface brightness would have.

how great the latter may be. The sun's temperature is far above the critical temperature for any known substance. Thus we conclude that the sun must be totally gaseous, although the gaseous form may be very different from that in our everyday experience.

When sunlight is analyzed with the spectrograph a dark-line absorption spectrum is observed against the background continuous emission spectrum (Figure 10-12). From preceding discussions we must conclude that the two types of spectra originate in different regions of the sun under substantially different physical conditions. It is beyond our scope to review the historical unraveling of solar radiation mysteries. We can, in light of physical principles established in this chapter, turn directly to the findings.

The Photosphere. The outermost region of the sun which presents the appearance of a solid surface is called the *photosphere;* this is the zone or region where the solar radiance appears to originate. Several interesting and important radiation phenomena are associated with the photosphere. Continuous emission and absorption are nearly complementary in strength. If no continuous absorption were present we could see to unlimited depth into the sun. If the continuous absorption and emission were exactly complementary, we could not see below the surface at all. That we can see to some depth is demonstrated in the next subsection. The absorptive ability of the solar gases is termed *opacity*. One of the great astrophysical mysteries, solved only as recently as 1938, was the extraordinarily high opacity of the photosphere. Astrophysicists had been unable to explain why essentially clear, transparent gases could in reality be so opaque. They are now pretty much in agreement that *negative hydrogen*, an ion (symbol H^-) is the culprit. Negative hydrogen is a hydrogen atom that has, temporarily at least, acquired an extra electron, unbalancing the net electrical charge on the atom. This ion is known to effectively absorb all radiation in the range from the far

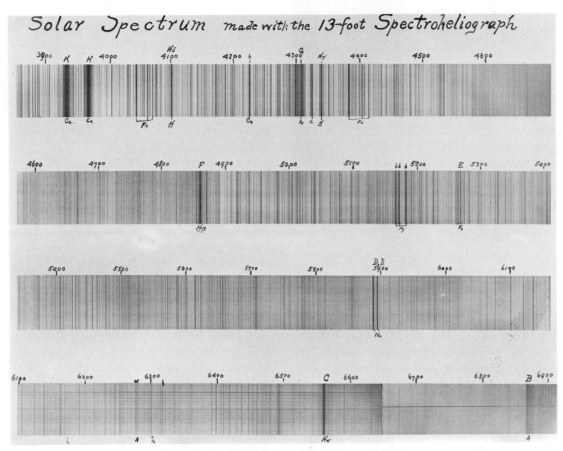

FIGURE 10-12 Solar absorption spectrum from 3900 Å to 6900 Å with lines of many chemical elements identified (Courtesy Mt. Wilson and Mt. Palomar Observatories).

infrared, through visible light, and well into the ultraviolet portions of the solar spectrum. Much of the difference between the continuous solar spectrum and the energy curve for a blackbody radiator at the same temperature is laid on the doorstep of the H$^-$ ion. Only about one such ion exists for every million neutral atoms. The extra electrons required are supplied by easily ionized metals. In spite of the relative abundances, there are sufficient H$^-$ ions to produce agreement between the predicted and the observed opacity.

<u>Limb Darkening.</u> Limb darkening (Figure 10-13), mentioned earlier in the chapter, is a direct consequence of opacity effects and has nothing to do with the sphericity of the sun. The concentration of negative hydrogen ions along the line of sight is dependent on, among other things, the density of the gaseous material. The upper limit of the photosphere is that zone where opacity ceases. The lower limit is, of course, the apparent visible surface. In between these limits, the depth to which we can see depends upon the angle of the path of the line of sight into the photosphere (Figure 10-14). We cannot see to as great a depth along a tangential path as we can along a direct path. Along a slant path opacity cuts off our vision in a higher, and therefore cooler,

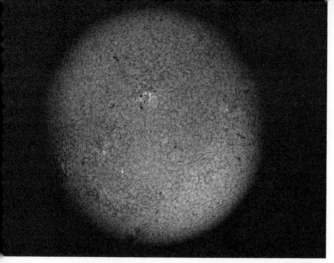

FIGURE 10-13 Limb darkening of the sun (Foothill College Photograph).

temperatures 200–300° higher. Radial velocity studies show vertical speeds of the granules of up to 1 km/sec. The obvious interpretation is that the lower photosphere where the granules originate is convectively unstable; it is a regime of furious seething and convulsive upheaval. The excess heat, reflected in the higher temperatures of some of the granules, is released in the recombination of the ionized hydrogen and is carried upward by the vertical circulation.

A physically analogous behavior can be simulated in the home. Fill a shallow cake pan with melted paraffin like that used to seal jars of preserves. Add a tablespoon or two of an oil-based silver or aluminum paint. Heat over a low heat source such as a hot plate or inverted flatiron. Vertical cellular motion that looks remarkably like solar granulation can be seen.

Sunspots. In 1610 Galileo risked ecclesiastical censure and possible physical punishment because of his announcement that the sun's surface contained dark spots. He was de-

zone of the photosphere. According to Stefan's law (p. 248) radiation intensity is less in the cooler region, making it appear less bright. The line of sight is most nearly tangent to the edge, or limb, of the sun; thus the limb appears darker. The fact that we can see to any depth at all is additional proof that the sun is not quite a perfect radiator.

The solar absorption spectrum is produced in this upper, quasitransparent zone of the photosphere, a zone that is probably between 100 and 800 km thick. Approximately 67 of the natural terrestrial elements can be identified in the solar spectrum; there are no unknown substances. From the relative line intensities, great abundances of hydrogen, calcium, iron, and their ions apparently are present. The other elements, some of which are ionized, and a few simple molecules are present in lesser concentrations.

Granulations. As mentioned earlier, photographs made of the sun in the light of a single wavelength when the earth's atmosphere is exceptionally tranquil show a polygonal network of fine, dark markings, called *granules* (Figure 10-2). The granules average several hundred kilometers in diameter, and they persist for only minutes. Some granules are brighter than the surrounding area, evidencing

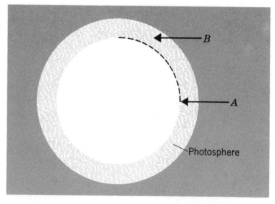

FIGURE 10-14 Explanation for limb darkening. Two lines of sight are shown: *A* is direct; *B* is a slant path. The penetration into the photosphere is governed by the sun's opacity and is the same for both lines of sight. Line *B* penetrates only as far as a higher, cooler layer compared to *A*, giving a darker appearance to that region of the solar disk.

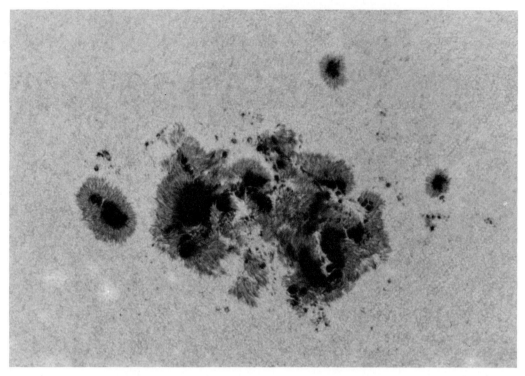

FIGURE 10-15 An exceptionally large group of sunspots, photographed May 17, 1951. Observe the excellent contrast between the penumbra and the darker umbra in most of the spots (Courtesy Hale Observatories).

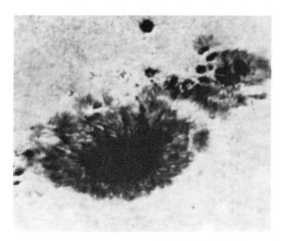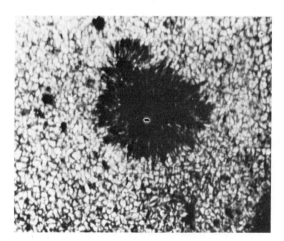

FIGURE 10-16 Two photographs of sunspots taken with comparable telescopes, from the earth's surface (left) and from a high-altitude balloon (right). Atmospheric turbulence clearly degrades the resolution as shown by the comparative detail in the solar granulations around the spots. Compare with Figure 10-2 (Courtesy NASA).

nounced for blasphemy; his telescope was pronounced an instrument of the devil since only *he* could "corrupt," in men's eyes, a perfect, incorruptible celestial body. Sunspots had in fact been reported for at least 15 centuries before Galileo; indeed some indication of the cyclical nature of their appearance had been chronicled. Galileo's only crime was to be in the wrong place at the wrong time.†

A sunspot begins as a *pore* resembling an enlarged dark area between granules, but it is longer lived. If the spot develops regularly the pore soon attains a diameter of over 1500 km; if it survives more than a day it is usually joined by others nearby to form a family or group of sunspots (Figure 10-15). A mature spot has a dark central region called the *umbra* and a less dark, occasionally striated region called the *penumbra.* (Compare the terminology with that of eclipses.) Both the umbra and penumbra have exceptionally sharp boundaries. The darkness of the spots is relative; both the sun's disk and the spots must be greatly reduced in brightness by optical devices in order to be seen in detail. Even the darker umbra is far brighter than any terrestrially produced phenomenon save a nuclear explosion. Studies give the average umbra a temperature some 2000° lower than the effective temperature of the photosphere.

An average sunspot is 37,000 km in diameter, although spots 245,000 km in diameter have been observed. It would be geometrically possible to drop a "string" of ten earths touching each other into the umbra of the largest spot without even touching its penumbra!

Bright patches, or *faculae,* appear as the forerunners of sunspots and they often persist for several days after the sunspots have disappeared. Several faculae can be seen in Figure 10-17. Sunspots themselves frequently last for days or even weeks. They do not occur anywhere at random on the sun; the vast majority occur within 30° latitude, but usually not within 5° of the solar equator. Most of the time sunspots occur in relatively equal numbers north and south of the equator. Sunspots wax and wane in count, with the maximums and minimums having a periodicity of about 11 years (Figure 10-18).

One of the clues as to the nature of sunspots is a whorl, or vortical appearance, of the penumbra. Vortices of a similar type are associated with large scale circulations in the earth's atmosphere; thus it is reasonable to assume these whorls are circulations. Spectrograms of sunspots show splitting of the absorption lines into two or three, or even more, components (Figure 10-19). The feat can be duplicated in the laboratory when absorption spectra are produced in a strong magnetic field. Complicated analysis of the line splitting, called the *Zeeman effect* after Pieter Zeeman (1865–1948), a Dutch physicist, shows that the sunspots are local centers of intense magnetic

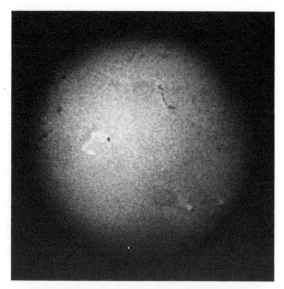

FIGURE 10-17 Irregular bright patches, called *faculae,* can be seen in association with three sunspots in this photograph (Foothill College Photograph).

† For an extraordinarily fine accounting of Galileo's tribulations read de Santillana, G., *The Crime of Galileo,* University of Chicago Press, Chicago, 1955.

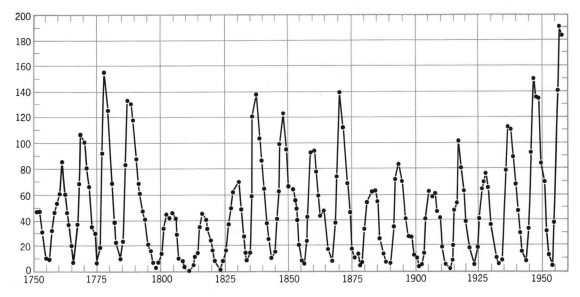

FIGURE 10-18 Eleven year sunspot cycle disclosed by the *Wolf* number for sunspot activity. The Wolf number is the sum of the individual spots observed plus ten times the number of sunspot groups.

disturbance. Curiously the magnetic storms have poles like an ordinary iron magnet. The magnetic polarity is reversed between a primary group of sunspots and a following secondary group. Following, in this sense, means to the westward in solar longitude; the following is produced by sun's rotation. Furthermore, the magnetic polarities of sunspot groups are exactly reversed in the opposite solar hemisphere. Finally, the polarities of the sunspot groups in both hemispheres are reversed at each successive period of the sunspot cycle.

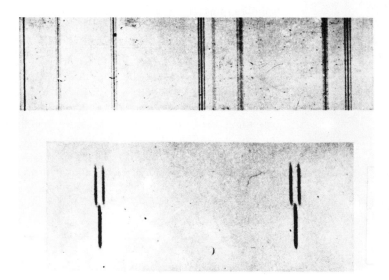

FIGURE 10-19 Spectral lines showing the Zeeman effect (Courtesy Hale Observatories).

Astrophysicists believe that sunspots represent relatively tranquil areas in the photosphere and that the magnetic forces seem to inhibit the usual convective turbulence seen elsewhere in the photosphere. Whatever the explanation, a phenomenon that can produce 2000° of cooling in areas extending in some cases over 45 billion square kilometers must be regarded as a truly gigantic refrigeration feat!

The Chromosphere. Just above the photosphere, indeed actually merging with its upper layers, is the *chromosphere*, the beginning of the true solar atmosphere. The chromosphere gets its name from its vivid red color. The chromosphere cannot be seen unless the photosphere is obscured during an eclipse, or

FIGURE 10-20 Solar filaments. The darker appearance indicates the eruptive gases have cooled expansively as they penetrate into the chromosphere. When the sun's rotation carries a persistent filament around to the limb it is seen as a prominence (Foothill College Photograph).

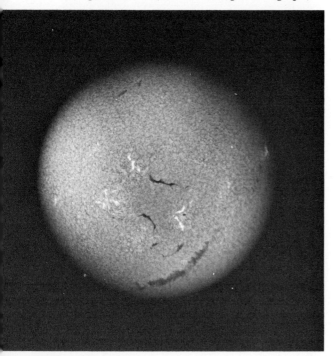

unless it is artificially obscured in a specialized telescope called a *coronagraph*. The bright red color is due to the dominance of an exceedingly strong emission line of hydrogen called the Hα line, although some 30 other emission lines are present. Among them are the lines of ionized calcium, helium, strontium, and magnesium. The hydrogen and calcium lines are the most prominent.

The chromosphere extends to perhaps 13,000 km above the photosphere, although specifying an upper boundry is somewhat arbitrary. In reality the chromosphere merges almost imperceptibly with the corona (to be discussed). Spectral studies disclose a decrease in the atmospheric density with height, but unexpectedly, a huge increase in the temperature, from 4500°K at the lower boundary to over 100,000°K at the upper one. The cause of this heating is discussed in connection with the corona. The upper boundary of the chromosphere is characterized by many small *spicules*, jetlike spikes of rising gas that reach heights of from 500 to 20,000 km above the photosphere. The gas often reaches speeds in excess of 30 km/sec in its upward motion. Individual spicules rarely last more than 10 minutes, most vanish in 1 or 2.

Specialized solar photographs called *spectroheliograms* often disclose dark, threadlike filaments against the brighter photospheric background (Figure 10-20). When the solar rotation carries some of the filaments around to the limb so that they can be seen in projection against the background sky, they appear as flamelike protuberances called *prominences*, extending well above the chromosphere into the lower corona. Indeed, time-lapse motion pictures made of such prominences show that some are really condensations occurring in the corona, and the material is descending into the solar surface. Prominences are often quiescent, not showing much change for hours or days on end (Figure 10-21). Others are definitely eruptive, the outrushing gases reaching speeds in excess of 10,000 km/sec,

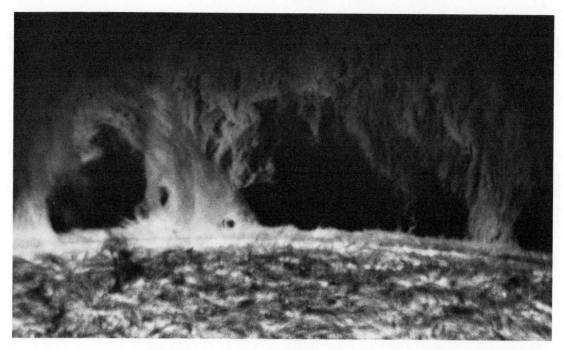

FIGURE 10-21 A quiescent solar prominence, often referred to as a *hedgerow,* extending several tens of thousands of kilometers above the solar limb. Smallest features visible in the photograph are on the order of 1000 km in size (Courtesy Big Bear Solar Observatory, California Institute of Technology).

well above the escape velocity of the sun (Figure 10-22).

Flares (Figure 10-23) are among the most remarkable of the chromospheric phenomena. A flare is a sudden outburst of very bright light of fairly short duration. Some flares last for only a few minutes, at most they dissipate in under 7 hours. Flares show emission spectra, and for that reason they are best seen when the photosphere is photographed in the light of a prominent hydrogen or calcium line (Figure 10-24). Flares are never centered on a sunspot, nor are they ever more than 100,000 km from the center of a group of spots. Their exact cause is a mystery, although it is certain that flares are not purely thermal in origin.

<u>The Corona.</u> The corona is a faint, pearly white extension of the sun's outer atmosphere. Unlike the chromosphere, the corona has been

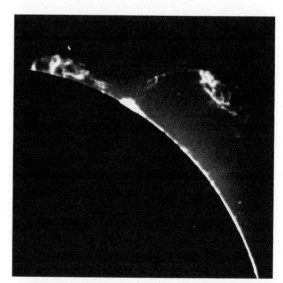

FIGURE 10-22 An eruptive prominence. The prominence has reached a height of about 230,000 km in this photograph (Foothill College Photograph).

Composition and Structure of the Sun   **259**

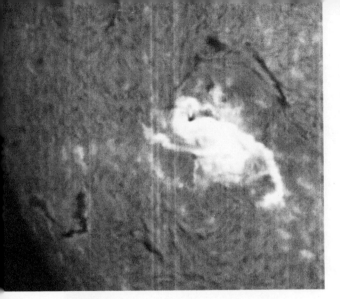

FIGURE 10-23 A solar flare (Courtesy Hale Observatory).

line broadening is interpreted to mean high particle velocity, 20 km/sec and more in a very tenuous, low-density gas. The emission lines are strange in that they arise only from iron atoms that have lost 13 of their 26 electrons, calcium atoms that have lost 14 of their 20 electrons, and nickel atoms that have been similarly ionized. The normal hold of the nucleus of these atoms on their electrons is so great that only under the extreme conditions of vanishingly low density, and temperatures in excess of 1 million °K can such high degrees of ionization occur. Thus the corona is more than 130 times hotter than the photosphere. This, however, does not mean that the corona contains 130 times more heat; the density is far too low. The best explanation for the excessive coronal and chromospheric heating is that shock waves originating in the very turbulent photosphere produce resonant, or sympathetic, vibrations in the atmospheric gases. The effect is analogous in some respects to the sonic boom produced by aircraft flying at supersonic speeds. Imagine what it would be like if a closely packed fleet of supersonic aircraft passed continuously overhead; the noise would be deafening. By analogy one might say that the sun's corona is plagued with extreme noise pollution!

The Internal Structure of the Sun. The radiation that ultimately makes its way outward through the photosphere originates deep within the sun. At first glance this does not seem to be a particularly erudite observation, except that when the laws of thermodynamics, which we shall investigate toward the end of the book, are invoked, it appears certain that the heat energy (and light) that we detect must be considerably different spectrally than the energy that is originated. Let us argue this point.

There can be no question that the *net* energy transfer is from source to sink, that is, from a hotter region ever toward a colder one. This fact is so commonplace in our experience that we cease to wonder at the self-evident-truth attitude with which the classical scientists

known for centuries. Plutarch mentioned it in one of his moralistic discourses, and Kepler described it in some detail. The corona extends many solar diameters above the photosphere; coronal rays or streamers have been traced to a distance of 22 million kilometers from the sun. It is possible that the solar atmosphere extends outward to the earth and beyond. The corona's innermost levels, 13,000 to 15,000 km from the surface are only one millionth as bright as the central portion of the solar disk. At a distance of one solar diameter, or 1.4 million kilometers, the brightness is less than 1½% of that of the inner levels. Over all, the corona is only half as bright as the full moon.

The faintness of the corona precludes its visibility except during total eclipses; the coronagraph under favorable conditions permits examination of the inner, brighter levels of the corona. The arrangement of the coronal structure (Figure 10-25) strongly suggests a close relationship with the solar magnetic field. The corona displays an emission spectrum that was not identified with known elements until 1942. Extremely broadened emission lines of highly ionized calcium, nickel, and iron are superimposed on a fainter continuous spectrum. The

FIGURE 10-24 Sequence showing the development of a solar flare. The elapsed time between the first and last photographs is 2½ minutes (Courtesy NOAA and Aerospace Corporation).

viewed it. But what happens along the way, from the bowels of the sun out to the surface? The energy generation processes within the sun appear to be relatively constant since no significant change in the quantity or quality of its energy output is apparent in all of geological history. Therefore, the sun must be hottest at its center. What implications does this logic carry with it as to the structure of the sun?

We can agree that the sun is totally gaseous, held together by the mutual gravitational attraction of all of the gaseous particles on each other. Since they are free to move, the atoms, ions, and molecules heterogeneously distribute themselves in a spherical configuration. Certainly, the sphericity is a matter of simple observation. According to Newton's gravitational law the net, or resulting gravity forces are directed inward. But gravity acting on mass is weight; this means that the ultimate crushing burden of the weight of all of the sun's matter is delivered to the exact center of the sun. In other words, the mobile gas particles are progressively compressed to higher and higher pressures in the depths of the sun. The values range from essentially zero at the surface to more than 1½ billion tons per square inch (the switch here to the more familiar English units is intentional), and the density of the solar sphere increases with depth from again practically zero at the surface to over 134 times the density of water at the sun's center. This latter value is nearly 10 times the density of mercury at the earth's surface, and 7 times the density

FIGURE 10-25 The sun's corona showing the streamer characteristics in the vicinity of the magnetic poles (Yerkes Observatory Photograph).

of gold. What prevents total collapse of the solar core under the enormous weight of the overlying matter? The pressure forces exerted by the gases themselves will just exactly compensate for the inward directed weight at every depth. But the magnitude of the pressure forces is profoundly influenced by another phenomenon to be discussed in the next section.

Solar Energy Sources. Gases heat up during compression; in the situation existing within the sun temperatures increase with depth to phenomenal values of around 14 million to 16 million °K. To be sure, these are determined theoretically, since it is impossible to probe instrumentally beneath the solar photosphere. The theory is derived from complex mathematical models in accordance with the known laws of physics of gases and plasmas. Support for the validity of the theory follows from observations of the energy emission rate through the photosphere, a process that involves multiple absorptions and emissions as well as reabsorptions and reemissions as the energy makes

its way from the bowels of the sun out through the surface.

What are the possible energy sources within the sun that support, and have supported for an estimated 5 to 10 billion years, an essentially continuous, unvarying rate of energy emission? Gravitational contraction with the associated conversion of gravitational potential energy to kinetic energy of moving particles can account for the internal thermal structure of the sun. But it can be shown that if this were the sole energy source, the sun's life expectancy in its presently observed form would be no more than a few tens of millions of years. Moreover, the loss of internal energy represented by surface radiation losses would necessarioy result in a gradual shrinking of the solar sphere, however slow, in the process of maintaining the existing thermal structure throughout the sun.

Both astronomical and terrestrial geological evidence support the conclusion that the sun has been radiating energy at a relatively constant rate for the 5 to 10 billion-year period as mentioned above. If the conclusion is valid, the thermal energy of the sun represents only about 1% of the radiative losses occurring during its presumed life span. Clearly the thermal supply is being replenished from processes other than gravitational contraction.

The most logical source of the replenishment is from energy released by nuclear reactions deep within the central core of the sun. Nuclear theory, supported by laboratory experimentation and the practical evidence provided by the successful detonation of the hydrogen bomb, shows that the central temperatures are sufficiently high to support several different kinds of nuclear reactions. Moreover, each type of mechanism is of particular importance according to the evolutionary stages through which the sun is presumed to have progressed, or through which it will progress in the future. Stellar evolutionary hypotheses are reviewed in more detail in Chapter 12.

We shall not here pursue the complicated mathematical argument that leads to an acceptable model of the solar interior. Nor shall we be concerned with the known details, including symbolic depiction, of the energy releasing processes that sustain the sun's unvarying luminosity and size. It is sufficient to note that the current nuclear process appears to be the fusion of four hydrogen nuclei into a single helium nucleus, a process that is in agreement with the existing central temperature. The relative mass of the helium product is about 0.07% less than the combined mass of the hydrogen nuclei entering the reaction. Thus the raw fuel, comprising at present about 75% of the solar mass, is hydrogen; the "ash" is mainly helium.

The mass loss, or *mass defect* as it is termed, appears as an equivalent amount of energy. Albert Einstein showed in his theory of special relativity that the equivalency could be represented by the relations $E = mc^2$, where E is the energy, m is the mass defect, and c is a proportionality constant equal to the speed of light. Thus c^2 in metric units is 9×10^{20} (cm/sec)^2, or 9×10^{10} (km/sec)^2. Even though we cannot fully comprehend such huge numbers, they do give us some feeling for the awesome magnitude of the energy released by a relatively small quantity of mass.

While it is clearly evident from the solar spectrum that substances other than hydrogen and helium exist in the solar structure, they cannot be accounted for by the hydrogen-helium fusion process. Presumably such (heavier) matter was present in the original dust and gas cloud out of which the sun evolved. We shall return to this topic in subsequent chapters on stellar evolution and cosmology.

Reaction Rates in the Sun. In view of the prodigious rate at which energy escapes through the solar photosphere, it is easy to grossly overestimate the rate at which individual nuclear reactions occur in the solar core. By extrapolating backward from the magnitude of the solar constant, taking into consideration the size and density of the sun, a value of about 4×10^{33} erg/sec can be postulated for the rate of energy release in the core of the sun. An erg is a miniscule amount of energy; 746×10^7 (7 billion 460 million) erg/sec is equivalent to only 1 hp. Now a *single* nuclear reaction in the sun typically releases a mere *few millionths of an erg*. Hence the number of reactions per second must be on the order of 10^{39}.

Even though this is an unimaginably large number (1 followed by 39 zeros), it is 10^{18} times smaller than the number of atomic nuclei present in the sun. And only about 10% of these are in the central zone where reactions can occur. Thus only one nucleus out of 10^{17} can be expected to enter into the reaction process in any given second. In other words, we can expect any given nucleus to exist for 10^{17} seconds, or 3 billion years, before undergoing a nuclear transformation.

While the above calculation is only a rough approximation, it does yield a realistic order of magnitude for the age of the sun. Moreover, the enormously long life expectancy of a hydrogen nucleus before losing its identity in the transformation into helium is indicative of just how truly continuous and long-term the solar furnace operation is.

Altogether the solar composition, the density distribution, the temperature, the mass distribution, the opacity, and the type of nuclear energy release operate nicely as a unit to keep the sun in gravitational-thermal equilibrium at every level from the center to the surface. The sun is losing mass at the rate of 4,700,000 tons every second. As astonishing as this quantity is, the sun is losing via radiation *less than 100-billionth of its total mass each year*. The mass loss is, of course, taking its toll on the sun, but we need fear no cataclysmic events, nor will the earth experience any appreciable change in its environment because of solar aging for another 5 billion years at least. This is indeed a comforting thought!

Solar Atmosphere and the Earth. Solar out bursts, notably solar flares originating in the sun's photosphere, are evidently accompanied by intense, irregular ultraviolet and x-ray emissions. The earth's upper atmosphere is normally ionized by the continuous background uV radiation from the sun; hence any disturbance in the uV emission levels produces corresponding disturbances in the ionized atmospheric layers. The most noticeable effect, in fact, the first indication of unusual solar activity, is in radio transmission and reception. Most of the broadcast band (not FM or TV) radio waves are reflected back to earth by the ionized layers. Thus disturbances in the ion layers of atmosphere produce corresponding changes in the reflection patterns. As a result, radio signal fading, echoes, noise, and degraded signal quality occurs. Often, radio blackouts last for many minutes. These disturbances are irregularly distributed over the earth's surface.

After about a day following the solar out-

bursts, erratic, intense fluctuations occur in the earth's magnetic field. These variations, called *magnetic storms*, often induce electrical currents in the surface of the earth that seriously interfere with telephone and telegraph communications. Storms of charged particles, electrons and protons principally, but including other ions as well, are believed to be responsible. The particles usually (and quaintly) referred to as *corpuscles* are ejected from the sun during a flare eruption with velocities of many millions of kilometers per hour. The stream of corpuscles is commonly called the *solar wind*.

Most of the earth's surface is protected from the "rain" of solar particles. Some are trapped by the earth's magnetic field in two, highlevel, toroidal (doughnut-shaped) belts known as the *Van Allen radiation zone*. James A. Van Allen, Iowa State University physicist for whom the

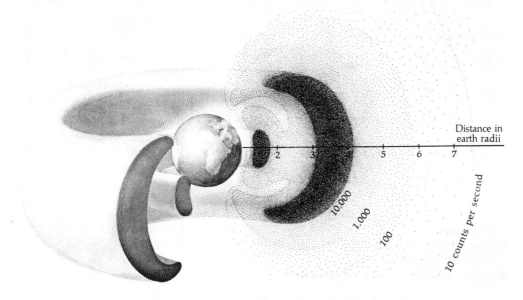

FIGURE 10-26 Van Allen radiation zones. Solid lines show the direction of the earth's magnetic lines of force. Two zones of high intensity surround the earth in the plane of the magnetic equator. The inner zone is about 4800 km thick, centered 3200 km above the earth. The outer zone is about 8000 km thick and is centered about 17,500 km above the surface (Courtesy Scientific American, Copyright © 1959 by Scientific American, Inc.).

belts were named, was the principal scientist in charge of the corpuscular detection experiments aboard the Army's Explorer I satellite in 1958. The geometry of the Van Allen zone is shown in Figure 10-26.

Others of the solar particles of low or intermediate energy are deflected toward the earth's magnetic poles and spiral in toward the earth over essentially uninhabited regions. The corpuscles emit energy in agreement with quantum theory as they pursue their helical orbits. The ions in the upper atmosphere interact with the electric and magnetic fields associated with the energy emission and in turn emit resonance "vibrations" at visible frequencies. We observe the effects in the form of the aurorae. The auroral rays and streamers are infrequently seen beyond 40° from the magnetic poles; the peak intensity occurs at about half this distance. The shape of the rays, streamers, and curtains follows the contours of the earth's magnetic field, strong evidence that the aurorae are indeed associated with the solar wind.

SUMMARY

The sun is the dominant gravitational member of the solar system; it holds all of its family in an invisible, tenuous, yet firm grip. The sun provides essentially all of the heat and light for the rest of the system.

The classical astronomers and physicists were certain that the source of the sun's radiant energy was not a simple chemical combustion process. Almost a revolution in physical thought was required before the true energy source was deduced.

As the star nearest us, the sun warrants detailed study as the best means for determining what the rest of the universe is like. The sun's distance from the earth is determined from Kepler's laws and from the measured parallax of closely approaching bodies such as certain planetoids. Radar echoes from the nearer

planets confirm a distance of 149.6 million kilometers; this is the length of 1 AU. With the distance known and the measured angular size we compute a solar diameter of 1.39 million kilometers, 109 times the earth's diameter. The sun's volume is 1,300,000 times the volume of the earth; its mass is 332,000 times the earth's mass. This latter quantity is obtained with the aid of Newton's laws applied to the measured angular acceleration of one of its planets, say, the earth.

The visible surface of the sun is extremely turbulent, and it is marked by a number of transient phenomena. Some of these, notably sunspots, are sufficiently long lasting to yield the information that the sun rotates differentially about an axis tilted 7° from the vertical with respect to the earth's orbit. The equatorial rotation period is 25 days; near the poles it is 34 days. As with Jupiter, the observed solar surface is not solid.

The rate of energy release from the photosphere of the sun is extrapolated by means of the inverse square law applied to the measured solar constant. Then from blackbody radiation theory, which applies to the sun within satisfactory limits, we calculate a surface temperature of nearly 6000°K.

Knowledge of the sun's composition and structure is obtained from its spectrum. The sun is totally gaseous. The opacity of the photospheric gases produces a phenomenon called limb darkening. Above the photosphere the chromosphere extends out to about 13,000 km. Above the chromosphere, for an indeterminate distance exceeding 22 million kilometers, is the faint, white, extremely high-temperatured corona.

From analysis of all of the physical evidence—mass, density, temperature, spectral characteristics, opacity, and energy emission rate—we can deduce the following: the sun contains no unknown elements, and roughly two-thirds of the elements known on earth are present in the sun. The upper 800 km of the photosphere are transparent, and this is the principal region for the production of the sun's absorption spectrum.

Photospheric phenomena include *granules* which are convective cells in the lower photosphere and *sunspots* which are cooler, tranquil regions in the photosphere. Sunspots appear to be associated in some manner with magnetic disturbances or variability in (or just below) the photosphere. Also seen are *faculae* (bright spots or regions near the sunspots) occasional solar *flares* of somewhat mysterious origin, and dark *filaments* that are frequently seen projected as bright prominences extending into the chromosphere and lower corona.

The sun appears to be in stable equilibrium; its size, shape, radiation rate, and physical processes all seem to be essentially unchanging, at least over the term of a few billion years. The stability is the result of the interaction of gravity forces, gas and radiation pressures, and nuclear energy release from the fusion of hydrogen into helium. Details of these processes and the evolutionary expectancy of the sun are reserved for discussion in a later chapter.

Questions for Review

1. How is the earth-sun distance in kilometers calculated, and what does the accuracy of the result primarily depend upon?

2. How is the angular acceleration of the earth employed in the determination of the sun's mass?

3. Why is the sun's disk darker around its limb?

4. Name, and briefly describe, the main subdivisions of the solar atmosphere.

5. What is meant by solar constant? What is its magnitude?

6. Can the sun's surface temperature be directly measured? If not, how is the temperature determined?

7. Define: (a) wavelength, (b) frequency, (c) speed, (d) velocity.

8. What is a blackbody radiator? How important is blackbody theory in astronomy?

9. Distinguish between absorption spectra and emission spectra.

10. Describe the appearance of the solar photosphere and its phenomena.

11. What is the chromosphere and why is it so named?

12. Discuss the appearance, properties, and significance of the solar corona.

13. What is the explanation for the extremely high coronal temperatures?

14. What is the solar wind? What effect does it have on the earth?

15. Describe the main properties of sunspots.

PART THREE
Structure and Properties of Stars and Galaxies

Prologue

We now turn our attention to the only other distinct population about which we have any knowledge: the stars. We shall find their statistics exceedingly varied. There are big stars and little stars, hot stars and cool stars. There may even be stars that are totally cold and dark. Certainly, some of the stars are young; some are very old. In spite of the appearance of random distribution on the celestial sphere, we shall find that stars are quite well organized socially. Three particulars are worthy of our attention. One is man's extraordinary ingenuity in securing or deducing information about the stars, some of which are so remote that their light has been traveling several billion years just to reach us. Second topic of interest is the similarities and differences among stars, qualities that will eventually enable us to construct a stellar family tree. A third matter of interest is stellar organizational schemes and the motions of the stars within these plans. But most important of all we shall discover obstructions that seem almost insurmountable which block the way toward understanding the universe. But all is not lost. The Game Plan changes dramatically and favorably in the next chapter.

Determination of Stellar Properties

11

The observant reader will by now have noticed that more than one-half of the text is devoted to the solar system, and that he is just now leaving the vicinity of the sun for the contemplation of the larger cosmos. Such preoccupation with the solar system, man's immediate environment, is both logical and necessary. It is here on earth that we develop the theories, acquire scientific sophistication, perfect experimental techniques, and improve our mechanical and intellectual skills, all of which are so very necessary to our further astronomical pursuits. We must rely on the universality of the laws of physics, laws that have been thoroughly tested and which so far have not been found wanting. Among these are the laws of motion, gravitation, optics, thermodynamics, and radiation theory. We had to thoroughly investigate the organization of the solar system in order to learn about celestial motions; we had to dissect the sun in order to prepare ourselves for the leap in space toward other stars.

But there are observational difficulties that will challenge our abilities to the utmost. In fact, many of the problems still await solution. The familiar measurement techniques are of no avail beyond about 100 light-years (LY);† there are no more than about 2000 stars within this proximity; there are billions beyond it. Because of the vast distances, because of internal and external impediments besetting our instrumentation, and because of the sheer weight of numbers of the stars and of stellar populations

† The unit of distance, *light-year*, will be defined presently.

we are forced to turn to statistical methods and to rely heavily on probability theory. We shall find that precision in measurement will have entirely different significance from what we are accustomed to in our laboratory environment. Yet in spite of the awesome obstacles that hamper our abilities to deal with distant objects and systems, we can learn a great deal about the universe, and we can certainly lay the groundwork for the future pursuit of that which is presently inscrutable. In the end, though, all of our present knowledge and skills become unavailing, and we are reduced to speculating about the ultimate questions in astronomy. But all is not lost even here; speculation is, to say the least, mentally invigorating.

Direct Measurement Techniques

Our attack upon the problems of the sun gave us direction toward the properties of stars that are of interest and that are pertinent to our investigations about the distribution of matter in the universe as a whole. We need to know *stellar distances, stellar luminosities,* and *stellar masses.* We want information about the *sizes of stars,* the *composition of stars,* and the *motions of stars.* These, at least, will do for a start. Let us begin the investigation of the stars beyond the sun by using techniques and methods that were adequate for the local solar system. When these no longer suffice we can try other, less direct methods, turning to statistical solutions only when all else fails.

Stellar Distances. In Chapter 5 the annual parallax of the near stars was employed in the proof for the revolution of the earth around the sun. In Chapter 7 we derived a geometrical technique for measuring the earth-moon distance (Equation 7.4 and Figure 7-13). We can employ the same geometry and mathematical relationships in the determination of stellar distances. The only modification we need is in the choice of a reference base line; even though we shall employ the diameter of the earth's orbit, we shall reduce the parallax angle to that subtended by the mean radius of the orbit.

Equation 7.4 states that $r = 206,265 \times b/p$. This relationship is readily applicable to our needs: r is the radial distance from observer to the star in question; b is the observational base line, the mean orbital radius; and p is the measured stellar parallax in seconds of arc (Figure 11-1). The constant 206,265 is, we recall, the number of seconds of arc in a full circle of 360° divided by 2π ($2\pi = 6.283$ . . .). The units in which r is expressed will be the same in which b is measured: kilometers, astronomical units, or more convenient derived units.

It is to our advantage to eliminate the numerical constant 206,265 from the computations. This can be done as follows. Let b be 1.0 AU and p be a parallax of 1″.0. Then r is just 206,265 AU. That is, a star whose parallax is 1″ is just 206,265 AU distant. Now, we define 1.0 *parsec* (pc) (coined from *parallax second*) as 206,265 AU. Here is the great advantage: by always using the earth's mean orbital radius as the base line, a convenience because by definition its length is 1.0 AU, then the distance r becomes simply $1/p$ in parsecs when p is measured in seconds of arc. For example, the star nearest the sun, *Proxima Centauri*, has

the largest known parallax, $p = 0″.785$. Then the star's distance is $r = 1/0.785 = 1.27$ pc. If we so choose we can express the distance in astronomical units by reintroducing the factor 206,265.

Another convenient distance unit in astronomy is the *light-year*, abbreviated LY. In the idealized vacuum of space light travels with a speed of 3×10^5 km/sec (in round figures). Since distance is speed × time, the distance a light photon will travel in 1 year is 3×10^5 km/sec × 32×10^6 sec/yr × 1.0 yr. This is 9.6×10^{12} (9.6 trillion) km. Now since 1 AU is 149.6 million kilometers, and there are 206,265 AU in 1 pc, we see at once that 1 pc is 30.9×10^{12} km. By direct comparison we find there are 3.258 LY in each parsec. These are useful equivalents. *Proxima Centauri* is thus 1.27 pc × 3.258 = 4.15 LY distant.

Stellar distances are so vast as to be unintelligible. Even solar-system distances are only vaguely comprehensible. Comparisons with the familiar are advantageous in attaining an appreciation for the enormous distances involved. Such comparisons are thus far from idle. Suppose that a light beam could be forced to circle the world, retracing its path on each lap. Suppose further that at the instant a high speed rifle bullet left a gun muzzle going in the same direction, an adjacent light beam sped off around the world. The beam would catch the bullet from behind and lap it before the bullet had traveled 100 m! The speed of light, 3×10^5 km/sec, amounts to 1.08 billion kilometers per hour or 670 million miles per hour.

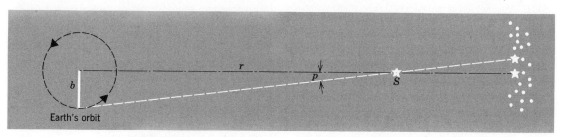

FIGURE 11-1 Parallax of stars. Parallax p is the apparent angular displacement of a star S when viewed from opposite ends of a baseline equal to the earth's orbital radius.

This is 10,000 times the speed of the earth in its orbit. Light travels the earth-sun distance in a little over 8 minutes, yet it still takes light 5¾ hours to reach Pluto from the sun, and, as we just saw, 4.15 years to reach the nearest star. Our best speeds are insignificant by comparison. Consider the following. We are accustomed to speeds of 60 mph. At this rate, traveling nonstop, ignoring the time it takes to accelerate or slow down, and with no speed variation enroute, it would take 100 years for us to reach Mars at its average opposition distance, 740 years to reach Jupiter, and 6800 years to arrive at Pluto. The travel time to the nearest star at the rate of 60 mph is 11 million years! At 1000 mph, about the fastest *sustained* speed, if we ignore orbiting satellites, thus far achieved by man, Mars is still 6.5 years away, Pluto is a respectable 400 years distant, and Proxima Centauri could not be reached in under 2.8 million years. If we could achieve constant speeds of 25,000 mph, the escape velocity from the earth, it would still take over 3 months to reach Mars and 21 months to reach Jupiter; the travel time to Proxima Centauri would still be 100,000 years.

Determining the stellar distances, as vast as they are, is simple and straightforward, except for one fatal flaw. The highest precision that can be attained, using the largest telescopes in the world and making repeated measurements, is less than 0".003 of arc. This is about the angular size of a penny viewed from a distance of 1450 km. Consider what an error of 0.003 either way means in a measured parallax of 0".010. We write it thus: 0".010 ± 0".003. The parallax has a most probable value of 0".010, but it could be as low as 0".007 or as great as 0".013 of arc. For the lower value the star's distance is 1/0".007 = 143 pc. For the upper value the distance is 1/0".013 = 77 pc. The most probable value is 1/0".010 = 100 pc. Thus, the lower limit is in error by 43%, the upper, by 23%. Both are insufferably large for such a small cosmic distance. If we are willing to accept a maximum error in distance of 10%

when parallax methods are employed, this puts a practical limit of about 35 pc or 100 LY on stellar distances for which parallax measurements are acceptable. As noted earlier, less than 2000 stars are closer in proximity to the sun than 100 LY; many billions of stars are beyond it. Accordingly we must develop other techniques for finding the vast majority of stellar distances.

Stellar Brightness. Properties such as a star's brightness, or better, its luminosity, its mass, its surface temperature, and its size are all intrinsic; that is, they are essential or inherent, belonging to the constitution, nature, or essence of the star. Distance does not belong in this category, although its determination may have a profound influence on the values we assign to certain of the inherent properties. Besides this, distance determinations are of fundamental importance in the larger problem of the distribution of the stars. In the interests of clarity we should reiterate the meaning of *brightness* or *luminosity*. *Absolute brightness* and *absolute luminosity* are synonymous. The terms refer to the rate of energy emission per unit area summed up over the entire area of the emitting surface. In other words these terms describe the total rate and quantity of energy emission. Stefan's law (p. 247) provides the relation between the rate of energy emission per unit area and the temperature of the emitting surface. Absolute brightness or absolute luminosity is, therefore, a measure of the total energy emitted per unit time by a body whose temperature can be evaluated. Note that a cool large star can have an absolute luminosity substantially larger than a smaller and hotter star; all that is required is that the cooler star be large enough.

It is essential that we distinguish between *absolute luminosity* (L_{abs}) and *apparent luminosity* (L_{app}). L_{abs} is an intrinsic property of the star; L_{app} is the brightness value we assign in terms of the star's appearance. The inverse square law is operational with respect to the apparent luminosity; L_{app} decreases as the square of the increase in distance. The following symbolic representations are both economical and convenient.

$$L_{abs} = E \times A \qquad (11.1)$$

where E is the radiation rate per unit area, and A is the area of the emitting surface. Also,

$$L_{app} = \frac{L_{abs}}{d^2} \qquad (11.2)$$

where d is the distance between source and observer. Equation 11.2 can be written alternatively as $L_{abs} = L_{app} \times d^2$, or $L_{abs} = L_{app}/p^2$. In the latter form d is replaced by $1/p$ (reciprocal of parallax) to which it is equivalent. We need not be concerned with the several physical systems of energy units in current usage. For our purposes, expressing stellar quantities as multiples or fractions of the corresponding solar quantities will give us the insight we need into the magnitudes of the stellar properties. We can accept one stipulation, however, and that is that all the properties, both solar and stellar, are evaluated as they would appear at a standard distance of 10 pc. The reason for moving the sun out to this distance is to diminish the effect of its overwhelming apparent brightness and to place it in proper perspective with the other stars.

The main problem in computing absolute luminosities from apparent luminosities is the determination of distance. Nothing in the relationship between the two luminosities themselves is of any help (yet) in the distance determination. On the other hand when the distance is known, as it is for the 2000 or so stars whose parallaxes are known, then the absolute luminosity for any one of them is just d^2 times the measured apparent luminosity. Apparent luminosities are measured with photoelectric devices or appropriately sophisticated light meters used in conjunction with telescopes.

Stellar Magnitudes. A system for designating apparent luminosities dating from Hipparchus (130 B.C.), while in common practice among astronomers, is really an anachronism by today's standards. A complete discussion of its intricacies and its several worthwhile variations serves no useful purpose here. Greater detail is provided for the insistent reader in Appendix 4. Briefly, the human eye differentially responds to light stimuli according to the apparent luminosity it perceives. Historically, the visible stars were classified in six magnitudes, the first magnitude being the brightest. The second magnitude was assigned to those stars that were just perceptibly fainter than those of first magnitude, and so on. Sixth magnitude stars are just at the limit of visibility. Because the eyes of no two persons respond to the same stimuli in exactly the same way, it finally became necessary to quantify the scale of magnitudes. In 1856 careful measurements determined that sixth magnitude stars were just 1/100 as luminous as first magnitude stars. Each magnitude, then, is just the fifth root of 100 ($\sqrt[5]{100}$) brighter or fainter than the next adjacent magnitude. The fifth root of 100 is 2.512; that is, $(2.512)^5 = 100$. Thus a first magnitude star is 2.512 times brighter than a second magnitude star; a fourth magnitude star is just 2.512 times fainter than a third magnitude star, and so on. *Apparent magnitude* is a measure of the luminosity of a star just as we see it. *Absolute magnitude* is a measure of the apparent luminosity the star would have if moved to the standard distance of 10 pc. Table 11.1 lists the magnitudes of several prominent stars.

Assigning magnitudes or apparent luminosities is beset with peculiar difficulties. Stars are of different colors; hence the luminosity assigned to a given stellar image depends upon the color sensitivity of the detector. For example, the human eye is most sensitive to yellow-green light; it is very unresponsive to blue light and to red light. The eye is insensitive (visually) to light stimuli in the ultraviolet or infrared. If two stars, one yellow, the other blue, both have the same true apparent luminosity, the yellow star will appear brighter to the human eye. Photosensitive materials such as films and plates are usually highly sensitive

TABLE 11.1 Apparent Magnitudes of Several of the Brightest Stars

Star	Apparent magnitude (for binary systems the primary component)	Distance (parsecs)	Right ascension (1950)		Declination (1950)	
			h	m	°	'
Sirius	−1.42	2.7	6	42.9	−16	39
Canopus	−0.72	30.	6	22.8	−52	40
α Centauri	−0.01	1.3	14	36.2	−60	38
Arcturus	−0.06	11.	14	13.4	+19	27
Vega	+0.04	8.	18	35.2	+38	44
Capella	+0.05	14	5	13.0	+45	57
Rigel	+0.14	250.	5	12.1	−8	15
Procyon	+0.38	3.5	7	36.7	+5	21
Betelgeuse	+0.41v†	150.	5	52.	+7	24
Aldebaran	+1.39	16.	4	33.0	+16	25
Spica	+0.86	80.	13	22.6	−10	54
Antares	+0.92†	120.	16	26.3	−26	19
Pollux	+1.16	12.	7	42.3	+28	09
Deneb	+1.26	430.	20	39.7	+45	06

† v indicates star varies in apparent brightness.
Distance to the more remote stars is approximate, based upon apparent magnitudes and spectral types.

in the blue and ultraviolet; they "see" red objects very poorly or not at all (Figure 11-2). On the other hand, specially prepared red-sensitive films or plates reverse the relative intensities registered. *Panchromatic* film is adjusted to respond uniformly throughout the visible range. By using special dyes and filters it is possible to adjust the sensitivity of photographic materials to a wide range of colors. Any system of apparent-brightness measure must take into consideration the color response of the detector employed.

Color Index. The overwhelming majority of stellar observations employ photographic and photoelectric techniques rather than visual inspection by the human eye. Before advent of photoelectric techniques, a useful relationship, still employed, between photographic color and visual color was devised. The color of photographic light, that to which standard astronomical photographic plates respond, extends from the violet to the blue-green, with an intensity maximum in the blue-violet at 4250 Å.† This is to say that the camera will see the 4250 Å wavelength as brightest, no matter what the true wavelength of the star's intensity maximum is. Similarly the color of visible light extends from blue to red with a maximum at 5280 Å in the yellow-green.

Now if the stars are radiating as blackbodies (a more realistic assumption for the very hot stars than for the cool ones) (Chapter 10), the relative spectral intensities at two separate wavelengths define the star's radiation curve. The curve, as we have learned, is unique for a given temperature, which means that the two relative intensities obviate the necessity of examining the whole stellar spectrum. This opens the door to rapid analysis of stellar temperatures and brightnesses. We compare the pho-

† The symbol Å stands for a unit of length called an *angstrom*, one one-hundred-millionth of a centimeter (Appendix 2). It is a unit commonly used in the wavelength measurement of light.

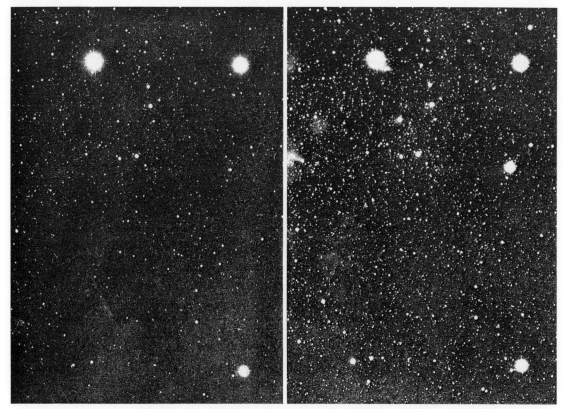

FIGURE 11-2 Apparent differences in magnitude between red and blue stars as recorded on a normal blue-sensitive star plate or negative. The larger images are those of the bluer star (Haute-Provence Observatory photograph).

tographic magnitude and the visual magnitude; the difference is defined as the *color index* (CI). These two magnitudes are taken to be equal for type A0 stars; this establishes a zero reference point for the color index. Modern practice substitutes photoelectric devices for the human eye; the resulting photoelectric color is called blue − visual (B − V). The numerical differences between the two systems, though important, are not large.

Our confidence in assigned CI and B − V values for standard comparison stars (Chapter 12) is based upon knowledge of their stellar temperatures obtained by more direct methods. Color index is an essential tool in the study of interstellar obscuration of the structure of the galaxy (Chapter 13).

Size of Stars. Except for highly specialized techniques employing the largest telescopes, and applicable to only a half-dozen or dozen stars, no star displays a visible disk; all star images are points.† At least they would be except for aberrations inherent in all telescopes (Appendix 3). Otherwise, the same principle employed in determining the angular size of the moon or sun could be successfully applied to the stars. Even if the stars did show an angu-

† The telescope is equipped with a *beam interferometer* (Appendix 3) that in effect greatly increases the resolving power of the telescope at the expense of image quality. The diameter of the star can be deduced from the interference pattern produced at the focal plane of the primary telescope mirror by two flat mirrors placed at the ends of the interferometer beam.

lar disk, determining the linear diameters would necessitate knowledge of the stars' distances. These are in short supply. We must defer the solution of the problem of stellar diameters until we find secondary, indirect measurement methods to be acceptable. We shall examine several alternative methods very shortly. Meanwhile we should not lose sight of the importance of the few stars whose diameters have been optically measured as indicated above. As few in number as they are, these stars are invaluable as controls in estimating the reliability of other, less direct, methods.

Composition of Stars. No star, not even the sun, is accessible to direct sampling of its matter. Stellar composition is determined by inference, using laboratory spectra techniques and quantum theory for comparison purposes. Stellar spectra show decided similarities as well as significant differences. All stars have a continuous background emission spectrum and all show line absorption spectra (Figure 11-3). In addition a few stars show selected bright line emission spectra as well. Most of the spectral lines have been identified. Moreover, from the intricate combinations of lines and the relative strength and width of the lines, much can be deduced as to the composition and distribution of matter in the stellar envelopes. All stars have a continuous emission spectrum in the

background of their line spectra. The continuous spectra display relative intensities over the wavelength distribution that can be correlated very closely to the intensity distribution of the blackbody radiation curves.

In Chapter 12 we shall examine in detail the significant similarities and differences in a variety of stellar spectra. These are invaluable in the study of population characteristics. In addition we shall find that the specifics of stellar composition as revealed by their spectra can be employed toward eliminating the distance roadblock that appears so troublesome now. For the present it is sufficient for our purposes to know that the spectra of stars are not formidably different from the sun's spectrum; hydrogen and helium are the principal constituents.

Stellar Temperatures. As with the sun, no direct temperature measurement of a star with a sensing instrument would be possible even if the stars were accessible. We must consider a resort to radiation laws as the equivalent of direct techniques. Even so, the problems are overwhelming. Equation 11.1, $L_{abs} = E \times A$ is of considerable assistance. If we employ solar units throughout Equation 11.1 takes the form $L = E \times D^2$. Here, L is in units of solar absolute luminosity; D^2 is in units of solar diameters; E is in units of radiation rate per unit area and is

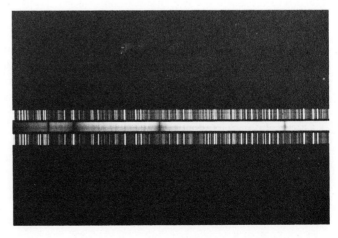

FIGURE 11-3
Representative spectrum showing superposition of comparison lines (Courtesy Lick Observatory).

independent of any given body. Solar diameters can be used instead of areas because, from geometry, the ratio of areas is directly proportional to the ratio of the squares of the respective diameters (Appendix 1). Thus if a star's diameter is D times the diameter of the sun, its surface area is D^2 times as great. Similarly, if the absolute luminosity of a star is L times as great as the sun's absolute luminosity, the radiation rate per unit area of the star is L/D^2 as great as the sun's.

From the Stefan-Boltzmann law, $E = kT^4$ (Chapter 10), we know that the sun's radiation rate per unit area is proportional to the fourth power of the temperature. In Chapter 10 we determined the sun's effective radiation temperature to be 5770°K. By the same argument a star's effective radiation rate is proportional to the fourth power of *its* temperature. It is also proportional to the temperature of the sun according to $k(5770)^4 \times L/D^2$. By the ordinary rules of arithmetic governing equal quantities, it is not difficult to find the expression for a star's effective temperature in terms of the temperature of the sun, Thus,

$$T_e^4 = (5770°K)^4 \times \frac{L}{D^2}. \qquad (11.3)$$

The factors that prevent unlimited application of this relationship to all observable stars are ignorance of the star's absolute luminosities and their diameters. The former requires knowledge of the stars' distances through application of the inverse square law; the latter requires determination of both angular size of a given star *and* its distance. Until these two problems are solved, this is as far as we can press the issue of stellar temperatures.

The Stellar Motions

We must be careful here to distinguish between the motion of groups or combinations of stars and the motion of a single member of a group or system. The former is a population statistic, and we shall attack that problem in

due time; the study of group motion is extremely important, but premature at this point.

There are a number of terms that apply to various aspects of a star's motion: *radial velocity*, which we have come to grips with already; *transverse velocity* which, with the radial velocity combines to produce a star's *space velocity*; *peculiar velocity*; and *proper motion*. *Radial velocity*, we already know, is the motion, or one component of total motion, that is directed along the line of sight. *Transverse velocity* is at right angles to radial velocity, thus it is directly across the line of sight. *Space velocity* is the true direction and speed of a star relative to the sun. The *peculiar velocity* (in this sense, singular, not shared in common, independent from others) is the space velocity corrected for the motion of the sun with respect to the neighboring stars. Finding the peculiar velocity allows analysis of the star's motion with reference to what amounts to a local standard of rest. Hence it represents a surgical procedure that cuts the umbilical tying a star's motion to motion of the sun. Finally, *proper motion* is the angular velocity in seconds of arc per year of the changing position of a star on the celestial sphere. Proper motion is not to be confused with the change in the stellar addresses caused by precession of the earth's axis (Chapter 5).

Radial Velocity Determinations. Figure 11-3 also shows a laboratory standard of reference, a comparison emission spectrum superimposed above and below the spectrum of the star. The comparison spectrum is photographed and optically projected on to the (photographic) star plate during exposure. Appendix 3 provides more detail than is warranted here. The comparison spectrum, often of iron vaporized in an electric arc, is well known; the wavelengths of the many emission lines are known to a fraction of an angstrom.

As we shall see in considerable detail in the next chapter, a star's spectrum is classified by comparison with that of a well known, inten-

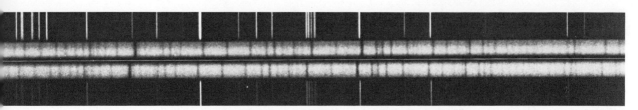

FIGURE 11-4 Doppler shift of spectral lines (Courtesy Lick Observatory).

sively studied reference star. This establishes the wavelengths of the star's absorption lines. Any measurable displacement (Doppler shift) of the absorption lines with reference to the adjacent iron arc comparison spectrum is associated with the radial velocity of the star (Figure 11-4). It is clear that possible radial velocity contributions arising from the earth's rotation, its motion around the sun, and the sun's motion through space may have to be corrected for.

Once the corrections are made to the apparent spectrum line displacement, the radial velocity is easily calculated. The ratio of the radial velocity to the speed of light is equal to the fractional change in the wavelength. The latter is the ratio of the observed apparent change in wavelength for a given line to the true wavelength associated with it under conditions of rest. If we represent the Doppler displacement as $\Delta\lambda$, the true wavelength as λ_o, and radial velocity as v_{rad}, $v_{rad} = 3 \times 10^5$ km/sec $\times \Delta\lambda/\lambda_o$. As an example, suppose the true wavelength of a certain spectral line of the star is 5000 Å, and its position is shifted to 5000.1 Å. The Doppler shift is $\Delta\lambda = 0.1$ Å. Then $v_{rad} = 3 \times 10^5$ km/sec $\times 0.1$ Å/5000 Å = 6 km/sec. The shift in this example is toward the longer wavelengths; as such it is called a *red shift* and indicates a velocity of recession. By convention the recession direction is taken as positive, thus we should write $v_{rad} = +6$ km/sec. Radial velocity of approach is considered negative.

The radial velocities of more than 10,000 stars have been measured: they range from about 0.2 km/sec, about the present lower limit of detection, to as high as 430 km/sec. Radial velocities in excess of 10 km/sec are quite reliably determined. The radial velocities of stars are independent of the stars' distances since only motion is involved. Distance hampers the determination only in rendering the spectral lines fuzzy so that the Doppler displacement may not be discernible.

Transverse Velocity and Proper Motion. Compared to radial velocity determinations, finding the *transverse velocity* of a star is a different ball game; the major stumbling block is readily evident through the following analogy. Suppose an object A moves across the line of sight (from a to b in Figure 11-5) in 1 second. Suppose further that another object B also moves across the line of sight (from a' to b' in the figure) in 1 second. Clearly object B covers a shorter distance during the time interval than object A and is therefore traveling slower. The vital question is, how much slower? More than that, what is the linear speed of *each* object?

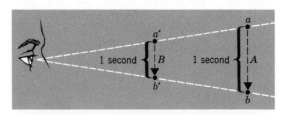

FIGURE 11-5 Transverse velocity paradox. During a given time interval objects B and A will travel the same angular distance. Object B, being closer to observer O, clearly travels slower. But if the distance to either object is not known, its speed cannot be uniquely determined.

The angular speed of the two objects is the same. There is no way to remove the ambiguities without knowing the distance to the objects.

As indicated earlier in this section the observed angular speed is the *proper motion* of a star. The proper motion does not tell us very much about a star's speed except as a population statistic. When proper motions are analyzed in connection with known space velocities a pattern emerges; large proper motions almost invariably place the stars near the sun, small proper motions tend to be ambiguous. The star Sirius has a moderately large proper motion of $1\overset{''}{.}32$/yr. At this rate it will change position on the star field by an amount equal to the sun's angular diameter in 1360 years. Bernard's star (q.v.) has the largest proper motion known: $10\overset{''}{.}25$/yr. This proper motion will cover the sun's angular diameter in only 175 years. Table 11.2 lists a number of stars with large proper motion.

To recap briefly, a star's proper motion alone is ambiguous as far as linear speed is concerned; the star could equally well be quite distant and moving rapidly, or quite close and moving much more slowly. Finding the distance to the star removes the ambiguity and is a task that is, unfortunately, frequently impossible to accomplish.

Suppose a star is close enough to display a measurable parallax. The distance is at once known; hence an observed proper motion can be found in distance units. A highly exaggerated example is shown in Figure 11-6. The observer is at O, and the path ab is the yearly displacement; the angular displacement is μ. The distance d in parsecs is numerically equal to $1/p$ as before. We have

$$\frac{\mu}{1,296,000} = \frac{ab}{2\pi/p}. \qquad (11.4)$$

Equation 11.1 is an old acquaintance in a new dress. The number 1,296,000 is 360° expressed in sec; ab (km/yr) must be expressed as km/sec $\times$ 32 $\times$ 10^6 sec/yr. Upon making these substitutions and evaluating,

$$v_t = 4.74 \, \frac{\mu}{p} \text{ km/sec.} \qquad (11.5)$$

This symbolic expression shows the relation between parallax and proper motion.

Some of the facts just developed are so important as to warrant reiteration. The proper motions of most stars suffer from the fatal defect of our ignorance of their true distances. As suggested earlier, the proper motions do have some statistical value. Assuming that all stars partake of some *secular* (nonperiodic) motion, the nearest stars will naturally have the largest

TABLE 11.2 Several Stars of Large Proper Motion

Name	Proper motion	Parallax	Tangential velocity, km/sec	Radial velocity, km/sec	Space velocity
Barnard's star	$10\overset{''}{.}30$	$\overset{''}{.}545$	90	−110	142
Kapteyn's star	$8\overset{''}{.}72$	$\overset{''}{.}251$	165	+242	293
Ross 619	$5\overset{''}{.}40$	$\overset{''}{.}154$	166	—	—
61 Cygni A	$5\overset{''}{.}02$	$\overset{''}{.}293$	81	−64	103
Wolf 369	$4\overset{''}{.}67$	$\overset{''}{.}425$	52	−90	104
Wolf 289	$3\overset{''}{.}89$	$\overset{''}{.}130$	144	—	—
α Centauri A	$3\overset{''}{.}67$	$\overset{''}{.}754$	23	−22	32
Washington 5583	$3\overset{''}{.}68$	$\overset{''}{.}034$	513	+307	598
82 Eridani	$3\overset{''}{.}25$	$\overset{''}{.}159$	97	+87	130

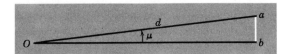

FIGURE 11-6 Geometry of proper motion computation (See text, especially Equations 11.3 and 11.4).

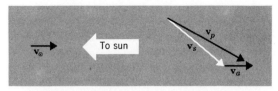

FIGURE 11-8 Peculiar velocity of a star. $\mathbf{v}_s$ is measured relative to the motion of the sun $\mathbf{v}_\odot$. This has the effect of foreshortening the star's velocity vector with respect to a stationary frame of reference. The correction $\mathbf{v}_a$ must be added to the star's space velocity in order to obtain the peculiar velocity $\mathbf{v}_p$ (Figure 11-20).

average proper motions; the more distant stars have little or no proper motion. Hence the search for nearby stars begins with the search for large proper motions. As unrewarding as the statistical treatment of proper motions is, it yields the only direct method of obtaining the distance to certain stellar populations. We shall see how very shortly.

<u>Space Velocity and Peculiar Velocity.</u> Let us assume that the transverse velocity, v_t, and the radial velocity reduced to the sun (corrected for the motions of the earth) are known. They can be vectorially diagrammed as in Figure 11-7 as the two mutually perpendicular sides of a right triangle. (Recall that the transverse motion is at right angles to the radial motion.) The space velocity, v_s, the resultant of the two component motions, can be computed by two methods: one from geometry (theorem of Pythagoras) and the other from elementary trigonometry.

As we shall discover in a following chapter, the sun has a motion through the star field

resulting from its revolution around the center of the galaxy. This motion will influence the value assigned to a star's space velocity only if we change reference systems. The space velocity, recall, was in reference to a local at-rest system that is fixed relative to the sun. For some statistical purposes it is desirable to consider a star's motion independently of the sun. This is easily done by merely correcting the space velocity for the *reflex velocity* (secular motion) of the sun. The resultant is a star's peculiar velocity (Figure 11-8). Again, this quantity will be useful later.

Indirect Methods of Measurement

Indirect measurement methods are those in which the desired quantities or properties of the stars or star systems are inferred from the laws of physics rather than being measured directly. Radial velocity measurements are in this category. It was, however, more logical to consider radial velocity in connection with the binary stars and with space velocity, to both of which it was vital. Another indirect measurement method we have already employed is the semiannual Doppler shift of the spectra of stars near the plane of the earth's orbit (used in one of the proofs for the revolution of the earth about the sun). Another concerns the manner in which the information from the spectroscopic binaries (discussion follows) can be

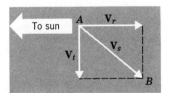

FIGURE 11-7 Space velocity diagram. Space velocity $\mathbf{v}_s$ can be readily computed if transverse velocity $\mathbf{v}_t$ and radial velocity $\mathbf{v}_r$ are both known.

used. Still others to be considered are the motions of star clusters, and the secular, or reflex, velocity of the sun.

The Masses of Stars.   Just as the mass of a planet cannot be determined unless its gravitational influence on another body can be observed, neither can the mass of a single star be evaluated. Fortunately, extremely fortunately for our investigations, there exists a class of stars that provides a breakthrough in this otherwise insoluble problem. These are the *binary stars,* pairs of stars that mutually revolve around a common center of mass, called the barycenter. The principle is the same one that governs the earth-moon motions. Binary systems are quite common; over 65,000 are known. On statistical grounds it is reasonable to suppose that anywhere from 20% to 50% of all stars are binary systems.

Binaries are classed according to an observational scheme; if the two components can be visually separated with the aid of telescopes the system is designated as a *visual binary.* If the components are too close for visual separation but periodic twinning of the spectra can be detected, or periodic Doppler oscillations of a single spectrum are detectable, the system is called a *spectroscopic binary.* Finally, if one or both of the components experience periodic eclipsing, they are classed as *eclipsing binaries.* Clearly the three classifications are not mutually exclusive.

Visual Binaries.   Compared to the other types, visual binaries are simpler to study; the components are visually separated, thus their individual orbits can be well determined. Furthermore their angular separations, and in some cases their physical distances from each other, are measurable. The combined mass can be determined from Kepler's harmonic law in the form of Equation 10.7, $(m_1 \times m_2)p^2 = a^3$, and more conveniently written $(m_1 + m_2) = a^3/p^2$. The masses of the primary and secondary components are m_1 and m_2 respectively. The primary component is always the

brighter one); a is the semimajor axis of their relative orbit, and p is the period of revolution. If a can be determined in astronomical units and p is measured in years, then $(m_1 + m_2)$ is in solar masses.

The period can be determined from five well spaced observations, although when the period is lengthy more observations are desirable. If the distance to the system is known, as it is in some cases through parallax determinations, then the physical separation can be evaluated in distance units. Otherwise, the angular separation is all that is measurable. In either case, the size of the orbits is inversely proportional to the masses of the two components. This fact permits determination of the relative distribution of mass between the two components, even if the magnitude of total mass is not known.

In practice, a best-fitting relative orbit for the system is graphed. The primary is considered stationary at one focus of the relative ellipse; all the motion is assigned to the secondary (Figure 11-9). Of course, we are unable to tell exactly how the orbit is oriented to our line of sight, although there are clues that eliminate some of the ambiguities. In general, the relative orbit, and of course the true orbits from which it is constructed, is not seen full-face, that is the orbital plane is in general *not* perpendicular to the line of sight (Figure 11-10). However, the *projection* of an ellipse from one plane into another will also be an ellipse but one with a different eccentricity. Neither focus of the true ellipse will project into the foci of the projected ellipse. Accordingly, the primary component of a binary system which is at one focus of the true relative ellipse will *not* be at a focus of the apparent relative orbit; it will be displaced from it by an amount we can measure, depending upon the angle between the orbital plane and the line of sight. The problem, one that is not particularly difficult, is to find the angle by which the apparent relative orbit must be projected in order to place the primary at a focus of the elliptical orbit. Said

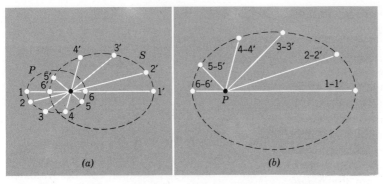

FIGURE 11-9 Relative ellipse with respect to the primary star of a binary system. (*a*) shows the apparent motions of both components, which will always be mutually opposite the common center of mass. (*b*) shows how an observer on the primary would describe the orbital motion of the secondary. Observe that the line segments 1–1'; 2–2'; . . . are the same length in both figures.

another way, the required angle is that at which the true relative orbit must be projected in order to produce the observed displacement of the primary from a focus of the observed relative orbit. Several techniques are available for securing this geometric solution which then yields the angular size of the true relative orbit, and with it the length of the semimajor axis required for use in Equation 10.7.

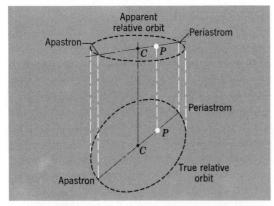

FIGURE 11-10 An apparent (observed) relative orbit is in general a projection of the true relative orbit (see text).

From the true relative orbit, then, the angular length of the semimajor axis is known. The length of the semimajor axis can be apportioned proportionately between the two true ellipses that go to make up the relative orbit inversely as the masses of the components. Thus their relative distances from the barycenter can be computed. If, in addition, the distance to the system is known, by parallax measurements, perhaps, the dimensions of the orbits can be obtained from our old friend (or nemesis), Equation 7.4. With all of the required information thus established, the masses of the components, collectively and also individually, are known. The distribution of mass is made with the assistance of Newton's law of gravitation. Here is a practical application of Newton's verification of Kepler's harmonic law (outlined in Chapter 10) as a natural consequence of the mutual attraction between astronomical objects. As an example *Sirius* (*α Canis Majoris*), the brightest star in the sky, is a visual binary. The parallax of the primary is 0".371 and the angular length of its true relative semimajor axis is 7".62. The period of the system is 50 years, and its distance from the sun is 2.67 pc. But from the definition of parsec the length of the semi-major axis in astronomical units is 2.67×7.62. Then from Equation 10.7, $(m_1 + m_2) = (2.67 \times 7.62)^3/(50)^2 = 3.4$ solar masses $(M_\odot)$. In addition the secondary is about twice as far from the barycenter as the primary, hence its mass, m_2 is about one-half the mass of m_1, or $m_1 = 2.2\ M_\odot$, $m_2 = 1.1\ M_\odot$, very nearly.

Astrometric Binaries. Occasionally one member of what would otherwise be a visual binary is too faint to be seen, but its presence can be deduced from the motion of the visible component. Such a system is called an *astrometric binary* (Gr., *astro*, star, and *metron*, to measure). The disturbance of the primary is periodic; it follows a 'wavy' path as it moves through space revolving around the common center of gravity (Figure 11-11). A very interesting example is Barnard's star which, besides being an astrometric binary, has the largest proper motion known (p. 279). In 1968 van de Kamp at Sproul University showed from an analysis of 51 years of data on its motion that Barnard's star displayed the tell-tale wavy oscillation. By using the analysis techniques described in this chapter, and a determination of mass by a method yet to be considered, van de Kamp deduced that Barnard's star has a mass of only 0.15 $M_\odot$; its dark companion has a mass that is a mere 0.0017 $M_\odot$. The secondary has an orbital distance from the primary of only 4.5 AU. The dark companion is only 80% more massive, yet it has a smaller orbit than Jupiter. It is able to produce an oscillatory displacement of its primary, an accomplishment Jupiter is unable to duplicate with the sun, because Barnard's star is such a lightweight compared to the sun. *The dark companion is in the range of planetary masses,* probably too small to be self-luminous. Hence, the object must be a planet, the first one outside the solar system to be discovered.

In April, 1969 van de Kamp announced the discovery of a *second* planet orbiting Barnard's star. The latest discovery has only about four-fifths Jupiter's mass, making it the least massive stellar companion yet discovered. Star Laland 21185, as it is listed in the catalogs, a near neighbor of the sun at a distance of only 8 LY, shows perturbations in its motion that are taken as evidence of the presence of a dark companion, very likely a planet. This unseen attendant is 10 times more massive than Jupiter, yet its mass is below what is considered to be the critical mass for a star to begin shining from nuclear fire initiated deep within its core. Thus the evidence mounts that as far as having planets in attendance, our sun is not *unique*.

Spectroscopic Binaries. In many of the binary systems the component stars are so close to each other that they cannot be visually separated. Moreover their orbits are so highly inclined that eclipses are not possible. Yet under favorable conditions one, or both, components show a periodic Doppler shift of their spectral lines. This periodicity is testimony to their binary character. What can we make of the visual evidence?

The Doppler shift, of course, yields directly the apparent radial velocities. And even though

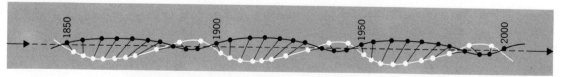

FIGURE 11-11 Space paths of Sirius A (black) and its white-dwarf companion Sirius B (white). Observe that both components oscillate about the mean space path of the bary center of the system. Diagram shows observed path (1850–1950) and predicted path (1950–2000).

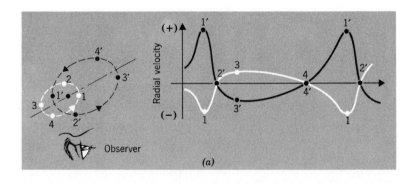

FIGURE 11-12 Radial velocity curve (right) for a hypothetical relative circular orbit oriented as shown with respect to observer.

the components cannot be visually separated there is a saving grace in the compactness of a spectroscopic binary system. The orbital periods are short, from a few hours in exceptional cases to a few months for most systems. Hence for the most part spectroscopic binaries complete many orbital revolutions in our life-

time; the data they yield can thus be highly refined.

In practice, the radial velocity of each component is plotted as a curve on a velocity versus time graph covering one period (Figure 11-12). This figure illustrates the idealized case of circular orbits. The manner in which eccen-

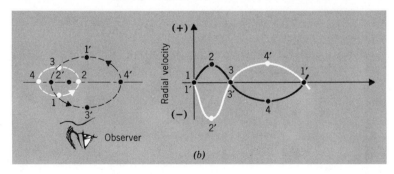

FIGURE 11-13 Asymmetry of radial velocity curves: for skewed elliptical orbit (a); for ellipticity only (b) (see text).

tricity and the oblique orientation of the semi-major axis (skewness) distorts the symmetry of the radial velocity curve is shown in Figure 11-13. In both figures the difference in amplitude between the two components results, of course, from their unequal distance from the barycenter. According to Newton's gravitational law the orbital speeds are affected in inverse proportion to their distances from the center of gravity.

If the spectroscopic binary system is moving with respect to the sun, the radial velocity components, if it exists, of the system's proper motion can be disentangled from the observed radial velocity without too much difficulty. If the radial velocity curves of both components can be plotted, they will intersect along a line that represents the radial velocity of the center of gravity relative to the sun.

Even if only one of the radial velocity curves can be plotted, as would be the case if the light from one component were so faint that its spectrum were undetectable, the radial velocity of the barycenter can still be deduced. No matter how skewed the orbit may be, during one-half of the orbit there is a radial velocity component of approach; during the other half there is one recession. Figure 11-14, the radial velocity curve of the star β Arietis, shows the *relative* radial velocity of the visible component. Now if a straight horizontal line is drawn through the curve such that two equal areas between the curve and the straight line are produced, the straight line will intersect the velocity axis of the graph; the velocity so indicated is the radial velocity of the center of gravity of the system.

How do we justify this procedure? Dividing the curve into two portions satisfies the requirement that the star is approaching during one-half of its orbital revolution and receding during the other half. Note that the *lengths* of the radial velocity curve segments are not equal. This is consistent with the variation in the speed of the star during its orbital period.

The problem of finding the size, the shape,

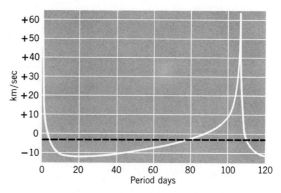

FIGURE 11-14 Radial velocity curve for β *Arietis*. This star has an orbital period of 106.9973 days and one of the largest eccentricities known, 0.89. Radial velocity of the center of mass of the system (the secondary is too faint to be seen) as shown by the dashed line is approximately 3 km/sec toward the solar system.

and the orientation of a spectroscopic binary orbit, of determining the separation of the components, and of determining their individual masses has a unique solution, but it is a solution of limited value. Not determinable in any degree are the diameters of the component stars and the inclination of the system's orbit. This last deficiency is the most damaging. We know that the orbit is not seen face-on, otherwise the radial velocity would be nonexistent.[†] We know that the orbit is not seen edge-on, or even nearly edge-on, otherwise eclipses would occur. But between these limits the inclination is unknown.

The radial velocity that we see is that associated with the apparent relative orbit. It is a component, fraction unknown, of the true relative velocity. Clearly the plotted radial velocity curves are for minimum values. Now the dis-

† The usual practice is to transform the data into that for the *relative orbit* with the primary at the focus. If only one component can be detected, of course, the true relative orbit cannot be deduced. In the discussions the relative orbit is implied.

tance *a* between the component stars is simply the radius of the apparent relative orbit; $2\pi a$ is the circumference of a circle whose radius is *a*.† The radius of the apparent relative binary orbit is equivalent to the speed of the star following the orbit times the period it takes to complete one round trip. We have $2\pi a = v_{rel} \times P$ where v_{rel} is the apparent relative velocity and P is the period. Solving for *a* we obtain $a = v_{rel} \times P/2\pi$. Again, the individual terms are modified for an elliptical orbit, but the *form* of the equation remains the same.

The relative velocity is necessarily a *lower limit;* how much faster than this the star is moving is unknown. As a result the separation must be the maximum possible. If the above value for *a* is inserted in Kepler's harmonic law $m_1 + m_2 = a^3/P^2$ we get the lower limit for the sum of the masses. If both radial velocity curves can be plotted, the ratio between the relative amplitudes reveals the relative distribution of mass between the pair; the least massive component has the slowest relative speed because it is the most distant from the barycenter. If, on the other hand, only one radial velocity curve can be plotted because only one component can be seen, we do not fare even this well. All we can obtain is a relation between the masses called the *mass function.* This is a ratio between the masses that depends upon the unknown inclination. As unsatisfactory as these results are, they do contribute to our over-all knowledge. As we shall see later, spectroscopic binaries yield valuable statistical information about close pairs of stars.

Eclipsing Binaries. If the components of a binary system revolve about the common center of gravity in a plane seen nearly edge-on from earth, and if the orbits are not excessively eccentric, each component will eclipse its companion once during each revolution about the barycenter. The qualification about the eccentricity is essential because of its effect on the geometry of the motions. If the orbits are eccessively eccentric, one or the other of the components may miss eclipsing its companion on some or all of the passes; the apparent path would be too high or too low to intercept the other star's disk. We shall not worry about the extremely eccentric systems because astronomy is eminently capable of dealing mathematically with the complex problems that arise in these cases. Since ours is a first course in astronomy, sufficient simplification in order to reveal the underlying principles is justified. Let us begin with the case of eclipsing binaries that are in nearly circular orbits, with two light minima per orbital period. The information that can be obtained in spite of this simplification is considerable.

Figure 11-15 illustrates the limiting case for the total eclipse of a smaller secondary component, and a partial (annular) eclipse of the primary. The figure clearly shows that if the orbit were even slightly more inclined beyond

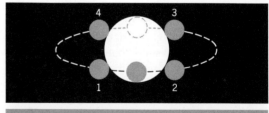

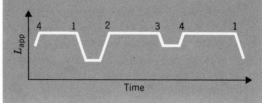

† If the orbit were truly circular, we would use *r* to indicate the radius. In this case *a* is the semimajor axis of an elliptical orbit; if the eccentricity of the ellipse is very small, *a* will not differ materially from *r*. The distinction between the two is important, however, and the use of *a* emphasizes this distinction.

FIGURE 11-15 Limiting orbital inclination for total eclipse of the secondary component of a binary star system. Schematic representation of the binary light curve is shown below the system.

that shown, both components would be only partially eclipsed. The graph at the bottom of the figure is a simplified light curve showing the variation in the apparent luminosity of the binary system. The actual change in brightness will not occur as abruptly as the straight-line segments suggest. If the components are of unequal size and their respective temperatures are not dramatically different, the *primary minimum* in the light curve occurs when the secondary component passes behind the primary; 100% of its contribution to the total luminosity is cut off. The *secondary minimum* coincides with the annular eclipse of the primary. Clearly, this minimum, except in unusual cases, will not be as deep as the primary minimum. Even though a portion of the light from the primary is cut off during the annular eclipse, the brightness of the secondary makes a compensating contribution. Of course, if both components were exactly the same size and of the same brightness, and if the orbit were seen exactly edge-on, the two light minima would be equal in duration and depth. The reader should ponder what happens when one or the other of the components is significantly hotter than the other.

Excluding the unusual cases, what can we learn from the analysis of the eclipsing binary system's light curve? At best the analysis is difficult, even with the simplifications imposed. In addition to the idealized circular orbits, let us assume that the orbits are seen exactly edge-on so that the eclipses are central, that total and annular eclipses occur, and that the stellar disks are uniform in brightness from limb to limb. In Figure 11-16 let us suppose that the secondary (smaller) component is one-half as bright as the primary, that is, the primary contributes two-thirds of the total apparent luminosity. In (a) the secondary is eclipsed by the primary. Positions 1, 2, 3, and 4, called first, second, third and fourth contacts, respectively, correspond to *grazing entry, first totality, last totality,* and *grazing exit.* Prior to the onset of the eclipse, the combined luminosities clearly produce the maximum apparent luminosity of the system. During the eclipse of the secondary between

positions 2 and 3, the light curve will show a constant minimum; the maximum apparent luminosity is reduced by one-third. In (b) the secondary transits the primary producing an annular eclipse of the latter. This minimum will not be as deep as the first one, although the duration of the minimum will be the same. Suppose that during this minimum the apparent luminosity of the system is reduced by one-twelfth of the maximum brightness. Then the relative brightness of *equal areas* of the two components is in the ratio of one-third to one-twelfth, or four to one. Since the primary component of the system that contributes two-thirds of the total luminosity is only one-fourth as luminous *per unit surface area,* the surface area of the primary component must be 8 times that of the secondary. But the ratio of the surface areas is proportional to the ratio of the square of the diameters; therefore, diameter of the primary must be $\sqrt{8} = 2.83$ times the diameter of the secondary. We now have the *relative* sizes of the primary and the secondary.

The foregoing constitutes only the barest details about the analysis techniques and the complications that beset them. Astronomers have learned to take in stride the complicated mathematics necessary when eccentricities are large, when limb darkening is of substantial consequence, and when the semimajor axis of the system is skewed relative to the line of sight. The mathematical treatment is beyond our scope and is unessential to our qualitative investigation. Even a discussion of the observational techniques is unavoidably complex. We can profit most from a simple summary of what can be learned from the binary stars.

The analysis of the light curves includes the relative depth of the minima, the relative duration of the minima with respect to the orbital period, and the time intervals between successive minima. It also considers the *shape* of the light curve at the onset and cessation of the eclipses. From the analysis we can find the inclination of the orbit and the size of the binary components relative to the orbital size. If the Doppler shift of the spectral lines of both

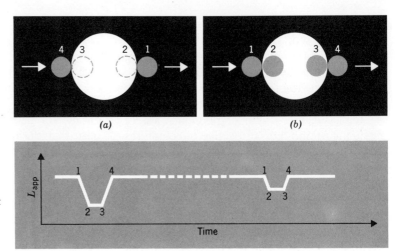

FIGURE 11-16 Analysis of the minima in the light curve for central eclipses of a binary star system (see text).

components can be determined, then the relative radial velocities can be computed. Then from the known orbital inclination, the effect of foreshortening on the observed speeds can be taken into account. This permits us to calculate the true orbital velocities; these, together with the period, pinpoint the real orbital dimensions.

Now, with the orbital size nailed down, the actual sizes of the components can be evaluated, and along with these the real distances of separation can be computed. And now we have the necessary data on the period and length of the semimajor axis for insertion into Kepler's harmonic law. The masses can now be determined. Knowledge of the masses, together with the size of the component stars, yields their mean densities.

If in addition, the angular size of the relative orbit can be measured, knowledge of the true size leads at once to the distance of the system. With the distance known the inverse square law can be applied to the measured total apparent luminosity to get the absolute luminosity of the system. Then by reverting to the relative depths of the eclipse minima, the relative surface brightness of each component can be deduced. These, together with the absolute luminosity, permit the effective temperature of

each component to be estimated. Truly the binaries, especially the eclipsing binaries, are most valuable aids in our assessment of the nature of the stellar population.

Temperature of the Stars. We have already found (p. 278) that the temperature of a star can be computed by means of the Stefan-Boltzmann law, $E = kT^4$, provided the radiation rate per unit area (E) can be measured. Often this cannot be done, but there is another way out of the dilemma. Let us assume that the spectrum of a star can be readily obtained, and that the wavelength corresponding with the maximum intensity can be determined. This is a critical task, requiring considerable experienced-based judgment, but it *can* be done. Then Wien's law (Chapter 10) comes to the rescue. Recall that the blackbody radiation curves for successively higher temperatures show a shift of the intensity peak toward shorter wavelengths. Wien's law states the relationship, $\lambda_{max} = k/T$; k is an experimentally determined constant, and the wavelength (λ_{max}) is read off the spectrogram. The *temperature* can then be computed. Our confidence in this procedure is governed primarily by our ability to find the intensity maximum in the spectrum. Experience shows that temperatures obtained in this way when compared to temperatures obtained by other means are within 10% of the true value.

Diameter of Stars. When the absolute luminosity of a star is known, as it is for all stars whose distance can be determined, the formula for computing effective temperature (Equation 11.3) can be rearranged and solved for the star's diameter. The effective temperature must be known, but Wien's law is available for the purpose. We need not be concerned with the algebraic manipulation that leads to the working formula for a star's diameter; the result is,

$$D = \left(\frac{5770}{T_e}\right)^2 \times \sqrt{L}. \qquad (11.6)$$

where D is the star's diameter, T_e is its effective temperature, and L is the star's absolute luminosity. Again, the confidence that can be placed in this method depends upon the assigned value for the effective temperature and the star's absolute luminosity. Diameters obtained in this way are accurate to within 25% or so of those found by more direct methods.

The plethora of alternatives for solving stellar problems is more apparent than real. Only in very few cases can a particular property or quantity be found by more than one method; in the majority of cases there is only one avenue of approach. The disquieting fact is that for the overwhelming majority of the stars none of the methods discussed here can be applied. Statistics is the court of last resort.

The Mass-Luminosity Relation

A most valuable tool for unlocking further mysteries of the stellar population concerns the binary stars. When the masses *and* the absolute luminosities of binary stars, for which both quantities are known, are plotted on a graph (Figure 11-17), a direct relation between the masses and luminosities is apparent. *The more massive stars are also the most luminous.* The

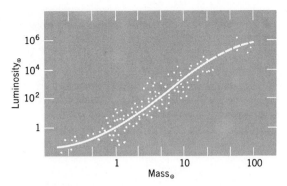

FIGURE 11-17 The mass-luminosity relation. Luminosities are in units of solar brightness; masses are in units of solar mass.

value of this relationship is substantial; it provides the means for estimating the masses of *single* stars (a property not otherwise obtainable) from a knowledge of their absolute luminosities. As we shall see in Chapter 12, the mass-luminosity relationship is not accidental; it arises from the fundamental structure and behavior of the stars. To be sure, not all stars fit the mass-luminosity relationship, but these deviants can usually be identified from peculiarities in their spectra.

Motions of Star Clusters

A prolonged study of the proper motion of the stars in certain clusters shows that they all have about the same space velocity (same speed, same direction). The open cluster, *the Hyades,* in the face of the constellation Taurus, the Bull, is an example. The motion of the members of the cluster is toward (or away from) a singular point on the celestial sphere. This is a perspective effect that yields the mean distance to the cluster. An analogy is helpful here. From a moving automobile on a straight stretch of road, fixed objects such as telephone or power poles, fence posts, and trees all appear to converge toward a point to the rear of the car. Conversely, similar objects all appear

FIGURE 11-18 Apparent convergence of proper motions of members of a star cluster. The convergence is a perspective effect.

to diverge from a point ahead of the moving vehicle. If we take the reasonable stance that the reference frame is fixed relative to the automobile, then all of the motion can be ascribed to the objects. Indeed, they all move toward the convergent point (or away from the divergent point) along parallel paths.

If a star cluster is close enough to the sun, the space motions of its stars can be measured. The space motions are all parallel, yet the stars appear to converge toward a point on the celestial sphere (Figure 11-18). Moreover, a line from the sun to the convergent point is parallel to the space motion of the cluster stars. Now we select a star in the cluster and determine its radial velocity (Figure 11-19). The angle α, between the line of sight to the star and the line of sight to the convergent point, is known from direct measurement. Also, the space motion of the cluster is parallel to the latter line,

and the radial motion is parallel to the former. Hence, the angle between the space velocity and the radial velocity is the same as that between the two lines of sight.

In the figure triangle SOP is a right triangle because the tangential velocity OP is, by definition, perpendicular to the radial velocity SO. We can solve for the tangential velocity OP since the other two velocities are known. Either the theorem of Pythagoras in plane geometry or the methods of trigonometry suffice. But we have a formula for determining tangential velocity (Equation 11.5; $v_t = 4.74\mu/p$ (km/sec) which can be rearranged so that $1/p$ (the distance) can be found, since all the other quantities are now known. The distance, of course, is in parsecs.

In three well known clusters, including the Hyades cluster and the cluster that contains most of the visible stars of the Big Dipper, there

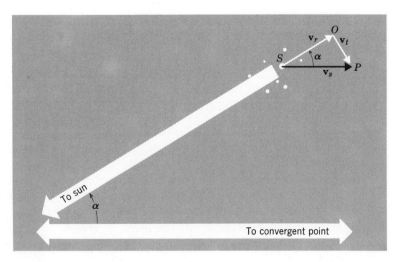

FIGURE 11-19 Space velocity diagram for stars in a moving cluster (see text).

are nearly 500 stars whose distances are thus established. This is almost as many stars as there are whose distances are known by parallax methods to an accuracy within 10% or better. Moving clusters have substantially expanded our distance horizons.

The Secular Motion of the Sun. We are now in a position to cast aside, without any reluctance, the notion of the fixed stars, since we have found so many that have detectable motions. But what should we expect to find on analysis of the proper motions (or better yet, the radial velocities) of a large number of stars in the vicinity of the sun? Should the motions of the stars be random, like people milling around in a crowd? Or should there be an organization pattern to the motions within the assemblage as a whole?

Let us imagine that we have at hand the radial velocities of many stars. Radial velocities are better to work with than proper motions because they are independent of distance; proper motions are not. Now if the star motions are entirely random, clearly the radial velocities should average to zero; there should be just as many stars approaching with an average velocity as there are receding with the same average velocity.

Suppose, however, that we find one direction in which the *average* radial velocity is one of approach; in the opposite direction the *average* radial velocity is one of recession. Perhaps this shows that the sun is moving through the starfield. How can we test the hypothesis? It is easy; if the radial velocities at right angles to the presumed line of motion average to zero, then the motion resides in the sun. It is called *secular* motion, meaning not periodic. This is a valid statistical conclusion, because the stars whose radial velocities are used in the sampling were selected at random. The probability that such a random selection would produce the observed distribution of radial velocities is essentially zero.

The sun, and its neighbors within a few hundred parsecs, are drifting approximately to-

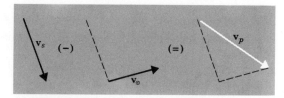

FIGURE 11-20 Space velocity $\mathbf{v}_s$ of a star when corrected for solar motion yields the star's peculiar velocity $\mathbf{v}_p$. Because of the sun's motion $\mathbf{v}_\odot$ and also because the star's space velocity $\mathbf{v}_s$ is measured with respect to the sun, the latter (in vector representation) is inclined too much toward solar antapex. Correction shows the star's true velocity $\mathbf{v}_p$ with respect to an "at rest" coordinate system.

ward the very bright star *Vega* in the constellation *Lyra*. This is a revised direction from that previously accepted, toward a point in the constellation *Hercules*. Actually the change in direction is very slight.

The direction toward which the sun is moving is called the *apex*; the opposite direction is called the *antapex*. Clearly, the *true velocity* of a star relative to a fixed local standard of rest will be different from the observed space velocity, since the latter is measured relative to the sun. Figure 11-20 illustrates diagrammatically the effect of the correction for the solar motion. The result is the *peculiar velocity* of the star (p. 282).

Proper motion studies yield approximately the same result for the secular motion of the sun. In the direction of the apex of solar motion, the proper motions display a relative drift radially outward from the apex, again a perspective effect. Toward the antapex, the net drift is radially inward toward the point (Figure 11-21). At right angles to the line connecting the antapex and the apex, the stars show an average drift in the direction of the antapex. Again, because of the randomness of the sample, the obvious conclusion is that the sun is moving. We will return to the analysis of the solar motion in our later search for the galactic motion of the stars.

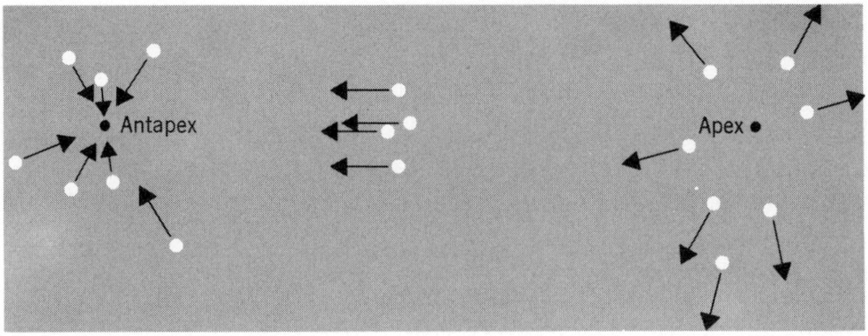

FIGURE 11-21 Proper motion *drift* of star field, a perspective effect that has all stars moving toward the antapex of solar motion.

SUMMARY

The properties of stars that are of interest in our attack upon the problem of the nature of the universe as a whole are: stellar *distances, luminosities,* and *masses.* We also need information concerning stellar *sizes,* stellar *composition,* and stellar *motions.*

Measurement methods can be broadly divided into two categories: *direct* determinations and *indirect* methods. Some quantities that can be directly determined, though only to a very limited extend, are: the distance to a star by direct parallax measurement, the apparent luminosity of a star by direct determination of the energy flux at the earth's distance, and the calculation of the star's absolute luminosity from knowledge of its apparent luminosity and its distance. Absolute luminosity can in principle be calculated alternatively from knowledge of the star's rate of energy production and its size; the sizes of a few stars can be measured by light interference methods using a beam interferometer.

The composition of a star is found by analysis of its spectrum; the star's temperature is calculated from the Stefan-Boltzmann law, E / kT^4, provided the radiation rate per unit area, E, can be directly measured. This method is weakened by the requirement that both the star's absolute luminosity and its size must be known in advance.

Stellar motions of interest are *radial velocity,* transverse (tangential) velocity, space velocity, peculiar velocity, proper motion, and, in the case of binary systems, orbital velocity. Of these, only radial velocity and proper motion are unaffected by the star's distance. Transverse velocity cannot be directly determined without prior knowledge of the distance to the star.

The masses of stars, the sizes of stars, the temperatures and distances of stars are most successfully determined indirectly. The binary stars provide the most complete information. The indicated quantities, with the exception of stellar sizes, can be learned from the *visual binaries;* the *eclipsing binaries* are even more valuable in that they yield information concerning the stars' sizes and the stars' temperatures. Analysis of the light curve is the technical procedure. *Spectroscopic binaries* make limited contributions except that the close association of the components produces short periods. Because of this, the available data can be continuously reviewed and refined, a practice usually not possible with the visual binaries because their periods are generally very long.

Stellar temperatures can be indirectly determined by resort to radiation theory and Wien's law (Chapter 10, and p. 279). The diameter of a star can be estimated from its effective temperature (usually found indirectly) and its absolute luminosity. The sun is usually employed as the reference standard; this simplifies the mathematical treatment considerably.

The *mass-luminosity* relation, derived from

the properties of binary stars, indicates that the more massive the star, the more luminous it is. This relationship between mass and intrinsic brightness can be used in the determination of the mass of single stars whose mass would otherwise be unaccessible.

Star clusters help in solving the distance problem. When the average space motion of the members of the cluster is parallel they all appear to converge toward a point, an effect of perspective. The direction of this convergence together with the radial velocity of a representative star in the group can be used to secure the transverse velocity. With the transverse velocity and proper motion known the mean distance to the cluster can be computed.

Questions for Review

1. What is the practical limit to stellar parallax in the determination of stellar distances?

2. Define parsec. How is it derived?

3. Differentiate between absolute luminosity (or brightness) and apparent luminosity (or brightness).

4. Star A and star B have the same apparent brightness. Star A is known to be twice as distant as star B. What is the ratio of the absolute brightnesses between stars A and B?

5. The apparent luminosity of a particular star is 6.3 times that of a second star. What is the difference between their apparent magnitudes?

6. One star is reddish, another is bluish in color. Which is the hotter star? Upon what evidence is this conclusion based?

7. Distinguish between a star's radial, its transverse, and its space velocities?

8. Define proper motion of a star.

9. Does the proper motion of a star give an accurate indication of its transverse velocity? Why?

10. Discuss how the mass of a star can be determined. (Hint: this applies to binary systems only.)

11. Is it possible for a given system to be simultaneously classed as a visual, a spectroscopic, and an eclipsing binary?

Prologue

Statistics comes to our rescue, and while we shall find that the fine structure of the stellar populations becomes indistinct, the larger picture begins to emerge more clearly.

We shall find a way to solve the problem of stellar distances that proved to be such a formidable roadblock in Chapter 11. And we shall discover that stars are born, grow old, and die in an evolutionary pattern that is significantly dramatic. We can say one thing about the stars: they are never at a loss for commanding eye-popping attention at each stage of their evolutionary life cycle. Some of the life styles of stars are exotic beyond imagination. At any one time we see normal, well-behaved stars, the "squares" as it were. We see bloated stars, shrunken, wizened stars, even stars that are pulsating in size for all the world like a beating heart. And there are stars that literally blow their tops, some to make a recovery of sorts, and others that vanish in one spectacular celestial puff. In this chapter we see how man's scientific skills have enabled him to account for these affairs.

Statistical Methods and Stellar Evolution

12

In Chapter 11 we reached the end of our resources in applying direct measurement techniques to the stars beyond the sun. In fact, some of the indirect methods we reviewed border on statistical treatment. The fact that some measured quantities were involved saved these methods from the ignominy of being little more than educated guesses. We saw that the determination of distance through parallax measurements becomes untrustworthy beyond about 100 LY. The 2000 or so stars within this distance, compared to the billions lying beyond it, bear mute testimony to the inadequacy of direct measurement techniques. Absolute luminosity determinations by application of the inverse square law to the stars' apparent brightnesses fail without knowledge of the distances to the stars.

Assigning apparent magnitudes as a measure of stellar luminosities is beset with a different kind of difficulty, namely, the effect of the stars' colors on the detection systems. The direct determinations of the diameters of stars require two generally unavailable bits of data: angular size and distance. Even resorting to a quasistatistical approach through stellar temperatures fails unless we know a priori the energy emission rate at the stars' surfaces. Most stars are too remote for direct photometric temperature measurements. In fact, determining stellar temperatures directly necessitates prior determination of absolute luminosities and diameters. But here we are again frustrated; determining absolute magnitudes and diameters

demands prior knowledge of the stellar distances. We appear to be dangerously near the quicksand of elliptic argument.

As for stellar motions, the transverse velocity, even though it is related to a given star's proper motion, cannot be determined unless the distance to the star is known. Proper motion by itself is almost totally ambiguous. The only bit of qualified information we can glean from proper motions that are unsupported by other data is that the largest annual angular displacements belong, in general, to the nearest stars. Here again, we are skating on the thin ice of statistical estimate.

Stellar masses are even more inscrutable. Were it not for the visual or eclipsing binary stars, we would be totally at a loss as to where to even begin mass determinations of the stars beyond the sun. Spectroscopic binaries, unless they happen to be also eclipsing or visual systems, are not of much help. The best that can be learned from these systems is a *lower limit*, a minimum value, for the masses of the stars in question. And, finally, as adequate as *Wien's law* is for deducing a star's temperature from the wavelength of the intensity maximum in the star's spectrum, this knowledge avails us nothing when the star is so remote that its spectrum is not obtainable. This is the prevailing situation.

The fact that the problems appeared insurmountable did not, however, deter astronomers from continuing their quest for knowledge about the distant stars. In the branch of physics known as *spectroscopy*, physcicists found the key that unlocked the mysteries of the properties of many stars beyond direct-measurement accessibility. That astronomers were able to breach the barrier at all is indeed a tribute to man's intellectual skill and agility.

The Spectral Sequence of Stars

A brief, historical review of the beginnings of spectroscopic analysis of starlight at this point is worthwhile in helping to set the stage for the statistical phase of our investigations. In 1802 William Wollaston called attention to the fact that when sunlight was passed through a narrow slit before the beam was dispersed by a prism, several dark lines could be seen crossing the continuous bright spectrum. Joseph Fraunhofer initiated the serious study of the solar absorption spectrum in 1814–1815 when he succeeded in determining the wavelength of some 324 spectral lines out of 600 or so that he found (Figure 10-17). The lines are still called *Fraunhofer lines* in his honor, and the alphabetical letter designation that he assigned to the more conspicuous lines is still in use.

A succession of laboratory experiments in the production of dark lines in the spectra of artificial light sources and in the production of the bright line emission spectra in glowing gases led Kirchhoff (p. 249) in 1859 to propose the *laws of spectral analysis* that bear his name in explanation of the phenomena.

By 1823 Fraunhofer had already discovered absorption lines in the spectra of several stars. In 1864 the English astronomer, Sir William Huggins, was the first to identify a number of terrestrial elements in the stellar spectra. As the studies multiplied, considerable differences between the spectra of individual stars were observed. The Jesuit astronomer, Angelo Secchi (p. 181), in Rome began a classification scheme for the spectra of stars according to general differences in the strength and arrangement of the spectral lines. Secchi's four classes have been augmented and modified over the years until at present we recognize seven principal, and four auxiliary, *spectral classes*. We will examine these presently.

Differences between the absorption spectra of stars arise from very complex relationships among a given star's radiation intensity, the effective thickness of its photosphere, and the depth and pressure of the stellar atmosphere. Pressure effects depend upon the total mass of the star and the star's radius according to Newton's universal law of gravitation.[†] Atmospheric effects vary according to the extent of the atmosphere, the pressure it exerts at the surface of the photosphere, and the intensity of the radiation passing through it. We have already seen in Chapter 11 that radiation intensity is directly proportional to a star's temperature. Also, we have seen that the absolute luminosity of a star depends upon both the radiation intensity and the surface area of the star, which in turn is a function of the star's radius. Thus we might anticipate that stellar masses, temperatures, and sizes will have their individual and collective signatures displayed in the stellar spectra.

We cannot here be too concerned with the details of absorption line production in a spectrum, even though we run a risk of oversimplifying in attempting a brief, qualitative explanation for the observed spectral differences. The electrons of different kinds of atoms are bound to their respective nuclei in varying degrees. Thus electrons are limited to varying degrees in their freedom to undergo transitions in energy levels, often referred to as changes in vibration modes. The more tightly the electrons are bound, the higher the temperature must be in order to excite them into changes from their normal vibration modes. If the radiation passing through the stellar atmosphere is intense enough, certain electrons may be torn away from their atomic nuclei leaving behind ionized atoms. The ionized atoms absorb and emit radiation differently from their un-ionized counterparts; the absorption lines produced are correspondingly different. The analysis of the spectral lines of the stars is exceedingly complex. We can accept temporarily

† In some cases we must turn aside from Newton's laws in favor of those of *general relativity*; this is not a serious defection, for Newton's laws are in reality simple, adequate approximations for the general relativistic laws. We shall pursue this topic in Chapter 14.

some broad generalizations about line formation, modifying these as we proceed and as the need arises. Differences in chemical abundances may show up as differences in line intensities, both within a given spectrum and between the spectra of individual stars. More important, differences in temperatures produce even greater influence on the line intensities than do the chemical abundances. Some lines are broad and weak; others are narrow and sharp. Moreover, theory predicts that the temperature of the emitting surface may either inhibit or enhance the production of absorption lines.

Astrophysicists now recognize that the principal spectral classes can be arranged in a *spectral sequence* of descending stellar temperatures. The modern classification system is based upon the Harvard Observatory scheme of the early 20th century in which letters of the alphabet in ascending order were assigned to the spectral classes. Subsequent studies, made after some one-quarter million stellar spectra were classified in and published in the Henry Draper catalog, necessitated the rearrangement of the letter sequence, and the addition of secondary classes in order to avoid a massive reclassification effort with its attendant confusion. Today, the scheme is,

$$(W)-O-B-A-F-G-K-M \begin{smallmatrix} \nearrow R-N \\ \searrow S \end{smallmatrix}$$

The original seven classes began with O and ended with M. Class W, and the branches R, N, and S are the later additions.† Figure 12-1 shows the seven principal stellar spectra. Astronomers frequently refer to stars in classes O and B as *early-type* stars and those in classes K and M as *late-type* stars.

The seven letters in the principal classifica-

tions indicate only broad characteristics. Miss Cannon of the Harvard Observatory, who did the yeoman's work in the original classifications, showed later that much greater refinement is possible. Accordingly each class is further divided into subclasses from 0 to 9 so that the spectral designations would appear as B0, A5, F8, G2, etc. To be sure, not all of the possible subclassifications are used; classification of the spectra is not yet that refined. As we shall see later, peculiarities called luminosity effects in the spectra of certain stars in a given classification necessitated the addition of a Roman numeral suffix, from I through VI, for proper identification. There are additional letter suffixes and prefixes that need not concern us here.

Some generalizations about the spectral classes will be helpful in subsequent arguments. Class O stars are blue, with temperatures in excess of 25,000°K. These stars are so hot that hydrogen is almost totally ionized; hence the hydrogen lines are very weak. Hydrogen has only a single electron; when it is removed, the hydrogen nucleus left has no vibration modes available and so cannot produce absorption lines. The dominant lines in class O stars are associated with ionized helium, an exceedingly difficult atom to ionize; doubly ionized nitrogen; and trebly ionized silicon. These latter elements require extraordinarily high temperatures in the liberation of the tightly-bound electrons. Class B stars are also blue with temperatures ranging between 11,000–25,000°K. Class A stars are blue to blue-white with a range in effective temperatures of 7500–11,000°K. Coinciding with the progressively lower temperatures, the helium lines rapidly weaken and vanish; hydrogen lines reach their peak intensity at about 10,000°K. Some metallic lines such as those produced by singly ionized magnesium, calcium, iron, and titanium become prominent. In the class A stars of lower temperature, lines associated with the more abundant of the neutral metals appear.

† A popular *mnemonic* still employed in assisting recall of the order of the stellar classification is, *Oh! Be A Fine Girl! Kiss Me! Right Now (Smack)!* Observe that class W is omitted—this class was unknown at the time the memory aid was devised.

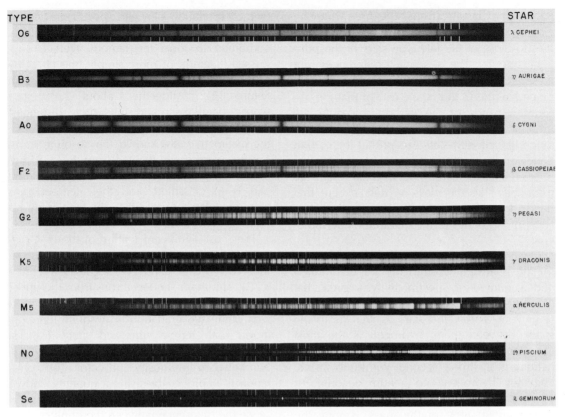

TYPE		STAR
O6		λ CEPHEI
B3		η AURIGAE
A0		δ CYGNI
F2		β CASSIOPEIAE
G2		η PEGASI
K5		γ DRACONIS
M5		α HERCULIS
N0		19 PISCIUM
Se		R GEMINORUM

FIGURE 12-1 Principal types of stellar spectra. The star spectra are the central light bands; above and below each are the bright line emission spectra, superimposed at the time of exposure for comparison purposes (Courtesy Hale Observatories).

Class F stars are white to yellow-white; they range in temperature from 6000–7500°K. The lines associated with ionized metals become strong; the Fraunhofer H and K lines of ionized calcium become very intense. Iron lines, peculiarly enough, appear to fade, but this is a phenomenon that can be attributed to the earth's atmosphere. The strongest iron absorption occurs in the ultraviolet region of the spectrum; but uV radiation is effectively absorbed in the upper atmosphere, so as a result the iron lines are only weakly seen.

In spectral class G, of which the sun is a member (class G2), the temperatures decrease from 6000 through 5000°K. Here the hydrogen lines rapidly fade; temperatures are insufficiently high to effectively ionize hydrogen. The lines of neutral metals become increasingly prominent except for calcium which absorbs most strongly at Class F5. Class G stars range in color from white to yellow.

In class K stars, the orange-to-red ones with 3500–5000°K effective temperatures, and in the class M stars, red in color and below 3500°K in temperature, the lines of neutral metals are the most prominent. Molecular bands appear, and since molecules cannot form except at relatively low temperatures, the band spectra provide major clues for the temperature classifications.

The Spectral Sequence of Stars   301

Secondary Classifications. The three "branch" classifications are spectra with strong similarities to those of the late-type stars of the principal group. The essential differences are these: class R stars are yellow-to-reddish, very much like the G to K stars, except that the absorption bands of the carbon molecule are present; class N stars are very red and are much like class M stars except that the carbon bands are very strong (Class N stars are often referred to as the carbon stars); Class S stars are essentially like the class M stars except that bands due to zirconium oxide are present instead of the bands of the titanium oxide molecule.

Once the effects of temperature and pressure on the relative line strengths are determined and accounted for, spectral analysis reveals the comparative chemical abundances in the stars. In general, within the stars as well as in the space between them, hydrogen is the most abundant of the elements. The abundance of helium is a strong second within the stars. Together, hydrogen and helium account for 95–99% of the mass of all stars; hydrogen alone makes up 50–80% of the mass. Among the "heavy" elements (when present) neon, oxygen, silicon, sulfur, and iron are the most abundant. Altogether, as we have noted earlier, 67 of the terrestrial elements have been identified in the spectrum of the sun.

The Hertzsprung-Russell Diagram

In 1911 the Danish astronomer, E. Hertzsprung, plotted the colors against the absolute luminosities of the stars within several clusters. The absolute luminosities were determined by methods that yield the parallaxes of stars in clusters, methods we considered in Chapter 11. Two years later an American, H. N. Russel, made a similar plot of the absolute luminosities of the stars in the vicinity of the sun against their spectral classes. Russell's diagram had the same general appearance of that of Hertzsprung, as of course it should because

of the correlation between spectra and color. The results of both studies were accordingly combined to form the famous Hertzsprung-Russell diagram, or more simply, the H-R diagram. A classical representation is shown in Figure 12-2. Altogether, about 2200 stars whose parallaxes, and thus their distances, were known were used in the studies. Since the distances were known the absolute luminosities were easily derived from the stars' apparent magnitudes. Moreover their spectra had been classified, and their temperatures were calculated by resort to the blackbody radiation laws.

The most significant feature of the H-R diagram is that the vast majority of the stars cluster into two broad bands or sequences (Figure 12-3). The main sequence runs from the upper left to lower right of the diagram, showing a well organized gradation in absolute luminosity in both temperature and spectral class. There is also a somewhat less reliable correlation between luminosity and color. The correlation is improved by using a quantity known as the color index, CI. The color index is in essence the difference between the photographic magnitude and the visual magnitude. The zero reference is the respective photographic and visual magnitudes of the star Sirius. The hottest stars are the bluest; these appear brightest on the normally blue-sensitive photographic plates. Yellow to yellow-white stars appear brightest to the human eye. Hence the early-type stars on the main sequence have large negative color indices; the late-type stars have positive color indices. The meaning of negative and positive here is the same as with the magnitude system. Among the main sequence stars the hottest are termed blue giants; the coolest are red dwarfs. These apellations refer to luminosities rather than size, since the names were assigned before anything was known about the relative stellar radii.

Above the main sequence is a rather sizable branch bearing slightly upward toward the right. This is the giant branch; it is broad, with

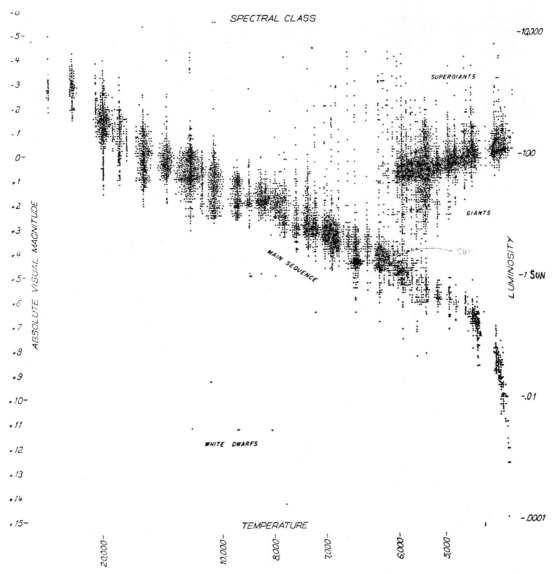

FIGURE 12-2 Hertzsprung-Russell diagram of stellar absolute magnitudes and spectral types (After Struve) (Yerkes Observatory Photograph).

ill-defined boundaries as is the case with the main sequence. The giant stars are highly luminous, yet they are relatively cool, and their spectra are distinctly late-type. It is clear that comparatively cool stars can be highly luminous only if they are very large in size. Hence,

the stars on this branch are referred to as *red giants*.

Above the giant branch are a few scattered stars of extremely high luminosity, whose spectral classes range from late F to M. The stars are called *supergiants*. Finally, neglecting for the

The Hertzsprung-Russell Diagram   303

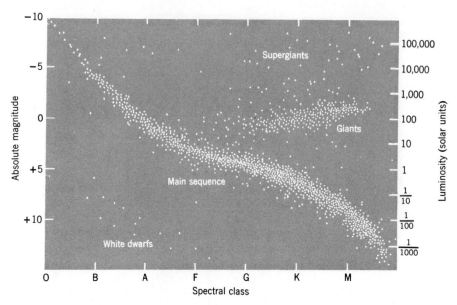

FIGURE 12-3 Schematic representation of the Hertzsprung-Russell diagram showing the two prominent bands or "sequences." Not all stars are on these main bands.

moment a very few isolated small groups, a dozen or so very near, very faint, white stars with luminosities on the order of 1/10,000 that of the sun are located in the lower left of the H-R diagram. These small, hot stars are called *white dwarfs*.

In examining the distribution pattern of the point-locations of the stars on the H-R diagram in Figure 12-2, we find rather well-defined vertical *columns* of stars. This graphical phenomenon arises from our inability to refine the spectra into all of the many subclasses within a spectral class. Thus many stars might be classified as B3, many others as B5, with few if any classified as B4 stars. Perhaps a significant number of stars in each of the adjacent classifications should have been classified as B4s. Another source of error that results in the broad vertical distributions arises in the random sampling methods among stars that differ markedly in their intrinsic characteristics. In all probability, the lack of precision in assigning absolute luminosities follows from uncertainties in the

stars' parallaxes. Nevertheless, the diagram shows that there are definite structural laws governing the behavior of stars. A reexamination of the spectra of the stars not on the main sequence bears out this contention.

Spectra of the Giant Stars. Since the parallaxes and therefore the distances of the stars on the H-R diagram are known, we can employ the methods of Chapter 11 in determining their absolute luminosities, their radii, their temperatures, and the intensity of the radiation through their photospheres. Thus there is no question but that the very luminous yet comparatively cool stars are truly giants compared to the sun. The star Antares, α *Scorpii,* has a diameter 450 times that of the sun. If placed at the sun's center, the photosphere of Antares would extend out well beyond Mars' orbit. Antares is a class M1 supergiant. Capella, α *Aurigae,* is a class G5 giant only 12.9 times the diameter of the sun. Betelgeuse, α *Orionis* is a class M2 supergiant whose radius is estimated to be 580 times that of the sun. If this supergiant were placed at the sun's center the orbit of Jupiter would be nearly 35 million kilometers below Betelgeuse's surface!

The luminosity effect manifests itself in a number of ways. We can gain an appreciation for the phenomenon by examining one or two examples. Let us consider Capella, a class G5 star whose surface temperature is about the same as that of the sun. Capella's radius is 16 times larger than that of the sun, and its mass, computed from the mass-luminosity ratio, is 4 times that of the sun. With the mass and size thus known, it is not difficult to determine that Capella's atmospheric pressure measured at the photosphere is only 1/100 that at the sun's surface, a consequence of Capella's lower surface gravity. As a result, its atmosphere is more attenuated, that is, less dense at all levels, than the sun's atmosphere. Individual atoms are farther apart on the average in Capella's atmosphere. Because of the wider separations of the atoms, the ones that become ionized take longer to reacquire the missing electrons. Hence there are in general *more* atoms ionized in Capella than in the sun during any given interval; this results in more intense ionization absorption lines than is the case for the solar atmosphere. Indeed, this is a general characteristic of the spectra of the giant stars (Figure 12-4). A similar line of reasoning accounts for more intense absorption lines for neutral atoms in the solar spectrum; the atoms are crowded so close together that only infrequently (by comparison) do they experience ionization. There are other spectral differences, with other factors playing their roles, but the above arguments serve to illustrate how observed differences between the spectra of the giant stars and the main-sequence dwarf stars can be accounted for.

Spectra of the White-Dwarf Stars. White-dwarf stars were originally thought to be restricted to colors ranging from blue-white to white, with spectral characteristics closely as-

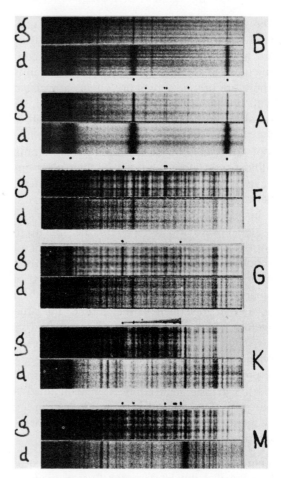

FIGURE 12-4 Spectra of the giant stars *g* compared with those of main-sequence stars *d* for several representative stellar classes (Yerkes Observatory Photograph).

sociated with the main-sequence classes A and F stars. Recent studies prove the name *white dwarfs* to be a misnomer, but the apellation is certain to remain. We now know the colors of these stars range down to the reds of the class M stars. All of these dwarfs are from 10,000 to 100,000 times less luminous than their main-sequence counterparts. The color temperatures and absolute luminosities imply radii in the range from 5 times to 1/3 that of the earth, yet their masses are on the order of the sun's mass!

The spectral lines of the white dwarfs are generally quite broad and somewhat fuzzy. They are always faint, sometimes being so inconspicuous as to be essentially undetectable. These are primarily pressure effects associated with unbelievably dense gaseous material. As we shall see later in this chapter, almost all of the electrons of the atmospheric atoms are free (not associated with any particular atom). Atmospheric atoms stripped of their electrons have no vibrational modes remaining to them; hence the absorption lines are very faint.

In addition, all of the absorption lines are shifted toward the red end of the spectrum, corresponding to a Doppler displacement of about 18 km/sec. Not until several white dwarfs were discovered to be members of binary systems, each in association with a normal main-sequence star, could the true average radial velocities be determined. This permitted isolation of the residual 18 km/sec phenomenon, which thus could only be interpreted as a *gravitational red shift*. We shall discuss the effect in connection with the life history of stars and stellar evolution later in the chapter. The spectra of several white dwarfs are shown in Figure 12-5.

<u>Value of the H-R Diagram.</u> There is no question but that the Hertzsprung-Russell diagram represents graphically the physical relationships among stellar sizes, luminosities, spectral classes, colors, and temperatures. Clearly this is true for the stars from which the diagram was derived. The only glaring weakness arises from a built-in bias in the sampling techniques by which the stars used in the H-R analysis were selected. Because of their greater luminosities, and hence higher visibilities, disproportionate number of bright stars is likely to be selected in any similar kind of random sampling. This flaw is present in the H-R diagram; as a result, the sampling contains a larger proportion of bright giant, supergiant, and highly luminous early-type main-sequence stars compared to the total number of main-sequence stars, many of which are fainter, late-

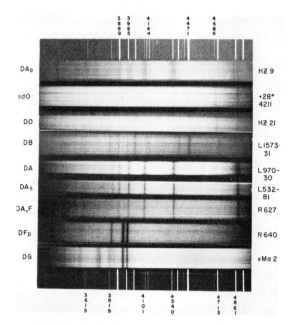

FIGURE 12-5 Spectra of white-dwarf stars. Since this photo mosaic was made, the second star from the top, +28°4211, has been reclassified according to modern ideas as a subdwarf type O star, not a hot white dwarf. This star is considerably more luminous than the typical white dwarfs as evidenced by weakness in the hydrogen lines. In addition, the relative sharpness of the lines is indicative of less surface pressure than is usual for white-dwarf stars (Courtesy Hale Observatories).

type stars. Similarly, there must be far too few white-dwarf stars in the diagram. How serious this imbalance is can only be conjectured. Probably the form of the H-R distribution would remain unchanged if the actual proportions of all stars could be considered. The greatest impact of the sampling bias is probably felt in connection with the analysis of stellar evolution, a topic which concludes this chapter.

Together with the mass-luminosity relation, the H-R diagram leads toward one inescapable conclusion. *On the main sequence,* the coolest stars are faintest, while the hottest stars are the most luminous. The cooler stars are the least massive, and the hottest stars are the most massive. The coolest stars are the smallest in size, and the hottest ones are likewise the biggest, although here, none of the stars have radii more than 4 or 5 times the solar radius. We could have anticipated the above relationships. The hottest stars are hottest *because* they are more massive. The larger the mass, sizes being otherwise comparable, the greater are the pressures deep within the stellar interiors. The gravitational effect of these greater pressures is to raise the temperatures drastically. Nuclear reactions are thus more vigorous, and proceed at a more furious rate, in these more massive stars. It is only when we try to apply these same arguments to the deviant giant stars and white dwarfs that our logic is dashed to bits on the shoals of reality. Evolutionary conjecture becomes our salvation, as we shall see.

Spectroscopic Parallax.   With most of the evidence in and favoring acceptance of the H-R diagram, we are now in a position to break the shackles imposed by the distance dilema that thwarted our investigations in Chapter 11. Only one further assumption is a necessary prerequisite. Since the H-R diagram was developed from a random sampling of stars, we can safely assume that all stars should fall into the same basic categories. All that is necessary then, in order to find a given star's distance, is to secure its spectrum, classify it on the H-R diagram, and read off the absolute luminosity. Then comparison between the deduced absolute brightness and the measured apparent brightness yields the distance to the star, or what amounts to the same thing, the star's parallax. This technique is appropriately called the method of *spectroscopic parallax*. With this new and powerful tool, the only limit to stellar distance determinations is in our ability to secure a decipherable spectrum. Securing the spectra becomes the paramount problem, one as yet not completely solved except for a few million stars nearest us. When applied to stars of known distance found by other methods, spectroscopic parallaxes correlate

quite well; distances are accurate within 25%. This is superior in achievement to measured parallaxes for stars beyond 80 LY distant.

Comparison Statistics of the Stars

Excluding for the moment the comparatively few and occasionally rare stars whose spectral peculiarities differentiate them from the broad categories of the H-R diagram, we find that among those stars whose spectra are obtainable the vast majority are not too different from the sun. Stellar luminosities range from the extremes of 1,000,000 times the sun's absolute luminosity to as little as 1/2,000,000 as bright. This lower limit is at the threshhold of detectability. Stellar radii range from extremes on the order of 1/100 to 1000 times the sun's radius. Excluding the exceptional cases making up the extremes, most stars fall within a range of 1/10 to 100 times the solar radius.

Stellar temperatures, as far as we can measure at the lower extreme, range from 1700°K to 200,000°K. This is a range of from 1/3 to 35 times the sun's effective temperature. Masses of the stars vary from 1/10 to 10 times the sun's mass in the extremes; the vast majority of stars have masses ranging from 0.4 to 4 or 5 times the sun's mass. Of all of these quantities, masses vary the least.

All stars whose spectra are obtainable show a continuous background emission spectrum; all have superimposed on it a dark line absorption spectrum. In truth, the absorption spectra differ among stars only in details, although it is true that these details are enormously important. We see spectral evidence, to varying degrees among the stars in general, for the presence of familiar terrestrial elements such as hydrogen, helium, nitrogen, carbon, calcium, and iron. In particular, no substance is found in the stars that is unknown on earth.

Thus, we are comfortably assured that the stars are all suns, or that the sun is actually a very ordinary, smallish, faint star. To be sure, there are great differences among individual stars. The task we now face is to break away from the statistical bonds and see what we can make of these individual differences. Can we find an organizational pattern in stellar structure? If so, what implications will it have for the total stellar population, for stellar distributions, for stellar evolution? To answer these questions we should first examine the deviant stars in more detail. We should also examine the composition of the local stellar population and the populations of the known star clusters in search of additional clues. We need all of the information that our ingenuity can marshal.

Variable Stars

For centuries astronomers have been aware that many stars exhibit variations in their apparent luminosities. Some of these variations are periodic, repeating with extraordinary regularity. Other luminosity changes are more or less irregular. Some stars display a meteoric rise from insignificance to extreme brilliancy, only to fade back into permanent, comparative oblivion.

During the past 2 centuries astrophysicists have discovered that variations in luminosity for some stars result from external influences. Transient obscurations, caused by clouds of gas and dust of variable density, produce irregular fluctuations in the apparent luminosities of some stars. For others, periodic changes in brightness can be ascribed to the phenomenon of eclipses. These stars, of course, are members of binary systems. Variations in brightness of the remainder of the variable stars are associated with physical processes occurring within the stars' photospheres, within the stars' or atmospheres, or within both.

The changes in luminosity caused by external effects are independent of the physical processes occurring within a star. Our interest here is not with these *extrinsic variables*. We are concerned only with the *intrinsic variables*, those stars whose luminosity variations are the consequence of dramatic, internal upheavals.

Besides being curiosities in their own right, these events occurring within the stars or their atmospheres are of statistical significance for the total stellar picture.

Over 15,000 intrinsic variables have been identified in the local collection of stars with which the sun is associated, the Milky Way Galaxy. Several thousand more have been found in the globular clusters, spherically symmetric groups of stars that are subsystems of our own galaxy. (*Galaxy* is a word of complex Greek origin meaning, literally, *milky circle*.) Many intrinsic variables have been discovered in other galaxies. Some variable stars display changes in color as well as in brightness — changes that are conspicuous to the unaided eye. With the aid of telescopes and a powerful array of sophisticated auxiliary instruments, astronomers have found that changes occur in stellar spectra along with changes in brightness and color. Even variations in radial velocities occur. Such stars are often referred to as *spectrum variables*. Still other stars show spectral changes that must be induced by fluctuations in their intrinsic magnetic fields; these stars are known as *magnetic variables*.

Most of the variable stars are discovered by comparing corresponding photographs of a given area of the sky made on different dates. A rather ingenious device, the *blink comparator*, makes variable stars on the plates or negatives figuratively jump out at the observer. Two otherwise identical plates made on different dates are placed in the comparator. The star images are projected on to a scanning table in such a way that the corresponding stellar images from each plate coincide. Then, the plates are viewed alternately at about 1 second intervals (Figure 12-6). The nonvariable star images remain constant in intensity; stars that vary in luminosity literally blink in rhythm with the period of alternation. Incidentally, the blink comparator is very useful in disclosing changes in position of astronomical objects that move too slowly to leave an image trace

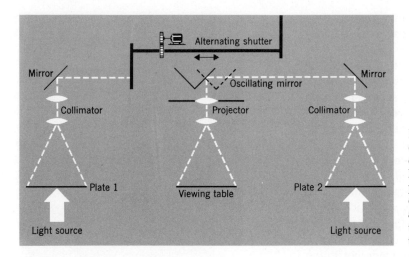

FIGURE 12-6 Schematic diagram of the operation of a blink comparator. Matching astronomical plates, differing in time or date of exposure only, are alternately presented to the observer's view (see text).

In the diagram: Alternating shutter, Oscillating mirror, Mirror (left), Mirror (right), Collimator (left), Projector, Collimator (right), Plate 1, Plate 2, Viewing table, Light source (left), Light source (right).

on a single plate. Tombaugh at the Lowell Observatory discovered the planet Pluto (Chapter 8) with the aid of a blink comparator.

Variable stars within each constellation are named with reference to that constellation and are designated by combinations of letters and numbers that indicate the order of discovery. The first variable discovered carries the prefix R thus: R Coronae Borealis. Subsequently discovered variables are designated S, T, . . . , Z; then RR, RS, . . . , ZZ. If additional designators are needed, the scheme continues with AA, AB, . . . , QZ, yielding a total of 334 unique designators. Beyond these, when required, the numbers 335, 336, 337, . . . , preceded by the letter V are used, thus: V 444 Cygni.

Periodic Variables. Most of the stars that exhibit systematic cyclical variations in brightness (hence, *periodic variables*) are giants or supergiants with periods ranging from 1 hour or so up to 3 years. A great variety of objects is included among the periodic variables. The stars with longest periods of variation invariably have late-type spectra; short period variables are of early-type, with basic spectra of class A or B. Astrophysicists have discovered a relationship of considerable statistical value between the period of luminosity variation and the mean density of a star. Red-giant stars are generally larger in diameter than bluer stars of the same luminosity; also, the mean densities of the red giants are lower. Studies show that the period of light variation is inversely proportional to the square root of the mean density.

$$P \propto \frac{1}{\sqrt{\rho}} \qquad (12.1)$$

where P is the period and ρ is the mean density. Thus, the variability must be indeed intrinsic, and the stars are pulsating, or vibrating in size, much like a beating heart!

Radial velocity changes accompany the variations in brightness, and this is attributed to a periodic rise and fall of the star's atmosphere; the range is about 10% or less of the star's mean size. Stars exhibiting this behavior are called *pulsating variables*. Changes in color and spectrum accompany the radial-velocity oscillations; the stars are bluest at maximum brightness and reddest at minimum. Analyses of a complex nature show that the radius of the star is about the same at both brightness maximum and minimum. Hence the temperature change associated with the change in color is indicative of a real variation in the intensity of the emitted radiation. Pulsating variables fall generally into four distinct groups.

Cepheid Variables. Cepheid variables, now often referred to as classical cepheids for reasons that will soon become clear, take their name from the naked-eye star, δ *Cephei*, the first such pulsating variable discovered. All cepheid variables are giants or supergiants with periods ranging from just over 1 day up to about 45 days. In our own galaxy, 7 days is the most commonplace period. Variations in brightness are most easily seen with the aid of a diagram called the star's *light curve*. Brightness variations are plotted against time, using appropriate scales. Frequently a composite diagram is constructed, consisting of the light curve, the radial velocity curve, a curve showing change in color, and another curve indicating the change in radius. Each curve is properly scaled against the other curves. Figure 12-7 illustrates such a composite diagram. The

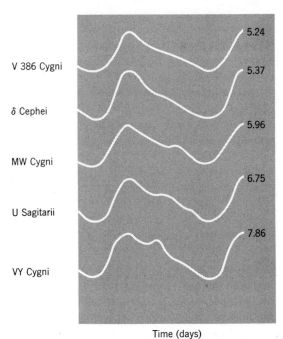

FIGURE 12-8 Light curves for several representative cepheid variables. The period in days for each star is shown at the right end of the curve. Note the similarity in the way each star rises to and subsides from maximum luminosity; the apparently minor irregularities are significant with respect to the period.

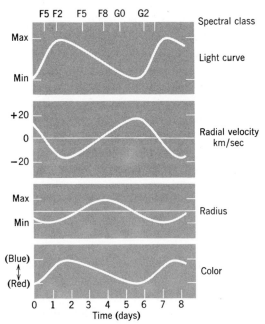

FIGURE 12-7 Composite diagram showing the corresponding variations in brightness (light curve), approach and recession of the surface (radial velocity curve), expansion and contraction in size (radius curve), and changes in color (color curve) for a representative cepheid variable.

radial velocity curve is seen to be a mirror image of the light curve. The star's surface is approaching us most rapidly when the brightness is at a maximum; it is receding most rapidly at minimum brightness.

Light curves of cepheids are not all the same, although great similarity does exist for stars of the same period (Figure 12-8). Irregular features such as humps in the light curve are thought to be associated with peculiarities in the atmospheric pulsations such as might be produced by shock waves. Usually the radial velocity curve shows irregularities corresponding to the humps in the light curves, lending strong support for the shock wave theory. Any given cepheid variable repeats its variation with extraordinary fidelity; most periods are constant to within 1/10,000 of 1%. The spectra of cepheids change with the variations in luminosity. At maximum brightness, a cepheid has a class F spectrum; at minimum it is class G for short-period variables, ranging to late class K for those with the longest periods.

Period Luminosity Relation. A remarkable and noteworthy feature of the cepheid variables is the relation between their periods and their mean luminosities. The longer the period of a cepheid variable, the more luminous it is. As important as the intrinsic brightness variation is in the study of stellar properties, the period-luminosity relation is far more significant to astronomy in general. It is one of the most powerful tools we possess for exploring the distance to remote stellar systems whose individual stars are otherwise inscrutable.

Two small, irregular galaxies near the south celestial pole, the large and small Magellanic clouds, each contain over 1000 observable variable stars. Studies of their light curves show most of them to be classical cepheid variables. Figure 12-9 is a schematic representation of the plot of the Magellanic cepheid's periods versus luminosities. All of the stars in each Magellanic cloud are at relatively the same distance from us. Thus the differences in luminosities, except for minor obscuration effects, are intrinsic.

Since the cepheids in the two galaxies are also at relatively the same distance from us, the period-luminosity relation can be validated once the mean distance is known. Moreover, the absolute luminosities of the cepheids are immediately available once the distance modulus, the constant difference between apparent and absolute luminosities for stars all at the same distance, is determined. The period-luminosity curves for cepheids in both Magellanic clouds practically coincide, showing that both galaxies are at essentially the same distance from earth. The problem is to find this distance.

Several hundred cepheids are known in our own galaxy. Unfortunately almost all are so remote that their parallaxes cannot be directly measured. Proper-motion studies yield only crude estimates; the rough values for the luminosities first determined in the 1920s and 1930s have since undergone considerable revision. Only in the early 1950s, with more powerful instruments available and with refined calibration techniques, could the luminosities of cepheid variables in our galaxy be reliably determined. With the assumption that cepheid variables of the same period are similar to each other wherever they are found, it was possible to establish a zero point, or reference brightness, for the period-luminosity relation. Imme-

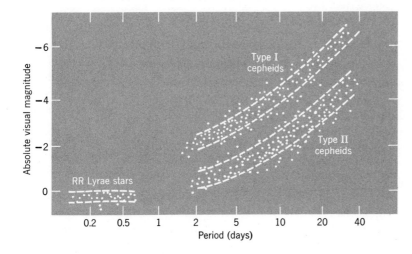

FIGURE 12-9 Period luminosity diagram.

diately, the distance to the Magellanic clouds was pegged at 170,000 LY. Since cepheid variables are extremely bright stars, they are visible for distances up to several million parsecs. Thus they serve as beacons and distance mileposts out beyond the Milky Way. All that is necessary in finding the distance to a remote stellar system is to identify a classical cepheid in its midst, determine its period, and obtain its absolute luminosity with the aid of the period-luminosity relation. The distance to the system is then computed by routine procedures.

<u>RR Lyrae Variables.</u> A group of short-period variables, extremely numerous in our own galaxy, takes its name from the prototype star, *RR Lyrae*. Large numbers of these stars are also found in globular clusters, but they are unknown elsewhere in the external galaxies save the two or three very nearest nonsymmetrical organizations. Even then, only a few RR Lyrae stars can be detected in the Magellanic clouds. In our own galaxy RR Lyrae variables are high-velocity stars; they are concentrated in toward the center of the galaxy and in the halo of stars surrounding the central portion.

The periods of the RR Lyrae stars range from a few hours to perhaps a day. Their light curves are not quite as regular as those of the classical cepheid variables. RR Lyrae stars increase in brightness quite abruptly; then there is a brief pause, or plateau, at maximum after which the decline to light minimum is quite slow. There is a decided hump, just before minimum is reached, in the light curve of many of the RR Lyrae stars (Figure 12-10). The overall light variation, or change in amplitude, is quite small averaging less than 1 magnitude, or only about a factor of 2. The spectral classes of the RR Lyrae range between A and F; all of them in a given globular cluster have about the same median apparent luminosity, and thus all have about the same absolute brightness.

The high velocities of the RR Lyrae variables facilitate studies of their proper motions. As a result their distances are more accurately deter-

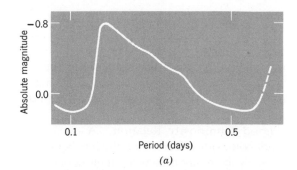

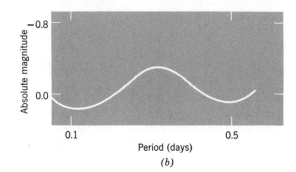

FIGURE 12-10 Light curves for *RR Lyrae* variables. (*a*) shows the light variation for a typical RR Lyrae star with a period ranging from one-half day to just over 1 day. RR Lyrae variables with periods of less than one-half day have more symmetrical light curves with smaller amplitudes (*b*).

mined than is the case for the cepheid variables. Their absolute luminosities exceed that of the sun by more than 100 times; yet they are some 100 times less luminous than the average for the cepheid variables. This explains why RR Lyrae stars are so rare in external systems; their intrinsic brightness is too feeble for them to be seen at great distances. Figure 12-9 shows the relationship between the luminosities of the RR Lyrae stars and the cepheid variables. The diagram also shows that there is no period-luminosity relation for the RR Lyrae stars.

<u>Type II Cepheids.</u> Another group of stars having periods comparable to the classical (often referred to as type I) cepheids, but whose spectra and light curves are distinctly different,

is that collectively referred to as type II cepheid variables. Two prototype stars lend their names to subclasses within the group: *RV Tauri* stars, with periods ranging from 30 to 100 days, and *W Virginis* stars with periods of from 10 to 30 days. Many of these stars occur in globular clusters in company with the RR Lyrae variables. Thus it is comparatively easy to find their absolute luminosities relative to the RR Lyrae stars. Type II cepheids average about 1.5 magnitudes, or 4 times fainter, period for period, than the classical cepheids. Accordingly these cepheids have their own period-luminosity relation, as shown in Figure 12-9, along with that for the type I cepheid variables.

The type II cepheids bear a superficial resemblance to the classical cepheids in that their spectra and colors are also those characteristic of stars of classes F and G. But there are significant differences. As the type II cepheids rise toward maximum luminosity, strong, bright hydrogen *emission* lines appear in their spectra. Moreover their radial velocity curves are discontinuous; during part of each period there are two distinct radial velocities. Even though the type II cepheids are pulsating stars, their atmospheres appear to be very extensive and violently turbulent, with the atmospheric gases simultaneously expanding outward and rushing inward.

The periods of the type II cepheids tend to be more irregular than those of type I, and their light curves are less precisely repeated in each cycle. Whereas all of the type I cepheids in our galaxy lie near its plane (Chapter 13), the galactic type II cepheids are all at large distances from the plane, forming a more or less spherical canopy or halo around the central portion, in much the same way as do the RR Lyrae stars. As we shall see in more detail in the next chapter, all of the type I cepheids are found in regions jointly occupied by substantial amounts of gas and dust. Type II cepheids exist only in globular clusters and in the regions of galaxies that are known to be dustfree. The discovery that cepheid variables fall into two distinct classes differing in luminosity by a factor of 4 had a profound effect on the estimates of the size of the universe, the distance to external groups of stars, and upon the theories concerning the origins and age of the universe. How the distance estimates were affected will become apparent in Chapter 13.

Long-Period Variables. The star, *Mira*, meaning *Wonderful One*, otherwise known as *o Ceti*, has been known to be a variable star for several centuries. Mira has a period of about 1 year, characteristic of the group known as *long-period variables*. These variable stars are easily recognized because they vary in light output from 16 to as much as 250 times in each cycle. A factor of 2½ is large for the other kinds of variables we have previously considered. Long-period variables have periods that range from 90 to 700 days and more. Their spectra resemble those of stellar classes M, S, R, and N. Many long-period variables display bright hydrogen emission lines during their cycle. This is a puzzling peculiarity because the production of the bright emission lines requires high temperatures, yet the molecular bands that appear in the late-type spectra are produced only at low temperatures. Clearly the spectra are composite, originating in two distinct zones in the stellar envelopes. Perhaps the molecular bands are atmospheric phenomena, and the hydrogen spectra are produced in an upper photospheric layer.

Other statistical differences exist between the long-period variables and the cepheids. The periods of the former are not as uniform nor do they repeat as regularly as do those of the cepheids. The light curves show distinct variations in amplitude from period to period. Radial velocity changes associated with the long-period variables are exceedingly small, and while there appears to be a period-luminosity relation for the long-period variables it is contrary to those for the cepheids. Within the long-period group the stars with shortest periods are brightest; those with longest periods are intrinsically fainter. The explanation for this apparent anomaly is interesting. The long-period variables are so cool (2000°K at light minimum and only 3000°K at maximum) that most of their energy output is in or near the infrared regions of the spectrum. The observed large fluctuations in luminosities are associated with the fraction of energy output that shifts between visible light and invisible infrared. Bolometric (rate of emission over the entire emission spectrum) measurements show that the *total* energy fluctuations are not much greater than those of the cepheids; it is just that the visible light variations are proportionally more noticeable.†

The space distribution of the long-period variables provides a clue as to the evolutionary role of all intrinsic variables. Because of their low mean luminosities, except for one or two individual stars in the Magellanic clouds, what we know about the long-period variables is gleaned from the galactic members. Those with longest periods occupy regions in space in association with type I cepheids. These are the dusty, nebulous regions of the galaxy. The long-period variables with shortest periods are typically associated with the RR Lyrae variables in the dust-free regions. Probably, then, the long-period variables represent both ends of a transition among all of the variables; that is, stellar variability may be symptomatic of a complete evolutionary cycle of stellar birth, death, and decay. These ideas will be explored in the last section of this chapter.

Other Variables. There are other miscellaneous types of intrinsic variables, all of which are no doubt of considerable importance in the studies of overall stellar behavior. Some are semiregular; others are somewhat periodic; still others are truly irregular in their light variations. We cannot dwell on the details here; those intrinsic variables with which we have

† A *bolometer* is an electronic instrument capable of measuring minute changes in radiant energy over the entire spectrum.

been preoccupied are essential in a stellar-life-history hypothesis which will culminate our investigation into the nature of individual stars. Yet, we should not leave the topic of intrinsic variables without mention of the following two important types of erratic variables.

Flare Stars. Flare stars belong to classes K or M and are main-sequence stars that suddenly brighten radically for a few minutes or so, then somewhat more slowly revert back to their normal luminosities. Presumably the outbursts are similar to solar flares: they are produced erratically; no periodicity is detectable. Solar flares would produce the same kind of variations in the sun's luminosity if the sun were not so brilliant. Statistical studies of the flare stars are impossible because there is no way that a systematic patrol of these transient outbursts can be established. The best we can do is initiate a program of *independent* surveillance for each suspected or known flare star. A representative light curve for a flare star is shown in Figure 12-11.

T Tauri stars exhibit erratic luminosity variations that may be partly extrinsic. These stars always tend to occur in groups, and they are always in regions where the concentration of dust and gas is heavy. Unquestionably the erratic behavior represents some sort of energy exchange between the stellar envelopes and the gas and dust in the space surrounding them. G. Herbig, O. Struve, and others have confirmed that the T Tauri stars are always in the midst of an unusually high concentration of stars. Possibly, as Herbig has indicated, the T Tauri variables are objects that have just completed, or are in the process of, condensing from the nebulosity in which they are immersed. If so, they are the youngest stars known.

Novae and Supernovae

A *nova* (L., *novus,* new) is a star, usually faint and insignificant, that suddenly and unpre-

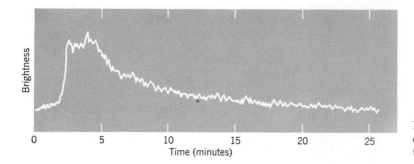

FIGURE 12-11 Light curve for flare star (see text).

dictably increases in luminosity by as much as 160,000 times in a matter of days. The drama is played out as the star gradually subsides somewhat irregularly, almost hesitantly, to its original luminosity. The decline stage may be completed in a few weeks, or it may last for months or years. Most novae are discovered on photographic plates where they appear almost as if by accident hence the apellation, new star. Most novae are one-time events in human experience, although a pair of groups, the *dwarf novae* and the *recurrent novae,* are known to repeat their

outbursts. The normal novae, to which we shall confine our attention, occur rather frequently in our galaxy; more than 100 have been discovered during the past century. Only a few novae (not to be confused with supernovae) are visible to the unaided eye. Novae are named after the constellation and the year in which they occur. Figure 12-12 shows Nova Herculis 1934 before and after the outburst.

We generally know little about the prenova star. Of the few whose existence have been verified on prenova plates, all are faint; no reli-

FIGURE 12-12 *Nova Herculis* of 1934 before and after outburst (Yerkes Observatory Photograph).

FIGURE 12-13 *Nova Persei* of 1901 photographed in 1949 (Courtesy Hale Observatories).

able spectra had been obtained for any of them. Most of the stars showed a slight tendency toward variability prior to the nova outburst. From what little we can learn from the color index of the prenova stars, all appear to be bluer, but with about the same luminosity, than the sun. Hence these stars appear on the H-R diagram *below* the main sequence. They are often called subdwarfs. On the average their luminosities, class for class, are from 2.5 to 6.0 times fainter than the corresponding main-sequence stars. Spectrally they cluster around class A in the H-R diagram. When a spectrum has been obtained after a nova outburst but before maximum luminosity is attained, the absorption lines are displaced strongly toward the blue end of the spectrum. Clearly, the atmospheric envelope is expanding rapidly. Astrophysicists are now generally agreed that the whole star does not expand; only a fairly thin layer of the photosphere, together with the atmosphere, is involved. In some cases a shell of material becomes detached from the star, expanding into nebulosity around the star's position (Figure 12-13). The expanding shell frequently displays a bright line emission spectrum; in many cases

the Doppler shift of the spectral lines reveals expansion velocities as great as 1000 km/sec, truly an explosive phenomenon. Except for an occasional shell of gas that may remain in the vicinity for up to several years, most novae revert to their original color and brightness. Little or no evidence remains to betray the paroxysm so recently suffered by the star (Figure 12-14).

The fact that normal novae have never been known to repeat does not preclude the possibility. If the period between repetitions exceeds the observational span of mankind, as seems likely, perhaps some recurrences among the normal novae may be a few thousands or tens of thousands of years in the offing. From an analysis of the light curve, the amount of energy released during a nova outburst can be readily computed. Then from the relation $E = mc^2$ (the Einstein mass-energy equivalence relation) we find that only about 1/100,000 of the total stellar mass need be converted to energy in order to account for the luminosity increase. If, as we are about to investigate, a white-dwarf star is the end process for the nova phase, then the average star must go through something like 10,000 nova type paroxysms in order to achieve the limiting mass.

Supernovae are stars that exhibit sudden, gigantic, cataclysmic increases in stellar luminosities, in connection with which a star will

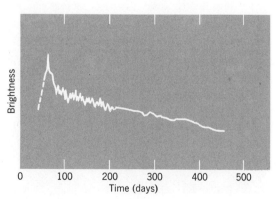

FIGURE 12-14 Light curve for *Nova Persei* of 1901 (Figure 12-13).

FIGURE 12-15 Supernova of 1959 in the galaxy NGC 7331 before and during the maximum (Courtesy Lick Observatory).

lose on the average between 1 and 10% of its mass. Some stars may possibly be annihilated in the catastrophe. Whereas a nova may increase in brilliance on the order of 150,000 times, a supernova ranges from 100 million to 1 billion times the star's normal brightness. Some observed supernovae at maximum exceed the collective luminosities of all of the stars in the galaxy of which they are members (Figure 12-15). According to one estimate, a supernova at maximum releases more energy in 1 second than the sun emits in a 10 year period!

One hundred or so supernovae have been observed; all but three of them occurred in galaxies external to our Milky Way. *SN Tauri 1054* (a supernova carries the designation SN followed by the Latin genitive of the constellation) was chronicled in China and Japan in 1054 A.D. Apparently it was overlooked in Western Civilization. SN Tauri is now the beautiful Crab nebula (Figure 12-16). The supervnova at maximum was reported to be 40 times brighter than Sirius, and it was visible in broad daylight. Tycho's star, *SN Cassiopeiae 1572*, was intensively studied by Tycho Brahe. This supernova was also a full daylight object, visually brighter than the planet Venus. The

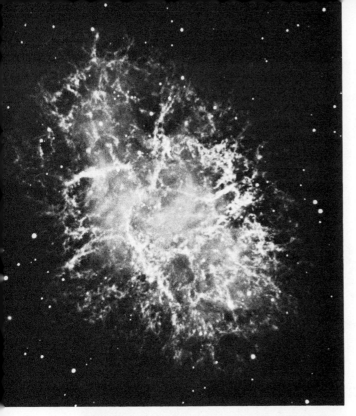

FIGURE 12-16 The Crab nebula (M1) in Taurus, remnant of the supernova of 1054 A.D. (Courtesy Lick Observatory).

Within the faint remaining wisps of nebulosity marking the locations of the supernovae of 1572 and 1604, as well as within the Crab nebula, radio emissions are still strong. In the Crab nebula, x-ray emissions accompany the radio noise. Rapid variations in both the light and the radio emissions from the core of the stellar remnant identify it as a *pulsar* (Chapter 13). Other sources of radio emissions—*radio stars,* or better, *radio galaxies,* but from which no light is emitted—together with other established pulsars, may all be remnants of ancient, otherwise undetected and unknown supernovae.

Stellar Evolution

When we marshal all of the evidence about stars that we have thus far generated, an intriguing possibility emerges; in terms of broad generalizations, probability may be a better word. One way of visualizing our effort is in terms of a stylized H-R diagram (Figure 12-18) showing all of the following: the main sequence and the giant branch; the supergiants, the T Tauri stars, and the RR Lyrae stars; the cepheid variables, the RV Tauri and W Virginis objects; the novae, supernovae, the white dwarfs, and the prenova subdwarfs. With the aid of the diagram let us speculate on the relationship between the individual and group characteristics of the stars, paying particular attention to associations between the various groups and to the role of the background interstellar dust and gas. When we do so, a definite pattern emerges; it appears that the H-R diagram graphically depicts the life history, the evolutionary cycle, of the stars.

Two approaches of analysis are open to us. We can take a purely theoretical tack which starts with certain hypotheses concerning the initial composition and structure of the stellar interiors. We then proceed with very complex calculations requiring the use of modern high speed computers, passing from stellar model to stellar model through all of the stages of a star's

third local supernova recorded in historical times, *SN Ophiuchi 1604,* is known as Kepler's star. Circumstances surrounding its discovery led Kepler to postulate that a supernova is a logical candidate for the explanation of the biblical Star of Bethlehem. Modern scholars, however, find no mention in the records outside of Christendom of a single nova (or supernova as well) during the period from 134 B.C. to 123 A.D. For other reasons, beyond the scope of this course, the supernova explanation seems highly unlikely.

Spectral studies of supernovae, in the few cases in which they are accessible, show extremely broadened emission lines. Evidently the ejected material is expanding at speeds up to 12,000 km/sec. After more than 900 years the Crab nebula is still expanding at a rate of 1300 km/sec in this remnant of a supernova.

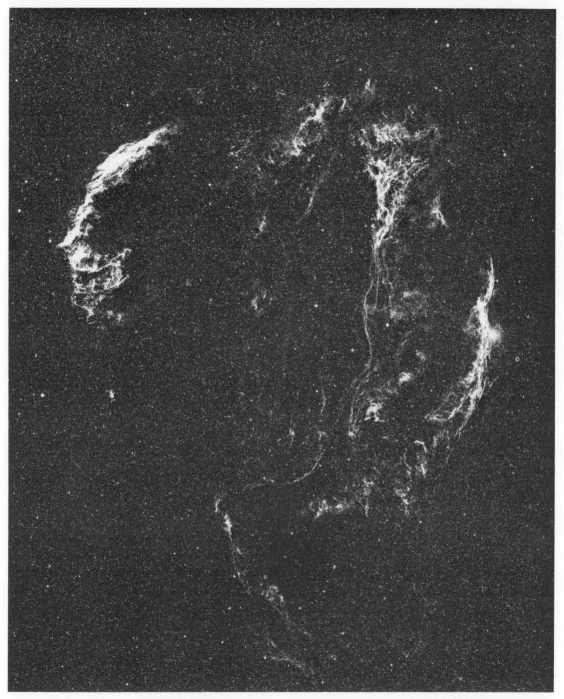

FIGURE 12-17 The Veil nebula in the constellation Cygnus. This filamentary nebula is believed to be the remnant of an ancient supernova (Courtesy Hale Observatories).

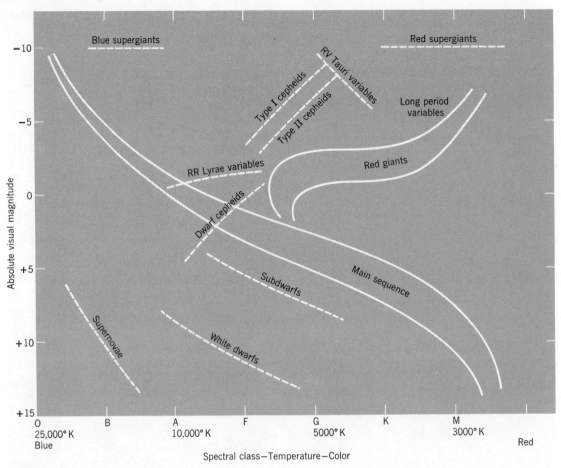

FIGURE 12-18 Schematic H-R diagram showing the relationships among all of the stellar types.

birth, life, decay, and death. The other alternative is an observational approach, based upon the assumptions that the variety of stellar objects we observe represent various stages in stellar evolution.

Neither approach is satisfactory by itself. Any theory needs observational evidence for its verification and support (or rejection). In astronomy the difficulty with a purely observational approach is the time scale, one that far exceeds the span of human experience. We never actually see a star evolving or, for that matter, dying. Our best approach is to combine

theory and observation, leaning first on one, then the other, for support, to see if the various kinds of stars we observe represent "snapshots" of the aging process. If this proves satisfactory, then the H-R diagram is indeed a composite picture of the whole life history of the stars. Let us see where the evidence takes us.

Like the sun, most other stars are in a state of equilibrium. In any one of them the surface temperature remains steady; the rate of energy output is constant. The star's luminosity is unchanging; this, together with the above facts, implies an invarying stellar radius. The energy release process and its development has already been detailed in Chapter 10 (p. 262 ff). A model of historical interest, attempting to account for the constancy of the sun's parameters and purporting to yield an accurate estimate of how long the sun has been shining at an undiminished rate, was put forward by Helmholtz and Lord Kelvin in 1854. Helmholtz showed that, for the sun, the amount of gravitational shrinking necessary for perpetuating the observed rate of solar emission would be so minute as to go undetected, at least over the span of recorded history. According to Helmholtz the sun has been shining thus far approximately 100 million years.

That the model was erroneous became obvious during the early part of this century, when geological evidence showed indisputably that the earth, and therefore the sun, was at least several billion years old. The Helmholtz contraction, while accurate for the initial phase of stellar radiation, was grossly inadequate over the demonstrated time scale. It had to be replaced by the currently accepted model that postulates nuclear transformations as the long-term energy source.

One other factor strongly influences the choice of a stellar model: the chemical composition of the prestar gas and dust cloud. The basic hydrogen-helium transformation is adequate in accounting for the observed energy release in the main-sequence stars. But the presence of absorption lines associated with metals and other heavy elements in the spectra of the main-sequence stars is paradoxical; the temperature in the stellar cores during the hydrogen-helium stage is simply not high enough for the formation of other substances. These apparently anomalous elements must have been present in the prestar nebulosity. If so, the solid matter in these clouds must represent debris from earlier stellar cataclysms: either material ejected from the stars or debris from total stellar annihilation. Such stars must have developed internal temperatures high enough for heavy element formation. As we shall presently see, this conclusion is consistent with

the evolution hypothesis. Thus, *the existing main-sequence stars must be second, or third, or later generation objects!*

Early Stages. No star can be infinitely old; the fuel supporting *any* conceivable energy release process must ultimately be exhausted. We can estimate from the observed mass and luminosity how long a star may expect to remain on the main sequence before the hydrogen fuel in the core is exhausted. For late-type stars similar to the sun, 10 billion years is not unreasonable; for the highly luminous class O and B stars, the life expectancy on the main sequence is a few tens, or hundreds, of millions of years. It is clear then, that we see millions of stars that were born long after the sun had settled into comfortable middle age. Probably the sequence begins in the following way.

A local condensation is initiated by random turbulence in one of the more dense portions of interstellar matter. The condensation is gravitationally induced and maintained. Depending upon the initial motions of the condensation, rotation may occur. As the gravitational collapse gains momentum, additional material along the boundaries of the condensation is attracted toward the gravitational center. How much is so attracted depends upon both the initial density of the dust cloud and the speed of the gravitational collapse.

Eventually, the heat generated with the build-up of internal pressure cannot escape from the condensing center as fast as the thermal energy is produced. As a result the internal temperature becomes high enough to vaporize solid matter and ionize the constituent gases. Ionization increases the opacity of the condensing region, and this further retards the rate of energy loss. With the accelerated temperature rise, the pressure and density in the local condensation rapidly increase and finally the gravitational collapse is halted. A *protostar* is formed; hydrostatic equilibrium is achieved. It is probable that T Tauri stars are objects of low mass that are in the final stages of achieving this steady state. The energy emission is a consequence solely of gravitational heating.

All is not ended, however. A slow, almost imperceptible gravitational shrinking of the protostar sets in. The process is so gradual that hydrostatic balance is not disturbed. Nevertheless, the slow release of gravitational potential energy is just enough to maintain the observed rate of energy emission. We have here a true Helmholtz contraction, but it is just an interim process, not an end state. If the protostar is below a critical value, believed to be in the neighborhood of 9% of the sun's mass, the central temperature never rises high enough to precipitate nuclear reactions. The protostar wastes slowly away with a slight rise in internal temperature until the stellar matter becomes so dense that electrons are totally stripped from their atoms, becoming what is known as a *degenerate gas*. The protostar has evolved to a white dwarf.

For protostars of larger mass, nuclear ignition finally occurs after a few million years or so of the Helmholtz contraction phase. The luminosity of the star and its size become stable; the star arrives on the main sequence at a position determined by its initial mass. Of course, the term arrives does not refer to a celestial journey. It means simply that the star now has radiation characteristics and a spectrum classifying it as a main-sequence object. It will remain on the main sequence until the hydrogen fuel in the stellar core is exhausted. As we have already learned, for a star of the sun's mass this will take several billion years; for the extremely luminous early-type stars, the hydrogen is exhausted after a few million years.

The Main-Sequence Stage. During the slow, even, hydrogen-helium-fusion stage in the core of the star, the star is extremely stable; no perceptible changes in luminosity of size or temperature occur. As the hydrogen in the core is gradually depleted and replaced by helium, the temperature rises slowly; along with it the star's luminosity increases slightly, but the star

remains on the main sequence. Finally, however, a helium core is all that remains. The relatively scant amount of hydrogen fuel that is consumed as the nuclear fires die out is insufficient to perpetuate the previous rate of energy release. Therefore, gravitational contraction once again commences. The energy source during this phase comes from hydrogen-helium conversion around the periphery of the core and from gravitational heating. The star begins a substantial readjustment in its internal structure; the temperature in the core rises dramatically, and the increased radiation pressure forces the outer layers of the star outward; that is, the star expands enormously. But the outer layers of the star are cooled by the expansion. Thus the star moves off the main sequence and becomes a red giant (or a supergiant as the case may be). The greatly increased pressure and temperature in the helium core initiate a new thermonuclear reaction in which helium is fused into a variety of heavier elements.

The Variable Stage. Considerable uncertainty exists concerning the details of the stellar evolution beyond the giant stage. Probably a variety of processes are available, each determined, among other things, by the mass of the star, the relative length of time it spent in previous evolutionary transitions, and the initial chemical composition. Theoretically, as the helium in the core becomes depleted and is replaced by carbon and other similar substances, there is still a surrounding shell in which helium fusion continues. It in turn is enclosed by a thin stellar layer where hydrogen fusion is still eating its way toward the surface. Complex contractions and expansions associated with variations in energy release begin as readjustment takes place in the various internal layers. Presumably the stars midway up the main sequence evolve to red giants; oscillations in the giant stars lead to the formation of RR Lyrae variables. Stars of higher luminosity evolve to supergiants; oscillations in these stars result in the cepheid variables. Differences in the initial composition and the locale of the prestar contraction may explain the differences between type I and type II cepheids. In any event, instability may become so pronounced that material is suddenly ejected from the star's outer shell. Mass and chemical composition may be the deciding factors as to whether a star becomes a dwarf nova, a recurrent nova, a normal nova, a supernova, or a shell star (Chapter 13).

The White-Dwarf Finale. Whatever the form of the internal readjustment, the nuclear and thermonuclear fuels are eventually exhausted. Since the beginning of the cycle, gravitational shrinking was first in the forefront, then in the wings; it now resumes center stage in the drama. The star collapses slowly at first, then more rapidly as the internal fires die out. Gravitational heating again becomes the principal energy source. By this time, however, the opacity of the outer layers has increased so much that the escape of energy into space is severely curtailed. Even though the temperature remains relatively high, the star's luminosity becomes feeble. As previously noted, the surviving atomic nuclei are embedded in an electron cloud called a degenerate gas. Each nucleus occupies vastly smaller space than it did before it was ionized. Also, limits exist as to how close electrons can be forced together. Even though the new white dwarf is extremely dense and compact, its interior is still largely empty space; the distance between the particles far exceeds their intrinsic diameters.

Stars of one, or less than one, solar mass will contract to about the earth's size. In this case the mean density may exceed 1 million times that of water. One cm³ of the material from the star's core would weigh 40 tons at the earth's surface! The stellar atmosphere is compressed under the influence of the high surface gravity to a thickness of only a few meters; from the photosphere to the top of the atmosphere the density differs by a factor of 1 billion.

Theories concerning the structure and behavior of white dwarfs are complex since they involve the postulates of general and spe-

cial relativity (Chapter 14). We can here do no more than acknowledge a few of the ideas. The structure of a white dwarf, including its radius, is determined ultimately by its mass. For example, consider a star that has achieved equilibrium between gravity and the internal pressure exerted by the degenerate gas. The radius of the star will be uniquely determined by the mass. Now suppose additional mass could be added. The increase in gravity assumes command, and the stellar material is crushed into a considerably smaller volume, more than offsetting the added bulk. According to S. Chandrasekhar, American astrophysicist, the upper mass limit for a stable electron-degenerate configuration is 1.2 solar masses. If the mass exceeds this value, the radius of the star theoretically vanishes; the star has a zero point volume (Figure 12-19). Moreover, the nearer a white

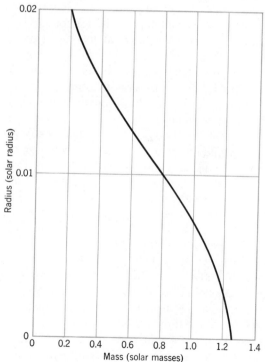

FIGURE 12-19 Theoretical relation between radius and mass of white-dwarf stars (After Abell).

dwarf is to the mass limit, the smaller is its radius, and the stronger is the surface gravity; the latter becomes infinite at the mass limit. Relativity theory predicts that the gravitational red shift will be so extreme that the whole spectrum will be shifted out of the visible range, meaning that the star becomes invisible. Obviously, stars at this extreme are not known.

But most of the known stars have masses exceeding that of the sun. All of these stars have comparatively short life expectancies. Somehow these stars with excess mass must get rid of the excess in order to complete their destiny in the white-dwarf stage. Otherwise the known, sizable white-dwarf population cannot be accounted for. Perhaps the novae, the supernovae, and the shell stars represent mechanisms for getting rid of the excess mass; we have no way of telling for sure.

Barring a mass in excess of the mass limit, a balance between the pressure exerted by the degenerate gas and gravity exists. Gravitational contraction is halted; all energy sources become extinct. The thermal energy loss to space, albeit slow because of the extreme opacity of the stellar envelope, is nevertheless inexorable. The star coasts to oblivion; it will eventually cease to shine, becoming a dark, cold, degenerate body, a "black hole" in space, with no hope of resurrection. Thus we have come full circle. We started with a gravitational collapse, and we end with one. One can look upon gravitational collapse as both the midwife and the undertaker in the stellar life cycle.

The Neutron Star Alternative. Before leaving the subject of stellar evolution, we should consider an alternative end product to gravitational collapse. This is the creation of a *neutron* star, at present only postulated and not yet observed directly. Neutron stars are discussed principally in connection with energy sources for pulsars and *quasistellar* objects (Chapter 14). Theoretically a neutron star is composed entirely of neutrons. Somewhere along the line in the final collapse of a star

whose mass exceeds 1.2 solar masses, the electrons, under tremendous pressure (such as could be produced by a supernova explosion), are forced *into* the atomic nuclei where they combine with protons on a one-to-one correspondence. The product is a neutron. Neutrons behave like degenerate electrons in that they too can become degenerate if crowded into a sufficiently small volume. Thus neutron stars are much smaller than white dwarfs because the space requirement for the now nonexistent electrons is abolished. A neutron star of 1 solar mass has a theoretical radius of only about 10 km.

The remnants of neutron stars after the nova phase are theoretically nearly invisible. Most of the radiation that can escape is in the form of X-rays; perhaps neutron stars are the as yet unexplained source of cosmic X-rays. Neutron stars are thus in general not observable from earth, except perhaps in terms of induced effects on surrounding systems. In any event their ultimate stage is a hyperdense chunk of cold, hard matter, with only one hope for reincarnation: an oscillating universe (Chapter 14).

Black Holes in Space. We should not leave the subject of neutron stars and pulsars without considering one of the consequences for theoretical physics of Einstein's theory of general relativity (Chapter 14). Under certain circumstances the radius of a neutron star may not stabilize at 10 km or so. Instead, the star may shrink in size in a prodigious implosion.

When the star's radius becomes about 3 km an extraordinary sequence of events is predicted by the theory of general relativity. We know from Newton's universal law of gravitation that the strength of the gravitational field of a star at any external point is governed both by the distance to the point and the mass of the star (Chapter 4). General relativity predicts that the propagation path of electromagnetic radiation is curved in the immediate vicinity of a large mass concentration. This phenomenon is the basis for describing the curvature of space, discussed in more detail in Chapter 14. Moreover, light photons emmanating from the surface of a neutron star are gravitationally retarded.

The combination of space curvature and photon retardation dominates the radiation characteristics of a neutron star whose mass exceeds 1.2 solar masses and which has shrunk to a radius of 3 km. The gravitational force increases without limit, and it is now stronger than any other known force in the universe, including the force that holds the nucleus of atoms together. Meanwhile, coinciding with the increase in gravity, the volume occupied by the star's mass shrinks to the vanishing point, literally squeezing the mass out of existence. Space becomes a closed curve around the point, called a *singular point;* literally, a hole in space! Both the prodigious gravitational field and the curvature of space operate to prevent any radiation from escaping from the hole. Thus a black hole in space is born.

Theoretical physics at present provides no other alternative to this hypothesis for black holes. But intuitively, such a demise for the enormous quantities of mass and radiant energy is repugnant to common sense. Singularities in theory, with their disagreeable consequences, have occurred before. Quantum theory was born when the atomic theory of Rutherford led to the singularity dilemma. Perhaps the black holes signal that we must discard, or modify, present theory in favor of a new one that deals with unknown forces over short distance that can withstand the crush of gravity, no matter how great it becomes.

Recently Igor Novikov at the Institute of Applied Mathematics in Moscow, and independently, James Bardeen at the University of Washington have presented a new mathematical example that does not require matter to vanish at a singular point. They propose a topological hole which does not trap matter, but which allows it to flow through, to emerge like a bubbling spring in some other region of our universe or in some other unsuspected universe. It is indeed an intriguing idea.

Addendum: The Evidence of Black Holes in Space.

In 1972 the first concrete evidence of the existence of black holes in space was discovered: the X-ray source, *Cygnus X-1.* According to the theories of gravitational collapse, consistent with Einstein's theory of general relativity, a sufficiently massive star (at least 3 times more massive than the sun) can shrink catastrophically through its *gravitational radius* without exploding as a supernova or halting in the pulsar or neutron star phases. The gravitational radius is a theoretically critical limit below which *no* conceivable forces could halt the crushing of matter out of existence. For example, the sun's gravitational radius is about 3 km; that for the universe (as we know it) is about 10 billion light-years.

There are in theory two critical stages through which a collapsing star passes on its way to permanent oblivion. First, at 1.5 times the gravitational radius, all photons leaving the surface in any direction except tangent to the surface (perpendicular to the star's radius at the point of departure) will escape forever. The others are trapped in a sperical cloud from which they slowly leak indefinitely. These photons could be the source of the radio noise.

The second critical stage is reached at the instant just prior to the time that the collapse passes through the gravitational radius. At this point all photons except those leaving the surface perpendicularly are pulled back into the imminent black hole; the others are trapped in a second cloud just immediately outside the gravitational radius. From here these photons in turn slowly leak away. The additional gravitational heating produced by the collapse from 1.5 to 1.0 times the gravitational radius could produce photons with X-ray energies. Clearly, once the limiting value is reached, no other photons can escape; thus observable evidence directly from the black hole is nonexistent.

Current theory suggests that the best way to observe a black hole in space is to examine close binary systems. The visible orbiting component provides information concerning the mass of the "extinguished" primary. If the mass is large enough, and if both weak radio signals and X-rays originate from the system, then probably the invisible component is a black hole.

If it can in the future be established that an observed blue supergiant star very near Cygnus X-1 is gravitationally linked to it as part of a binary system, then in all probability Cygnus X-1 *is* a black hole in space.

The information about Cygnus X-1 cited above was made from data secured from the *Uhuru* earth-orbiting satellite which is equipped with an X-ray telescope. Perhaps, just perhaps, by the time this textbook reaches the market, the proof of the existence of black holes in space will be conclusive.

SUMMARY

Among the most important of the astronomical tools are stellar spectra. Spectra can be differentiated into several distinct classes even though there are subtle differences within the classes. Each class relates to the mass of the star, the rotation of the star, the gravitational field and the radius of the star, and the specific temperature of the star's photosphere.

Not only does a star's spectrum give information about the radiation conditions at the surface, but it also yields information concerning the nuclear processes occurring deep within the star. The spectrum also yields valuable information about the chemical composition of and the relative abundances of the elements or compounds in the star's atmosphere.

The Hertzsprung-Russell diagram is a graphical depiction of the relationships between absolute brightness, color, temperature, and spectral class of stars. As such it is a statistical instrument derived from the characteristics of stars at known distances that are radiating essentially like blackbodies. The stars fall statistically into two broad bands on the H-R diagram; each band marks similarities in radiation characteristics. The *main sequence* iden-

tifies stars in their first-phase radiation state. The *giant branch* statistically lists stars in second and subsequent radiation phases. The great value of the H-R diagram rests in its ability to classify stellar luminosities and temperatures when the stars' distances are not directly determinable. From the absolute luminosity determination made with the aid of the H-R diagram, a star's distance can be calculated from the measured apparent brightness. The method is known as *spectroscopic parallax*.

Statistically, all stars have similarities. All show continuous background emission spectra; all have atmospheres; all have photospheres that are essentially opaque. The principal differences between stars are in mass, in temperature, in intrinsic brightness, and in size. Presumably they also differ in age and in maturity relative to their radiation life history.

Many stars are intrinsically variable in their radiation output. Some are regularly periodic, others vary more or less irregularly, although they are still periodic in their energy output. Still other stars vary cataclysmically, usually only once during historical intervals.

The cepheid variables are most valuable as a stellar resource. The relationship between their absolute brightnesses and their periods can be shown through a graphical device known as the *period-luminosity* relation. Cepheids are visible to great distances; knowing their periods allows us to calculate distances to star systems that contain them and are otherwise inscrutable. Cepheid variables occur in two luminosity classes that differ by 1.5 magnitudes. Star for star, type I cepheids are 4 times brighter than their type II counterparts.

There are several other types of variable stars, each having unique characteristics. Each type provides valuable information in its own way. RR Lyrae variables, for example, occur mainly in globular clusters and in the galactic halo. Because of their extremely high luminosity they can be seen from great distances. Thus they make excellent beacons in the proper-motion studies of the globular clusters.

The evidence, provided by the differences in and similarities among stellar properties, suggests that we are viewing stars collectively at different stages in a general evolutionary pattern. Stars appear to condense from galactic nebulosity, radiate for a time as main-sequence stars, evolve through one of several patterns to giant stars, and then evolve to regular or irregular periodic variables. Eventually they reach a terminal phase as a white dwarfs.

Questions for Review

1. What is the relationship among the spectral class, the temperature, and the absolute luminosity of a main-sequence star on the Hertzsprung-Russell diagram?

2. Answer *true* or *false*: Star A is redder than star B. Then it is not possible for the absolute luminosity of star A to exceed that of star B. Explain your answer.

3. Given: Star A; distance, 16 LY; apparent luminosity, 1 unit; color, red. Star B; distance, 4 LY; apparent luminosity, 1 unit; color, white. (a) Which star has the greatest absolute luminosity? (b) Which is the hotter star? (c) Which is the largest star?

4. Sketch from memory a simple line drawing of an H-R diagram, with spectral class as the horizontal coordinate and absolute luminosity in solar units as the vertical coordinate. Show the relative placement of main-sequence A, G, and M stars as well as white dwarfs, red giants, and red supergiants. Practice until you can do it on demand.

5. How can a red-giant star be distinguished from a main-sequence star of nominally the same spectral class?

6. What is there about the spectra of white-dwarf stars that identifies them as such?

7. Discuss the value of the H-R diagram in astrophysical research.

8. What limits the determination of stellar distances by the method of spectroscopic parallax?

9. What method is routinely employed in the discovery of intrinsic variables (stars)? Briefly describe the method.

10. What contribution do cepheid variable stars as a class make toward determination of distances in the universe?

11. What is the relationship among the period, the absolute luminosity, and the mass of cepheid variables?

12. What evidence suggests an evolutionary cycle in the life history of stars?

Prologue

There are a few exotic stellar types that do not seem to fit into a definite evolutionary scheme. Peculiarities in their spectra give most of them away, although some stars with detached atmospheres can be seen and identified visually with the aid of a telescope. Of all of the exotics, the pulsars are the most enigmatic and astonishing.

We must turn our attention beyond these few irrascibles to the details of the organized star groups: the clusters, the galaxies, and the clusters of galaxies. This is the last step in preparing for our glimpse of the ultimate nature of the universe. How and why stars congregate in crowds as they do, how these crowds mill about and surge forth into unoccupied regions of space, and how separate and distinct crowds can collide and pass through each other without losing their identity, yes, without even a single bruising encounter between stars, these facts command part of our attention in this chapter.

Content and Organization of the Universe

13

The evolutionary hypothesis for the life history of a star appears to adequately explain most of a bewildering variety of stellar objects. It is neither complete nor self-contained, however. There are a few extraordinary objects (to be considered next) that do not seem to fit into a stellar life-cycle scheme. Neither does such a scheme provide an explanation for observed groupings of stars. Nor can the life cycle hypothesis be put to an observational test; the time scale is too vast. In all likelihood man will have made his appearance on earth, played out his role, and have long vanished from the scene in much less time than it takes a local nebular condensation to evolve into a star, complete its appointed time on the main sequence, and emigrate to the giant stage.

In spite of this insurmountable handicap, we are entitled to a reasonably high degree of confidence in our astronomical surmisings. Unlike that of the ancient astrologers, our work represents the healthy, vigorous offspring of a successful marriage between astrophysical theory and laboratory experiment. There seems no reason (at present) to reject the theory that stars *are* born, reach maturity, grow old, and finally die. Indeed, the accumulating evidence is progressively strengthening our position.

Let us, then, turn to the final stage of our investigation of the physical nature of the universe and see what we can learn about the organization of stars and of interstellar material, of clusters and associations and galaxies, and of clusters of galaxies that populate the universe as far out as we can see in space, and time. In the process we can acknowledge the few remaining astronomical objects for which we could find no certain place in the evolutionary scheme. Some of these objects are very peculiar; some are completely mysterious; all are exceedingly interesting.

Stars with Peculiar Spectra

Almost all of what we know about the peculiarities of unusual stars is deduced from their spectra. Fortunately, the components of some of the eclipsing binary systems display these same peculiarities. Thus knowledge of the systematic behavior of eclipsing binaries can be employed in verifying the truth of the deduced peculiar behavior. In the sense that we employ the term, *peculiar* refers to intrinsic stellar behavior that is statistically rare among the stars in general.

Among the binaries we find rapidly rotating stars and stars with deep, extensive atmospheres. We find shell stars, stars that have detached or semidetached atmospheres. Some components of binary systems appear to be exchanging atmospheric mass with each other. A few class O and B stars, having otherwise normal spectra, show considerably broadened and weakened spectral absorption lines. How do we interpret this phenomenon? We can rule out gravitational effects. These would produce mainly Doppler displacement of the spectral lines toward longer wavelengths, not broadening of the lines. This leaves two other possibilities: either the stellar atmosphere is in a state of violent convective turbulence, or else the star is in rapid rotation. Here eclipsing binaries come to our rescue; at least, where Doppler broadening is observed in binary stars,

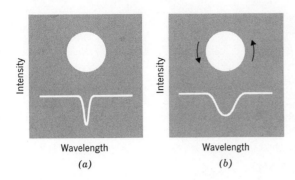

FIGURE 13-1 Rotational broadening of an absorption line. Normal intensity profile (*a*); rotationally broadened profile (*b*). The reference intensity (horizontal portion of trace) corresponds to background continuous emission spectrum.

rapid rotation is the most likely explanation. Let us consider a rapidly rotating star oriented in such a way that we are not looking directly down on one of its poles. Then part of the stellar surface is being carried rapidly toward us; this portion will produce spectral lines shifted toward the blue end of the spectrum. Another portion is moving directly across the line of sight and will show no Doppler displacement. Finally, the portion of the surface being carried away from us will produce lines displaced toward the red end of the spectrum. Now since the star's surface is not discontinuous, the observed spectral lines will result from the integrated effect of all of the discrete surface motion. Figure 13-1 shows how the lines are broadened from the combined Doppler displacements. Analyses of the amount of broadening show that some stars have equatorial surface speeds in excess of 177 km/sec, nearly 90 times that of the sun.

Let us now consider a fast rotating star that is a member of an eclipsing binary system. We can construct a radial velocity curve and a correlated light curve for the component in question. The radial velocity curve is related to the motion of the center of the star. When the star enters eclipse, the light from that portion of the surface having a rotational velocity of approach toward the observer is first to be cut off. The spectral lines then show an exaggerated velocity of recession, composed of the rotational recession of that part of the surface not yet obscured and the recessional motion of the

star itself. As the star emerges from eclipse the effect is reversed, producing a spurious velocity of approach. Ideally the radial velocity curve is distorted as shown in Figure 13-2. This distortion establishes with certainty the fact that the star is rapidly rotating. Some intriguing questions arise from the observed phenomenon. Why do some stars have such extremely high rotational speeds? And why does high-speed rotation appear to be confined to main-sequence O and B stars? We do not yet have answers to these questions.

Extensive stellar atmospheres can be detected from progressive changes in the quality and composition of the absorption lines as the light from the binary component being eclipsed has to pass through successively deeper and denser layers of its companion's atmosphere. In some cases it is possible to work out in some detail the composition and structure of the atmospheric layers, giving us additional insight into the stellar natures. One interesting, peculiar system deserves mention. The periodicity of the light curve shows the star *ε Aurigae* to be a member of an eclipsing binary system. But there is only *one* light minimum per cycle. After consideration of all of the possibilities revealed by analysis of the light curve and changes in the spectrum, there is only one possible answer for the enigma. The fainter component of the system is so cool that it makes no measurable contribution to the total luminosity of the system. Furthermore, the eclipse minimum comes from reduction in the

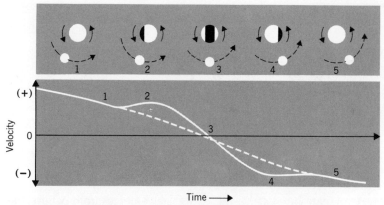

FIGURE 13-2 Radial velocity curve for primary component of an eclipsing binary system showing rotational distortion effect. In 2 the companion star eclipses the approaching surface of the primary, thereby producing an augmented velocity of recession. In 3, corresponding to the primary minimum, rotational effects cancel. In 4 the companion star eclipses the receding surface of the primary; the velocity of approach is exaggerated. Broken line 1–5 represents a portion of the radial velocity curve in the absence of rotation. Velocities of recession are taken as positive; velocities of approach are negative.

light of the primary as it passes behind and *shines through the secondary!* The secondary component is over 2000 times the diameter of the sun. From knowledge of its size and mass, we deduce an average density for it that is astonishingly low. Compared to laboratory standards, one might describe the star as being a red-hot vacuum!

Some binary systems have components with gaseous streams and rings. Evidently such detached material originated during atmospheric convulsions associated with evolutionary expansion in the giant stage. If the binary components are very close to each other, a stellar diameter or so, theory predicts and spectral evidence substantiates the existence of atmospheric mass exchange between the components. In the case of β *Lyrae*, a close binary system consisting of a giant B9 component and a lesser giant F component, atmospheric gases flow from the B9 star to the F component. Spectral evidence together with a regular annual decrease in the period of the system

suggests strongly that some of the atmospheric material is lost to the system and spirals outward to form a shell that encloses both components (Figure 13-3).

Shell Stars and Planetary Nebulae. Some class O and B stars appear to have detached

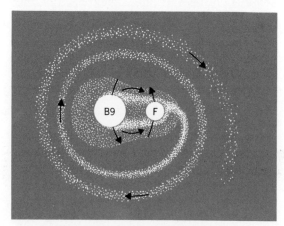

FIGURE 13-3 Deduced atmospheric configuration attending the close binary system β *Lyrae* (see text) (After Struve).

atmospheres, called shells, that are rapidly expanding into space. Several hundred representatives, called *Wolf-Rayet* stars, are known. Wolf-Rayet stars are extremely hot O-type stars that are ejecting shells at speeds from 1000 to 3000 km/sec. Obviously we cannot see the shells as such; evidence for their existence comes solely from peculiarities in the spectra of the parent stars. In general the spectra have weak, broadened absorption lines, upon which are superimposed less broadened emission lines. In turn, the emission lines are bisected by sharp, narrow absorption lines. We shall find it instructive to see how the existence of shells is postulated from the spectral evidence.

First, the general character of the spectra and the positions of the intensity maxima show that the shell stars have surface temperatures ranging from 80,000°K to as much as 200,000°K. These stars radiate very intensely in the ultraviolet. The emitted uV photons are absorbed by gaseous atoms that are too cool to give off visible thermal radiation of their own. Reemission of the absorbed energy occurs in the form of lower energy photons, some of which are in the visible light range. Thus the gas is said to *fluoresce,* that is, give off light which is not associated with the thermodynamic properties of the gas. We popularly refer to this type of illumination as cold light, an apellation not strictly correct. The only way an extended body of gas associated with a star could be relatively cold, yet subject to fluorescence by uV radiation, is to be far removed from the stellar photosphere.

Now let us consider the situation illustrated in Figure 13-4. The upper part of the illustration consists of a plan view of the star and its shell (not drawn to scale). The lower part of the figure is a *profile* in intensity of one of the spectral lines. The weak, broadened absorption line, a–a', is produced by the rapidly rotating star. The somewhat narrower emission line, e–e', is produced by the fluorescent shell of gas. In agreement with Kepler's laws, the shell is rotating more slowly than the central star,

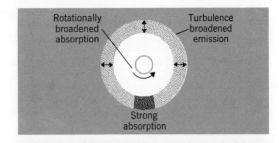

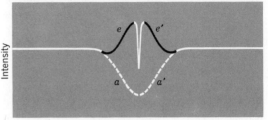

FIGURE 13-4   Spectrum line profile for a typical shell star. a–a': rotational broadened absorption line profile produced by star; e–e': turbulence broadened emission line produced by detached atmosphere (see text).

hence its emission lines are less broadened. The narrow central absorption line is produced by that portion of the shell directly between the observer and the central star. This line feature is not rotationally broadened since the portions of the system producing it have no radial velocity component. Moreover, the atmospheric shell is cooler than the photosphere of the central star, hence absorption occurs as with any cooler atmosphere. Thus we can account for all of the spectral features of shell stars. However, one problem is left unsolved. Why should shells be formed almost exclusively in Class O and B stars? One explanation, as yet only tentative, states that because of the extremely high temperatures, the stars are not in thermal equilibrium with their atmospheres. Large convective motions must exist. Also, rapid rotations can theoretically be expected to produce large magnetic forces. Thus, radiation pressure, convective instability, and high rotation speeds all contribute toward shell formation.

Figure 13-5 is a photograph of the well-known Ring nebula in the constellation Lyra, NGC 6720, also known as M57. † This object is the most famous of a group known as *planetary nebulae*. The misnomer arose early in telescopic astronomy from its superficial resemblance to planets. Planetary nebulae vary in apparent size from one-half the moon's diameter down to pin points, indistinguishable, except for spectra, from the stars. The largest

† M57 refers to the *Messier Catalogue of Nebulae and Star Clusters*, completely listed in Appendix 8. NGC 6720 refers to the 6720th entry in the *New General Catalogue*, Dreyer's revision of Sir John Herschel's *General Catalogue* (1864) published between 1888 and 1908.

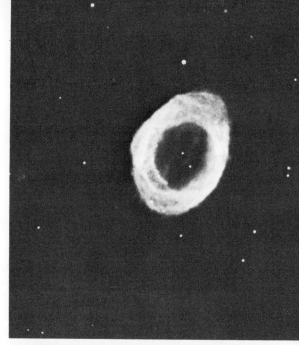

FIGURE 13-5 The Ring nebula in Lyra (M57; NGC 6720) (Courtesy Lick Observatory).

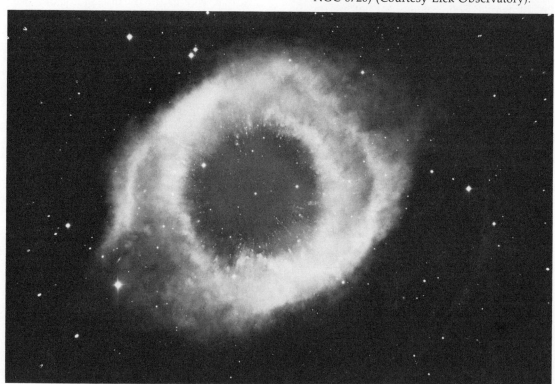

FIGURE 13-6 Planetary nebula NGC 7293 in Aquarius (Courtesy Hale Observatories).

Content and Organization of the Universe

known planetary nebula is the rather faint, NGC 7293 in the constellation Aquarius (Figure 13-6).

Planetary nebulae are slowly expanding shells of gas surrounding extremely hot, faint class O or W stars. Radial velocities place the expansion rate between 20 and 30 km/sec. This is a very slow expansion compared to the rates of the nova shells or the shells of the Wolf-Rayet stars.

Even though the central stars are extremely hot, invariably they are very faint and thus appear to be white-dwarf stars. If this is the case, then the shell of gas represents an appreciable fraction, 10–20%, of the star's total mass. The ringlike appearance of planetary nebulae results from looking through the shell of material along a central axis and also from looking into greater optical thickness around the periphery. One can simulate the optical effect by looking down on the rim of a transparent tumbler or bowl.

Typical planetaries are estimated to be from $\frac{1}{2}$ to 1 LY in diameter. If this estimate is correct, then the objects are short-lived transients with an expectancy of about 100,000 years. The current age of the known planetaries, some 1000 objects, is estimated from their size and rate of expansion. The mean value is around 20,000 years. Presumably the planetary shells will thin out from expansion until by the end of 100,000 years they will cease to be visible. We are able to conclude from this life expectancy, together with the presently known quantity, that planetary nebulae are very common in the celestial scheme.

The significance of the planetary nebulae is still quite speculative. The nature of the central star suggests advanced age; perhaps it is an object like a nova, engaged in the process of reaching mass stability in the decline toward quiescent white-dwarf status. Yet the ejection of shell material is at a rate far less than that occurring in the nova process. At best, we can tentatively classify the planetaries as very slow novae.

Neutron Stars and Pulsars. In Chapter 12 we touched upon the subject of neutron stars as the end product of a gravitational collapse when the stellar mass exceeds 1.2 solar masses. As noted such objects are at present purely speculative; considerable theoretical difficulties have yet to be resolved. In principle neutron stars can exist but we do not know how stars could ever achieve the necessary density for existence. For stars with less than 1.2 solar mass, gravitational collapse is successfully resisted by electron pressure; it is halted at the white-dwarf stage. For masses in excess of 1.2 solar mass two degrees of further stellar evolution have been postulated. The electron crush described in Chapter 12 is one of them. The other carries the collapse hypothesis further. During the early stages in the formation of neutron stars, *neutrinos* are released which readily escape through the outer layers. † But as the collapse rushes headlong to completion, the density throughout the star increases tremendously. At some given stage the energy carried by the neutrinos is absorbed in the outer layers in the form of heat. The heating occurs at such a cataclysmic rate that a shock wave is produced, resulting in the explosion of the outer layers in the form of a supernova. The explosion produces a reaction implosion of the central material, creating the destructive collapse of the core into a superdense neutron star. This hypothesis accounts for the pulsar in the central remnants of the Crab nebula (p. 318).

The first of several tens of known pulsars (*pulsar* is an acronym for *pulsating radio star*) was discovered in the constellation Vulpecula in late 1967 by J. Bell and A. Hewish at Cambridge, England. It is designated CP 1919 for Cambridge Pulsar at right ascension 19^h 19^m (declination $+21°$). CP 1919 has defied all efforts to identify it with an optical source. This

---

† Neutrinos are hypothetical particles without mass or electric charge. They represent the mechanism for carrying energy away during certain nuclear transformations.

pulsar emits radio energy in the form of a sharp pulse of very short duration every 1.33728 seconds. Its regularity is so exact that it could very well be used for precision timekeeping. About 20 pulsars have been investigated; pulse periods for those so far studied range between $1/30$ second to 2 seconds.

Most pulsars emit complex radio waves with several subpeaks within the periodic sharp pulse; the radiation extends over a wide range of radio frequencies (Figure 10-14). The lower frequencies are retarded as they journey through space in proportion to the increase in wavelength, presumably through interaction with interstellar electrons. The effect is exactly analogous to the interaction of visible light with the transparent media through which it passes, producing refraction and dispersion effects. If we knew the exact electron density in space, we could determine the distance to the pulsar from the measured delay time for the lower-frequency components of the pulses. Unfortunately we do not have this information; based on best current estimates of the number of electrons in free space, most pulsars appear to be a few hundred parsecs distant; a few may be over 1000 pc away. Considering these vast distances, the strength of the radio pulses and the sharpness of their spectra indicate that an incredible amount of radio energy originates within a region of no more than a few hundred kilometers for each pulsar. NP 0532, with a period of only 0.03309 second, is the fastest pulsar known. It was discovered in October, 1968, near the center of the Crab nebula. In January, 1969 the radio source was positively identified with an extremely faint, optically visible star; this star is emitting light pulses in exact synchronization with the radio emissions. The evidence suggests that this star is the supernova remnant, possibly a collapsed neutron star, in very rapid rotation or oscillation. The pulsar is perceptibly slowing down, that is, the period between pulses is lengthening. On theoretical grounds, this should be the case since energy must be conserved; the energy lost through radio emissions must occur at the expense of rotational energy of the star. This is the only conceivable source for the radio energy, at least according to the present state of our knowledge.

The nature of the pulsars, just exactly what they really are, poses one of the most complex and perplexing problems of modern astronomy. Of the various hypotheses—that pulsars are pulsating white dwarfs; that they are rotating neutron stars; that they are local oscillations in the degenerate gases in white-dwarf star atmospheres; or even that they are communication signals from distant, advanced civilizations searching for other centers of life in the universe—not one has gained general acceptance. We must classify pulsars as unfinished business and await discoveries produced by future research in the hopes that the mysteries will in time be unraveled.

Galactic Clusters and Associations

As we turn our attention away from the singularities of individual stars (binary systems included) and begin to assess their distribution in space, a significant fact quickly becomes apparent. Practically *all* stars are members of stellar systems of one form or another. Some of these are subsystems within larger groupings. And in connection with each system and subsystem, we find equally significant differences in types, or *populations,* of stars. This statistic, of course, excludes those specialty stars with prominent individual differences that we have already investigated. The collections of stars commanding our immediate attention are the *galactic clusters* and the *associations.*

Several hundred galactic clusters are directly accessible to visual inspection (by telescope); a few are just barely naked-eye objects. All of these collections have been thoroughly studied spectrally. Galactic clusters are so named because they are all confined to the plane of our galaxy, in the vicinity of the Milky Way. Because they are irregular in organization and appear loosely knit, they are often referred to as *open clusters.* This is in contrast to *globular clusters,* a very different kind of collection that we shall investigate next. Among the best known open clusters, and easily accessible to the amateur with even a modest telescope, are the *Hyades* in Taurus, the *Pleiades* (often called the Sisters) in Taurus (Figure 13-7), *Praesepe* in Cancer (often referred to as the Beehive) (Figure 13-8), and the Double Cluster (h and χ Persei) in Perseus (Figure 13-9).

Several hundred galactic clusters have been cataloged; the Messier Catalogue lists 27 that are all easily accessible with a small telescope. Thousands more as yet undiscovered clusters must exist; all are either too remote or too obscured by interstellar matter. Indeed, one characteristic of the galactic clusters is their location in the vicinity of galactic nebulosity, even though the clusters themselves may be free of the material.

We have previously made use of the motions of galactic clusters in the study of the proper motions of stars in general. In the process, we found the distance to individual clusters. With information available concerning the distances and the apparent luminosities of the individual stars, we can construct H-R diagrams for individual clusters. This ability enables us to reach out further in space than is possible from the evidence obtained from single stars alone. Most open clusters have relatively few members; at most they number in the hundreds. Most have red-giant members; many have sizable numbers of T Tauri stars. This information assists us in reconstructing from the cluster diagram an approximate age for a given group of stars (Figure 13-10). In the illustration, the age scale on the right side of the diagram corresponds to the points where the clusters leave the main sequence. Thus, from the diagram, the Pleiades cluster is about 150 million years old.

Open clusters must be comparatively young, even though some of the individual member stars have reached advanced evolutionary age.

The phenomenon of clustering suggests that, save an occasional transient interloper, all stars in a given cluster evolved from protostars at about the same time. Because of the looseness of their association, gravitational bonds holding them together are not nearly as strong as is the case in tighter, more compact clusters. We should expect an individual star to escape from the grasp of its companions through gravitational interaction with a chance close passerby. Also, as we shall see later in the chapter, stars of open clusters are susceptible to differential revolution rates about the center of the galaxy. The reason is that the stars nearer the galactic center have higher orbital space velocities than those farther removed from the center. This motion will inevitably disrupt the cluster. That open clusters do exist bears testimony to the fact that disruptive effects of different space velocities and interference from external stars have not had time to wreak destruction of the cluster by scattering the individual members.

Associations are loosely knit groups of stars, generally very few in number, all having essentially the same spectral characteristics. Conspicuous examples appear in the great nebula in Orion, M42, and in the constellations of Perseus and Scorpio. Class O and B stars tend to be grouped in associations; members of a given group may be spread out over several tens of parsecs. Associations should be broken up by scatter even faster than open clusters, usually within a few million years. Because of the short time scale, associations represent probably the youngest groupings known. As with ordinary galactic clusters, associations are found in galactic regions also occupied by interstellar matter. The significance of this will be reviewed presently.

Globular Clusters

Seeing a globular cluster such as M13 (NGC 6205) in Hercules in the field of view of a telescope for the first time is an awe inspiring and

FIGURE 13-7 The *Pleiades* cluster in Taurus (Courtesy Lick Observatory).

unforgettable event (Figure 13-11). The grand Hercules cluster, high overhead in summer in northern midlatitudes, roughly resembles a popcorn ball. Myriads of stars are concentrated with increasing density toward a central nucleus, so much so that individual stars in the central condensation are impossible to resolve. The outer reaches of the cluster appear to merge with the surrounding star field, but this is just an illusion. The cluster is far removed from the local population of stars. Nevertheless it is difficult to determine where the cluster begins and what its apparent angular diameter is. A commonly accepted figure is 18' of arc, roughly one-half the diameter of the full moon.

Under favorable conditions M13 can be seen with the naked eye, but of course all of the detail is lost.

The Hercules cluster is typical of globular clusters in general, although not all of them are as compact. Star counts vary from hundreds to thousands of time those in the galactic clusters. Thirty thousand stars have been counted in the Hercules cluster, excluding the unresolved central portion. Because it is so vast the total number of stars is estimated in the range of 100,000 to 200,000. The concentration of stars in the outer portions is more than 40 times that in the sun's immediate vicinity. Nevertheless the distance between individual stars in this

FIGURE 13-8 *Praesepe* (The Beehive), an open cluster in Cancer (Yerkes Observatory Photograph).

FIGURE 13-9 The double cluster in Perseus, *h* and *χ Persei* (Courtesy Lick Observatory).

part of the cluster is still over 1 LY. Even in the dense central population, individual stars are many thousands of astronomical units apart.

The velocities of individual stars in globular clusters are not very great; thus the odds against stellar collisions are quite high. Even so, because the number of globular clusters (with such numbers of stars within them) is so large, an occasional collision should occur somewhere. We assume that such a collision would result in an explosive cataclysm of supernova proportions; it should thus be extremely visible. Yet in all of the photographic evidence, only one supernova, occurring in the outskirts of M80 in Scorpio, has been observed.

The number of known globular clusters in the vicinity of our own galaxy is slightly in excess of 100; probably there are others that are obscured by the galactic nucleus. The color-brightness diagram for a globular cluster differs radically from those of most galactic clusters (Figure 13-12), and the diagrams differ substantially among globular clusters although the resemblance among all is unmistakable.

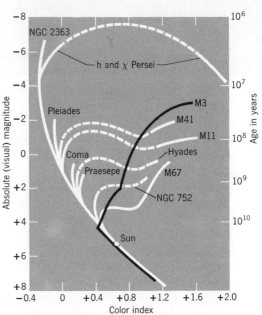

FIGURE 13-10 Composite color-luminosity diagram for several clusters. M3 is a globular cluster; the rest are galactic clusters (After A. R. Sandage).

FIGURE 13-11   The Globular cluster M13 in Hercules (NGC 6205) (Courtesy
Hale Observatories).

There are no bright, bluish main-sequence stars in globular clusters. Moreover, the giant branch of the diagram joins the main sequence much lower down toward the later yellow stars, and it is inclined upward toward the right more steeply. Thus the statistical characteristics of globular-cluster stars differ substantially from those of main-sequence stars in general (Figure 13-13; compare with Figure 12-3). Unlike the H-R diagram, which was constructed for stars near the sun plus those of a few nearby galactic clusters, the color-brightness diagram (which is equivalent to an H-R diagram) for globular clusters has a distinctive horizontal branch. Midway in this branch, we find cluster variables, the RR Lyrae stars. Curiously enough, no stars other than the variables occupy the same position on the diagram. For that reason it is often referred to as the "cluster variable gap." The interpretation from the color-brightness diagram is that globular clusters are very old systems that have been quarantined from outside influences for their entire history. Most of these clusters are estimated to be from 10 to 20 billion years old; in all of that time no new stars have been formed since the initial condensations. There is no interstellar matter between the member stars of a cluster from which a descendant could be conceived and nurtured.

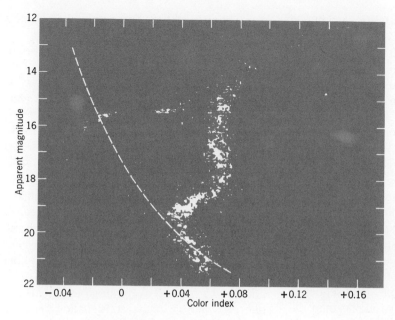

FIGURE 13-12
Color-brightness diagram
for the globular cluster
M3 (NGC 5272) in *Canes
Venatici* (After A. R.
Sandage and H. L.
Johnson).

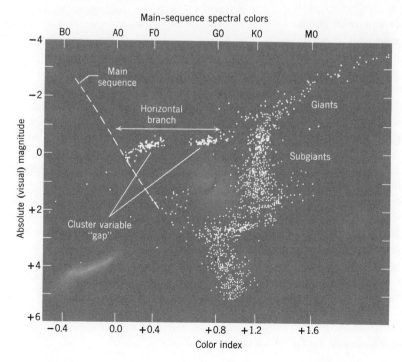

FIGURE 13-13 Schematic
H-R diagram for the
globular cluster M3
(Figure 13-12).

Distances to globular clusters can only be statistically determined. Our confidence in the estimates is subject to a ±20% range in accuracy. Several methods are employed in the distance determinations. The most reliable one involves the RR Lyrae variables whose absolute magnitudes are, as we have seen, all about the same wherever they are found. Because of their luminosity variations, the variables are quite prominent in the clusters. Also, their apparent brightnesses can be observed. Then, in the familiar manner, the ratio of their mean apparent luminosity and their absolute brightnesses (estimated from their periods) yield the distance to the cluster. Another distance-determination method makes use of main-sequence stars and their position on the H-R diagram. But the absence of early-type main-sequence stars makes this method less reliable than the one just described. When all else fails, the brightness of the cluster as a whole, or the apparent diameter of the cluster, is used. If we assume that all globular clusters have about the same intrinsic brightness, then the apparent brightness of a given cluster is a measure of its distance. Similarly, if we assume that all globular clusters are about the same size, then the apparent diameter of a given cluster is a measure of its true size. Both of these methods rely heavily on statistical averages derived from the preceding methods. The degree of accuracy resulting from these techniques leaves much to be desired, but they are better than no method at all.

It is important to understand that, in dealing with globular clusters, color-brightness relationships must prevail over techniques depending upon identification of spectral types. The reason is that the spectra of *all* globular-cluster stars show deficiencies of metals in the stellar atmospheres. Thus the spectra of these stars cannot be compared exactly with the standard stellar spectra; to do so would yield spurious values for their absolute luminosities. Incidentally, the absence of metals in the stellar compositions lends added weight to the speculation that the stars in globular clusters are first-generation objects (p. 321).

The work of Harlow Shapely at Mount Wilson during 1915–1920 resulted in a scale of distances and dimensions for globular clusters. Subsequent research showed that interstellar matter adversely affects distance estimates by dimming and reddening the light from distant stars in a manner to be considered in a following section. The original estimates had to be revised, placing many of the clusters much closer to us than originally calculated. Even so, all known globular clusters are many thousands of light-years distant. The brightest, ω Centauri, is 15,000 LY away. M13 in Hercules is 25,000 LY distant. The most remote cluster known to belong to our galaxy is 200,000 LY away. Two other globular clusters, one of which is in the constellation Leo, are 400,000 LY distant. This latter cluster may be a remote member of the Milky Way system, or it may now be a fugitive object, having made good its celestial escape from the galaxy. The diameters of globular clusters range between 80 and 400 LY with an average of about 150 LY. The mean diameter of the dense, unresolved centers is about 20 LY.

As we shall demonstrate in a later section, the distances of the globular clusters and their directions from our vantage point give the impression that they cluster about the galactic nucleus as shown in Figure 13-14. Most of the clusters are within 10,000 LY of the center of the Milky Way Galaxy, but a substantial number are in the neighborhood of 30,000 LY distant.† Then also there are a few far ranging clusters whose status we have just considered.

The motions of the globular clusters suggest that they represent a closed system not par-

---

† We have not yet argued formally the shape of the Milky Way Galaxy, nor have we considered its motions or its composition. This deficiency will be remedied very shortly. We are justified in departing from a strict developmental sequence in order to dispose of the problem of the globular clusters in one section.

FIGURE 13-14 Schematic edge-on view of a typical spiral galaxy showing the distribution of globular clusters around the nucleus.

taking of the rotational motion of the galaxy, but that they are gravitational captives nevertheless. The clusters seem to have highly inclined, randomly oriented, elliptic orbits of high eccentricity, resembling those of some of the comets. As with comets, the clusters will spend most of their time at the outer extremes of their orbits. Their brief plunges into the nucleus of the galaxy are widely separated in time. Here again, the danger of stellar collisions is almost nonexistent; the spacial separation of stars is just too great for them to occur.

Stellar Populations

The concept of a normal stellar population consisting of a hierarchy of stars in different stages of evolution, and one that is summarized in the indispensable Hertzsprung-Russell diagram was, we recall, formulated from the studies of stars in the immediate solar vicinity. We also reviewed the evidence that points to very young stars in the process of evolving, or having only recently evolved, from the stellar nebulosity in which they were immersed. We will find that in the Milky Way

Galaxy, and in other similar galaxies, spiral arms are seemingly wrapped around a central galactic nucleus; it is only in these arms that nebulosity is found. Thus, galactic clusters are invariably located in the spiral arms of the galaxies. The situation is radically different with respect to globular clusters and the central nucleus of the galaxies. Here, the early-type main-sequence stars have disappeared through aging. The occasional excursions of the globular clusters into the dense central regions of the galaxy accomplish at best a house-cleaning function. Whatever gas and dust that might have accumulated in the clusters from ejection from the stars in the aging process are swept away and lost to the dominant gravitational mass through these orbital incursions. Thus in the stellar sense the globular clusters are sterile regions. They represent an aged population with no youngsters to brighten and invigorate the scene. The evolutionary process in globular clusters is approaching a dead end. The scarcity of metallic lines in the spectra of the stars of globular clusters is consistent with these observations. As we have seen earlier, metals are produced in the later stages of the initial core burning. In the evolutionary process, they are introduced as debris into the surrounding regions as the result of mass ejection. If such matter is allowed to accumulate, subsequent generations of stars are successively enriched in metals. But this process is denied to the globular clusters.

Thus two very different *populations* of stars can be recognized: population I consisting of the normal, metal-rich stars found in spiral arms of galaxies; and population II, the metal-poor stars found in globular clusters, as isolated stars in the galactic halo, and in the central galactic nucleus. Actually, the distinction made here between just two populations is forced. A sharp distinction between two populations can be made only for stars of the globular clusters (extreme population II), and the very young associations of Class O and B stars (extreme population I). Elsewhere there is a gradation in population properties that need not concern us here. The significance of the existence of two major stellar populations will become more evident as we proceed. For example, we found earlier that two types of cepheid variables are recognized; on the average type II cepheids are 1½ magnitudes fainter than the type I cepheids. Now type II cepheids are found only in association with population II stars. Type I cepheids belong to population I and are found only in association with these stars. How this distinction affected the determination of intergalactic distances is a fascinating chapter in the history of astronomical growth. We shall encounter it soon.

The Milky Way Galaxy

For those of us who are living in population centers having an unavoidable nighttime glare from artificial illumination and the usual urban level of atmospheric pollution, the celestial sphere presents a somewhat bleak appearance. Only a few of the brightest stars and the brighter planets can overcome these handicaps to stellar visibility. But the viewer who has the advantage of clear, moonless nights far from population centers sees a breathtaking display of stars. In every direction he looks he can see countless numbers of the celestial populace. Some stars are bright; some are faint; more are fainter still; the whole vista creates a feeling of looking so far into the vast reaches of space that all objects ultimately disappear from view.

Depending upon the time and season, the fortunate observer will see the Milky Way, an irregular, mysterious appearing band of light encircling the whole celestial sphere. With the aid of binoculars or a telescope the viewer at once discovers that the Milky Way is not a faint, luminous nebulosity, it is an incredible concentration of stars (Figure 13-15). The concentrations appear uneven. Some regions are aptly described as star clouds; other regions are nearly black, almost devoid of stars, as though they were great tunnels through the Milky Way, favorably oriented so that we are looking into the blackness of space beyond.

The presence of the Milky Way attests to the extraordinary differences in the spatial distribution of the stars. These differences lead to such logical questions as: Is there any symmetry, pattern, or plan to the arrangement of the stars? If so, from our position *within* the system can we ascertain what the pattern is? The answer to this last question is no, not precisely, although certain broad characteristics are discernible. Let us retrace the historial arguments concerning the stellar distribution with an eye toward evaluating the relative strengths of our successes (and failures) in discovering the shape of our local sidereal organization.

Galileo is thought to be the first to ascertain that the Milky Way is a vast collection of discrete stars. In 1750 Thomas Wright, in his *Theories of the Universe*, wrote that the sun is a member of a natural star system that is limited in extent, and that the Milky Way somehow delineates its boundaries. Other observers, including Immanuel Kant, the German philosopher, reached similar conclusions. The first to undertake a scientific investigation of the distribution of the stars, however, was Sir

FIGURE 13-15 The Milky Way, a mosaic of several photographs of the region from Cassiopeia to Saggittarius (see map, Figure 13-16) (Courtesy Hale Observatories).

William Herschel (1738–1822), the great English astronomer. Herschel's method of approach, the results of which were published in 1785, consisted of a sampling procedure he called *star gauging*. A systematic gauge, or count, of stars visible in Herschel's telescope was recorded for 683 selected regions scattered over the sky. The counts ranged from a single star up to nearly 600 stars per region. To Herschel the variation could mean only one thing: the sidereal system was disk-shaped, much like a solid wheel. The Milky Way was its rim, the sun was at the hub. Herschel's conclusion was only partially correct.

At the beginning of the 20th Century J. C. Kapteyn of the Kapteyn Astronomical Laboratory in Groningen, Netherlands planned a sophisticated mapping of the Milky Way together with a uniformly distributed sampling of 206 selected areas. The sampling included determination of stellar magnitudes and colors, classification of spectra, measurement of radial velocities, and star counts. In this effort Kapteyn was assisted by astronomers in many nations in all parts of the world. The resulting sidereal model, called the *Kapteyn Universe,* was published in 1922. The model preserved the slablike character of Herschel's concept. It still placed the sun at the center, but it was improbably small, only 10,000 LY in diameter. For the excellent reason that he was unaware of the real extent and magnitude of obscuration, Kapteyn failed to make the necessary allowances for its effect. We now know that the Milky Way is much larger than he suggested and that it extends over much greater distances in one preferred direction.

Broad Outlines of the Galaxy. Let us tentatively assume, subject to further verification, that the galaxy resembles a circular sheet of stars, and that the mean center of the Milky Way defines the galactic plane. This coordinate is inclined 63° to the plane of the celestial equator, and this inclination is unaffected by precession of the earth's orbit. From the brightest region, located in the constellation Sagittarius, we can trace the Milky Way northward through the constellation Cygnus, Cepheus, Cassiopeia, and Auriga. The galactic plane continues southward through Gemini and Canis Major (Figure 13-16). In the southern celestial hemisphere the Milky Way extends from Sagittarius southward past Centaurus and the Southern Cross to the constellation Carina. From Carina northward to Canis Major, the Milky Way is very faint.

The poles of the galaxy, each of which, of

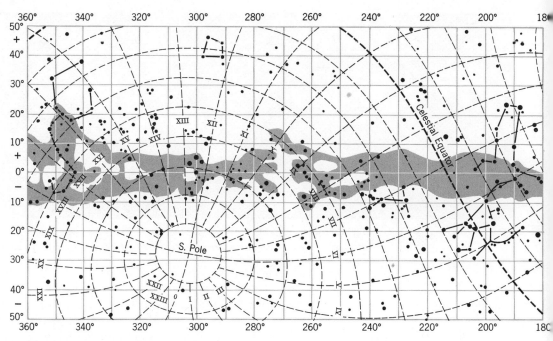

FIGURE 13-16 Galactic chart in rectangular projection, 0–360° longitude and −50–+50° latitude, showing the extent of the Milky Way. A few prominent constellations are shown for reference (Courtesy Sky Publishing Corp.).

course, is at 90° latitude from the galactic plane, are in the constellations Coma Bernices (north) and Sculptor (south). Originally, 0° galactic longitude was placed at the ascending node in Aquila. In 1958 the International Astronomical Union redefined the galactic coordinates, placing 0° galactic longitude at RA 17ʰ42ᵐ.4 (1950) in Sagittarius. This position coincides with a strong radio source, presumed to be the center of the galaxy. The International Astronomical Union also redefined the location of the galactic poles, which location in turn governs the position of the galactic plane. The new latitude coordinates do not differ substantially from the former when precession is taken into account.† Now if the galaxy is truly

slablike, the concentration of the stars toward the galactic poles ought to decrease with the decreasing luminosity of the stars at a rate much faster than it does in the Milky Way. This is indeed the case. However, the star counts in the galactic plane do not follow the statistically predicted quantities (p. 345) in anything like reasonable precision. Moreover, the star counts in the Milky Way at galactic longitude 180°, in the constellation Auriga, are much smaller than in the opposite direction in Sagittarius. This statistic does nothing to diminish our confidence in the location of the galactic center; it does reinforce the conviction that the sun is *not* at the center of the galaxy.

What additional support can we muster for the selection of the region of Sagittarius as the center of the galaxy? The counts of faint stars are more than 10 times greater in this direction than toward the (hypothetical) anticenter in Auriga. The same holds true (the percentages

† While the *angle* between the celestial equator and the galactic plane is unaffected by precession, the coordinates in right ascension and declination do change with time (Chapter 2).

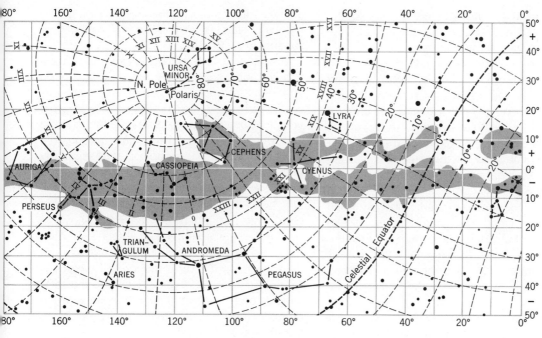

FIGURE 13-16 *(Continued)*

differ somewhat) for the counts of planetary nebulae, cepheid variables, and cluster cepheids in the direction of the supposed galactic center but removed from the galactic plane. Novae and globular clusters above and below the galactic plane also show a marked preference for galactic longitudes in the direction of Sagittarius.

In 1918 Harlow Shapely proposed a model of the galaxy based upon the observed distribution of globular clusters and their distances from the earth. This model eventually prevailed over the Kapteyn Universe. As we have seen previously (p. 342), globular clusters are distributed spherically around the galactic nucleus, which thus marks the galactic center. Since we know the distances to the clusters within reasonable limits, the size of the galaxy is readily obtainable. Shapely computed a diameter of 80,000 LY for the galaxy. With the discovery of interstellar absorption in the

1930s the value was corrected to 100,000 LY. The solar system is in an outlying part of the galactic plane some 30,000 LY from the center. The galactic plane is some 3000–5000 LY thick in the sun's vicinity; the galaxy is estimated to be some 15,000 LY thick at the central nucleus which is some 40,000 LY in diameter. We shall investigate the details of the fine structure of the galaxy in a subsequent section.

Galactic Motion

At the level of our investigation in this text we necessarily lack sophistication in the dynamics of moving systems, most of which are extraordinarily complex. But this need not be a serious obstacle; we can certainly appreciate the qualitative aspects without resort to higher mathematics. Clearly, the galaxy must be rotating around an axis passing through the center of the galactic nucleus and perpendicular to the galactic plane. Were this not so, the galaxy could not maintain a flattened, disklike shape. The stars are not attached to each other except by gravitational bonds. Gravity can restrict stellar motion, but it cannot *a priori* prevent it. In the absence of rotation, gravity would pull every star in toward a common center of mass that would result in a catastrophic crush. Just as the planets in their continual fall around the sun are carried around it by their orbital velocities, so are the stars on the average carried around the galactic center. The motion, to be sure, must be much more complicated than in a simple system where essentially all of the mass is concentrated in one body. In the galaxy the mass is *not* concentrated at the center of a homogeneous sphere. Moreover, the gravitational forces link billions of stars, not just a handfull or so. Therefore, the Milky Way *must* represent the mean plane containing the vast majority of the stellar orbits; the galactic poles are thus on the axis of rotation. It is in the galactic plane that centrifugal effects will be the strongest, resulting in the flattened, disklike configuration.

How can we verify the rotational motion and find its magnitude? We recall that Kepler's laws describe the motions of celestial bodies. The laws may not yield precise information about the rotation of the galaxy because the mass distribution is much more complex than it was in the simple Keplerian systems. They can, however, give us direction for our search. Kepler's harmonic law asserts that among objects orbiting a common center of mass, those with the smallest orbits have the highest orbital speeds. Now if the sun is neither at the center nor at the remotest outskirts of the galaxy, then we should be able to detect speed differences between stars closer to the galactic center and those farther away from it than the sun. In particular, the speed differences should be reflected in the star's radial velocities, even though all except the high-velocity stars are moving along parallel paths. Let us therefore analyze the radial velocities of particular groups of stars, in all directions from, but reasonably near, the sun.

We know that we are being carried toward a relative point called the apex of solar motion and away from a point known as the antapex (Chapter 10). We first find the average radial velocities of a group of stars a little to the left of and toward the apex of the sun's motion. We repeat this with a group of stars similarly placed to the right of the sun's path. We find on the average that the group of stars on the left shows a residual Doppler shift toward the blue, and the group on the right shows a similar shift toward the red end of the spectrum. Next, we analyze the motions of the stars at right angles to the sun's motion; all we see is an average transverse velocity toward the solar antapex on the left and toward the apex on the right. The net average *radial* velocity relative to the sun for both groups is zero.

Finally, we look toward the antapex and analyze the group radial velocities as before. This time we find the stars on the left side of the sun's space path show a Doppler shift toward the blue; those on the right show a net displacement toward the red end of the spectrum. The situation becomes clearer with the aid of Figure 13-17. On one side of the sun's path, say the left side when looking toward the apex, the stars nearer the apex show a blue shift; those to the rear toward the antapex show a red shift. Again, on the right the Doppler shifts are reversed. The only rational explanation for the reversal of the Doppler shifts on either side of the sun's path is that it is a perspective effect. The stars on the right of the sun's motion are overtaking, passing, and running away from the sun; those on the left are being overtaken by the sun, passed, and left behind. *Only rotation of the stars and sun about a common galactic center could produce such an effect!*

Having thus verified the rotational motion, and with it, justified the arguments for a disk-shaped galaxy, we next try to determine the

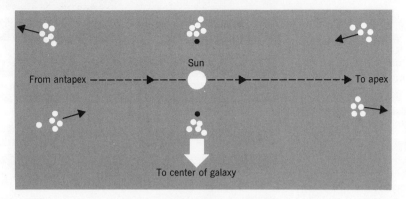

speed of the solar system around the center. The easiest, and most logical, way to do this is to select a frame of reference that is stationary with respect to the center, or if moving, is doing so at a very slow rate. Immediately, the apparent radial velocities of the globular clusters (with respect to the sun) and high-velocity stars command our attention. The radial velocities of the globular clusters range from −135 km/sec to +290 km/sec. After making allowances for a mean residual velocity of the clusters about the galactic nucleus of 75 km/sec (the average speed necessary if they are not to be caught in a gravitational trap that would cause them to rush in toward the center), we conclude that the sun is traveling around the center of the galaxy at about 250 km/sec. At this rate the sun completes one round trip every 200 million years or so. If our estimates of the age of the sun are correct, has completed about 25 revolutions around the galactic center since its birth as a star. The speed of the high-velocity stars is thus accounted for as well. They are stars of the galactic halo having minimal velocities of their own about the galactic center. It is we, not they, who are moving at high velocity, along with the "slower" orbital stars in our own neighborhood.

With a minimum of one additional observation of the motion of a group of stars at a different distance from the galactic center than the sun, say 1000 pc farther out, we can apply Kepler's laws to arrive at an estimate of the mass of the galaxy. It turns out to be on the order of 100 billion solar masses, Some of this mass is in the form of interstellar matter, to be considered next. In passing, it is of interest to contemplate that if the density of the sidereal population were a million times as great as it is, there would still be little or no risk of stellar collisions. The galaxy is mostly empty space.

Interstellar Matter

In arriving at the estimates for the mass of the galaxy and its size, we avoided consideration of the way interstellar matter interfered with the early star gaugings, thus influencing the distance determinations. Indeed, we have paid lip service to its presence and its effects without reviewing the arguments for its existance and what its density distribution is. Let us backtrack briefly in order to see how the search for interstellar matter was stimulated. As was discussed earlier (and as discussed in more detail in Appendix 4), a difference of 1 magnitude in the apparent brightness of stars corresponds to a factor of 2.512. We also know from the inverse square law that if a given star is removed to twice its original distance, its apparent brightness will decrease by factor of 4. Suppose, however, the distance is changed by a factor of only 1.585, not 2. Then the star's apparent brightness will decrease by a factor of $(1.585)^2$ or 2.512. More succinctly, a star of any magnitude will appear 1 magnitude fainter if its distance is increased by a factor of $\sqrt{2.512}$, or 1.585.

From geometry, we know that if the radius of a sphere is doubled, its volume is increased by a factor of $(2)^3$ or 8. Similarly, if the radius increase is 1.585 times, the volume increase will be $(1.585)^3$ or 3.98 (very nearly 4). Consider how this applies to star gauging. Main-sequence stars of the same spectral class will all have essentially the same absolute luminosities. For these stars, differences in apparent luminosity represent differences in distance from the sun. Thus if we select any given magnitude as indicative of the volume of the celestial sphere containing such stars, then 1 additional magnitude fainter will represent a volume increase of 4 times. If the stars are *uniformly distributed* in space, then the count of stars at each successive magnitude fainter ought to be 4 times that of the previous magnitude.

The actual numbers of stars show no such rate of increase at successively greater (fainter) magnitudes. The maximum ratio is less than 3; this is in the direction of galactic latitude 0°. Toward the galactic poles the ratio falls below 2. That the expected ratio of 4 is not even approached anywhere leads us to two possible conclusions. One is that the stars thin out in all directions from the sun, with thinning occurring more rapidly toward the galactic poles. (Now we see how a heliocentric, disk-shaped galactic model can be argued.) Another alternative is that the stars at fainter magnitudes appear fainter than they should because of the presence of some kind of obscuring matter.

Historically, both effects were found to be operating. The obscuration is not uniform; the greatest amount occurs toward dark rifts or dark nebulosity such as seems to split the Milky Way from Cygnus to Sagittarius.

Interstellar Gas. In the year 1904 attention was focused on the spectrum of the star δ *Orionis* because of the presence of two sharp lines of ionized calcium, the H and K lines of

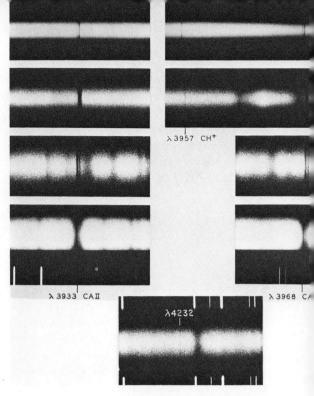

FIGURE 13-18 The H and K lines of interstellar calcium (Courtesy Hale Observatories).

Fraunhofer's nomenclature (Figure 13-18). The presence of ionized calcium lines was not singular; other stars show ionized calcium in their spectra. But in this case all the other spectral lines were diffuse and weak; only the calcium lines were sharp and clear. Moreover, δ Orionis is a spectroscopic binary showing periodic Doppler shifts of the spectral lines. However, the calcium lines do not partake of the Doppler displacements. There is only one conclusion that can be drawn from these phenomena; the calcium lines do not originate in the atmosphere of the star. The gas that produces them must be *interstellar*.

Subsequent investigation disclosed the presence of other absorption lines associated with interstellar gas, among them being the lines of sodium, potassium, iron, titanium, together with a few molecular compounds containing carbon, oxygen, or nitrogen. From detailed studies of the strengths of the lines in much the same manner as is employed for stellar atmospheres, we find that the various chemical abundances are not too different in interstellar gas from those of the atmospheres of many stars. Hydrogen, understandably, accounts for 75% by mass of the interstellar gas, and helium accounts for another 24%. The remaining gaseous substances have concentrations more on the order of impurities in the hydrogen-helium mixture.

Analysis also yields the density of the interstellar gas. On the average there are less than 10 atoms/cm³, although within a cloud the concentration may run from 100 to 1000 times this amount. In terms of earth-bound laboratory standards, even these higher concentrations represent a near-perfect vacuum. The gaseous clouds themselves are so tenuous that the mass contained in a volume the size of the earth is only about 3 kg. Even so, the amount of gas in the galaxy staggers the imagination. Perhaps 5 or 6% of the total galactic mass, equivalent to 5 or 6 billion solar masses, is in the form of interstellar gas.

Most of the gas is cold and nonluminous. But in the regions near very hot stars such as stellar classes O and B1 the intense concentration of ultraviolet radiation ionizes the gases, particularly hydrogen. The ionization may be complete within a sphere of radius 500 LY. The temporary recombination of atomic nuclei and electrons releases energy in the form of fluorescence, producing emission spectra. We shall consider these ionized regions presently.

In 1932 Karl Jansky of Bell Telephone Laboratories announced the discovery of radio signals in the microwave frequency range that originated from various regions in the Milky Way. In 1944 the Dutch astronomer H. C. Van de Hulst predicted that a radio emission line from neutral hydrogen at a wavelength of 21 cm should be observable. When equipment sensitive enough to detect the weak signals became available in 1951, the prediction was realized; radio astronomy came to full flower. Radio telescopes are discussed briefly in Appendix 3.

Radio astronomy is not confined to the 21-cm hydrogen emission line, but a major effort is devoted to its study. We find the 21-cm line the most prominent in the radio spectrum, and the most extensively disbursed in the galaxy. It is beyond the scope of this text to investigate the nature of the radio emissions, except to say that neutral hydrogen has a double ground state (lowest energy level). The energy separation is extremely minute. As a result of collisions, the atoms of neutral hydrogen are excited to the uppermost of the bilevel ground state. Eventually the excited atom will emit a 21-cm photon and return to the lower level. On the average, a given atom will emit a 21-cm photon only once every 11 million years. That this radio noise can be detected at all attests to the fact that the infrequency of individual emissions is offset by an enormous number of hydrogen atoms partaking of the transitions. It has been estimated that the intensity of the 21-cm radiation over the entire earth's surface is less than the power available from a transistor radio cell.

The neutral hydrogen in the Milky Way Galaxy is confined to a thin layer, less than 100 pc thick, all in the plane of the galaxy. The emission line often shows multiple Doppler displacements indicating that much of the hydrogen exists in vast clouds having independent motions around the galactic center. Fortunately, radio waves can penetrate obscuration caused by particulate matter, something that light waves cannot do. We shall make use of this vital behavior shortly.

Interstellar Dust. The presence of dark lanes and patches in otherwise bright star clouds (Figure 13-19) betrays the presence of particulate matter that produces a variety of obscuration effects. These seemingly vacant regions are in reality silhouettes of clouds of interstellar dust (Figures 13-20 and 13-21). Some of the stars visible against the dark background are between us and the cloud. Others are embedded in the cloud at depths less than that which produces complete extinction. In some

FIGURE 13-19 The small star cloud in Sagittarius, showing obscuration by
dust in the Milky Way (Courtesy Lick Observatory).

FIGURE 13-20   Cluster and nebulosity M16 (NGC 6611) in Serpens (Courtesy Lick Observatory).

FIGURE 13-21 The Horsehead nebula (NGC 2024) in Orion (Courtesy Hale Observatories).

striking cases the dark nebulosity is outlined by stars behind the edges of the cloud. The effect is similar to the silver lining produced by the sun when it is behind the edges of broken cumulus clouds.

The outlines of only the nearest dust clouds are readily seen. The more distant clouds show less contrast against the background stars because of increasing numbers of foreground stars. On the other hand, some dark clouds are illuminated from within by the embedded stars (Figure 13-22), producing what is known as *reflection nebulae* (p. 357). Under proper conditions, and depending upon the particle size

and composition, interstellar matter absorbs light, reflects light, and scatters light. Often all these effects are combined. If particles are large compared to the wavelength of light, all wavelengths are absorbed with equal efficiency. The net effect is dimming of the light; just how much light is dimmed depends upon the concentration of the particles. For particles of molecular size selective scattering of light is the dominant phenomenon (Chapter 6). Just as in the earth's atmosphere, the amount of scattering is inversely proportional to the fourth power of the wavelength. Thus, blue light is scattered much more effectively by molecular

FIGURE 13-22 Gaseous nebula, the "Cone" nebula, in Monoceros. The dark nebulosity is in the outer regions of the cluster which contains 29 or 30 identified T Tauri stars (Courtesy Hale Observatories).

sized particles. For larger particles, with diameters less than about 10 wavelengths of the light in question, scattering obeys a lesser power law. In both cases a beam of light is reddened because of the blue components that are scattered out of the direct path.†

Finally, particulate matter can selectively absorb light of certain wavelengths, provided the particle size is sufficiently small. In particular, red light can penetrate obscuration without being absorbed more effectively than light of shorter wavelengths. Most of the particulate matter in interstellar space is mixed with gases, and the particle sizes, though very minute, do vary through a moderate range. As a result, interstellar obscuration is complex, with usually more than one of the above effects operating in any given situation. The concentration of solid particles averages about 1000/km³, or one for every billion quarts of space.

The physical composition of the interstellar dust grains has not yet been clearly established. Most likely they are ices of various gases, such as water, ammonia, and methane; perhaps there are some particles of heavier matter, including some of the metals. Dust grains must be growing from slow molecular accretion where electrostatic forces predominate. Subsequent collisions between particles may terminate their growth. These phenomena would account for both the presence of interstellar grains and the limited range of sizes. If growth by accretion is the source of the grains, the average concentration in space of the constituents suggests that a given particle reaches equilibrium size about 30 million years after the first atomic interaction.

Interstellar Reddening. In 1930 at Lick Observatory, R. J. Trumpler first recognized that diffuse interstellar obscuration existed. This is general obscuration, distinct from nebulosity. Trumpler was comparing distance determina-

tions for certain galactic clusters made by two independent methods. One, employing the familiar luminosity measurements and the inverse square law, yielded systematically *larger* distances than the other method, which was based on apparent size measurements. Trumpler correctly concluded that interstellar obscuration was the culprit. Recent determinations indicate that diffuse obscuration creates a dimming factor of 0.8 magnitude, representing a twofold reduction in apparent luminosities at right angles to the galactic plane; along the plane, dimming is much more severe; extinction is complete beyond 10,000 pc.

In addition to dimming, starlight is reddened in a manner just reviewed. Because of the definite relationship between the color index of a star (Chapter 11) and its spectral class, a star's color index must increase if interstellar reddening occurs.††

Now, if we can determine a star's spectral class, we know what its color index should be in the absence of obscuration. If its measured color index is greater, then the difference between the two values is called *color excess*. Color excess is a measure of the *differential* absorption of light between blue and yellow (visual) wavelengths. If in addition we can obtain the star's distance by some reliable independent method, then we can estimate the total light absorption from the reduction in apparent luminosity below the predicted value. The ratio of total absorption to differential absorption (the ratio of dimming to reddening) has a value of 3. The ratio seems relatively constant from region to region in our galaxy and in others. This statistic shows that particle size is relatively uniform from region to region. As soon as we review the characteristics of nebular obscuration, we shall use this

†† The color index values are increasingly negative from 0.0 at spectral class A0 toward type B and type O stars. It is increasingly positive from type A0 toward later-type (cooler) stars. An increase in color index refers to a change in the direction of increasing positive (or decreasing negative) values.

† A review of the section on the earth's atmosphere in Chapter 6 may be helpful here.

knowledge in the arguments for the fine detail of the structure of the galaxy.

Galactic Nebulae

The great clouds of interstellar matter, both dust grains and gas, give us some of the most spectacular vistas in astronomy. Most of the interstellar dust clouds are cold and dark, the gas clouds are transparent. A few dust clouds shine by reflected light, and some gas clouds give off light of their own by the process of fluorescence. A review of the broad characteristics of interstellar nebulosity is in order at this time.

Dark Nebulae. As we have seen, some dark clouds are very conspicuous when seen against the star clouds of the Milky Way. They also appear as rifts and obscurations in connection with other kinds of nebulosity (Figures 13-19 through 13-21). These obscurations are produced by such high concentrations of particles that extinction of light is complete. From our knowledge of the extremely sparse concentrations of particles in even the densest clouds, we have to conclude that they are vast in size in order to produce complete obscuration. Their dimensions are measured in parsecs, not in kilometers or even in astronomical units.

In addition to the dark curtains of matter that produce the rifts in the Milky Way and in some of the bright diffuse (emission) nebulae, small dark patches are often silhouetted against the background (Figure 13-22). Many of these globules are hundreds of thousands of astronomical units across, and they are so dense that they dim background objects a hundredfold or more. Some of these patches are oval or spherical in shape, suggesting initial condensations of matter that will ultimately become stars.

If it were not for the total extinction of light from the galactic nucleus in the region of Sagittarius, the night sky illumination (at the appropriate season) would be brighter than that produced by the full moon. Even the opposite regions of the Milky Way would produce enough light to read by were it not for the general diffuse obscuration.

Reflection Nebulae. Besides being splendid viewing objects, reflection nebulae give us valuable information about particle size and concentrations. When enough light from a nearby star is scattered by the dust grains, the cloud itself is illuminated. If it is not so dense as to obscure the illuminating star (yet sufficiently dense to provide a high concentration of scattered and reflected light), the cloud can be seen visually, or better yet, photographically (Figure 13-23). Reflection nebulae are always bluer than the light from the stars that illuminate them. It takes a very bright star to provide sufficient illumination to render the clouds visible; in the Milky Way regions where dust clouds are found, the brightest stars are the blue main-sequence class O and B stars. In some cases these intrinsically blue stars, identified as such from their spectra, appear as red as late class M stars as a result of interstellar absorption and reddening. The spectra of reflection nebulae are essentially those of the stars that illuminate them.

Emission Nebulae. Two kinds of nebulae that shine by emitted light in the form of fluorescence, rather than by reflected light, have already come under our scrutiny. They are the planetary nebulae and the remnants of supernova cataclysms. Another form is called diffuse emission nebulae, of which the Orion nebula, M42, is the finest example visible in northern latitudes (Figure 13-24). M42 is visible to the unaided eye as a hazy appearing star in the middle of the sword at Orion's belt. Telescopically the nebula is about the diameter of the full moon. Photographically the nebulosity extends much further. A group of hot, class O stars known as the *Trapezium* is embedded in a mottled appearing, bluish-green centrally located haze. Dark dust clouds are also present as are several detached globules of dark nebulosity in the bright emission region. The *Trifid* nebula, M20, in Sagittarius (Figure 13-25) is

FIGURE 13-23 The star *Merope* in the Pleiades cluster in Taurus, showing reflection nebulosity (Courtesy Hale Observatories).

another fine example of emission nebulae. The sharply outlined dark clouds that appear to divide M20 into three parts are well in front of the nebula along our line of sight.

The Orion nebula is 1600 LY distant and about 30 LY in diameter. In our galaxy, other emission nebulae of the same type are about the same size. Vastly larger emission nebulae are known outside our galaxy; an excellent example is in M33, a spiral galaxy in Triangulum (Figure 13-26).

The spectra of diffuse emission nebulae resemble those of the planetaries, but they are not so highly excited. In addition to bright line spectra, diffuse emission nebulae invariably show absorption spectra that match their central stars.

The Fine Structure of the Milky Way

As we shall see in the next chapter, the existence of various types of galaxies far from the plane of our galaxy, and originally thought to be forms of (mostly) symmetrical nebulae, assuredly marked them as external to, and independent from, the Milky Way. The intensified efforts of astronomers during the first third of the 20th Century to discover the fine structure of the Milky Way Galaxy culminated with the advent of radio telescopes in 1951. The particular problem that intrigued most investigators was to determine whether or not the Milky Way is a spiral galaxy.

In other galaxies the gas, the dust, and most luminous stars are found in the spiral arms. In our galaxy, the early-type main-sequence O

FIGURE 13-24 The Orion nebula M42 (NGC 1976) and below it NGC 1977
(Courtesy Lick Observatory).

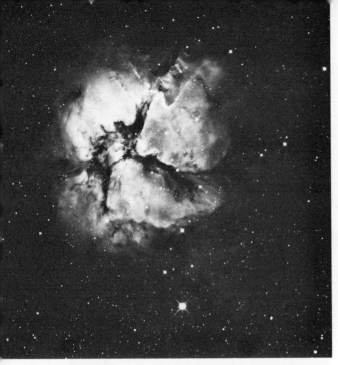

FIGURE 13-25   The Trifid nebula M20 (NGC 6514) in Sagittarius (Courtesy Lick Observatory).

FIGURE 13-26   The spiral galaxy M33 (NGC 598) in Triangulum, showing emission nebulosity in the galactic nucleus (Courtesy Lick Observatory).

and B stars are found exclusively associated with gaseous and dust nebulae. They are believed to be among the youngest stars, only recently condensed out of the galactic nebulosity. These stars, then, particularly those forming loose associations and clusters, for which distance determinations are more reliable through analyses of group proper motions, can be used as survey markers in tracing the galactic outlines. Figure 13-27 is a diagram showing the relative positions of such associations and galactic clusters relative to the sun and the galactic center. Plots of thier positions definitely allied the groups into distinct bands. The center of one band, called the Orion arm because it contains the Orion nebula, is about 1000 LY outward from the sun; the sun is on the inner edge of this arm. It extends from the bright star cloud in the constellation Cygnus all the way along the Milky Way to Carina where the early-type blue supergiants are as much as 4000 pc from us. The dark rift in the Milky Way in the direction of Sagittarius is in this arm.

A second band, called the Perseus arm because it contains the double cluster in Perseus, is about 3000 pc outward from us. Bright clusters have been mapped in this arm for a distance exceeding 10,000 LY. At both ends, the arm disappears into dark nebulosity.

A third, less distinct band, called the Sagittarius arm, has been located about 3000 pc nearer the galactic center from the sun. This arm contains objects such as the Trifid nebula. The mapping suggests that all three arms are winding inward around the nucleus of the galaxy in a clockwise direction when viewed from the galactic north pole. This is consistent with the stellar motions around the galaxy in the vicinity of the sun.

If this were all that we had to go on, the fine structure of the galaxy would still be an enigma; the data are too limited upon which to build the gross conclusion that ours is a spiral galaxy. But radio astronomy provides the missing clues. The 21-cm radio waves of neutral hydrogen can penetrate the galactic obscuration. Thus they are detectable on both the near

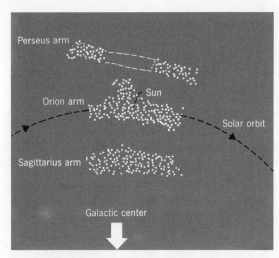

Perseus arm

Orion arm

Sun

Solar orbit

Sagittarius arm

Galactic center

FIGURE 13-27 Optical evidence for the spiral shape of the Milky Way Galaxy. Numerous clusters and associations (exaggerated for clarity) mark the locations of portions of three spiral arms in the vicinity of the sun.

and far side of the galaxy except for a narrow, cone-shaped shadow zone caused by the nucleus of the galaxy (Figure 13-28). Maps of the distribution of neutral hydrogen, combined with the optical evidence, strongly suggest a spiral shape. In the schematic diagram (Figure 13-29), the location of the spiral arms and the dimensions of the galactic system have been adjusted to the distribution of neutral hydrogen.

Strong correlations for the spiral hypothesis are provided by the characteristics of the stellar population. Population I stars: (the bright, superluminous stars of the main sequence), the Wolf-Rayet stars, type I cepheids, and the young open clusters are found exclusively in the regions containing gas and dust. In external galaxies, these are in spiral arms. Population II stars are confined largely to the galactic halo, the globular clusters, and the nucleus of the galaxy. On the other hand main-sequence stars of classes F through M, together with red giants and white dwarfs, are found in all parts of the galaxy. True, only the nearest white dwarfs can be seen because of their faint luminosities.

Certain other classes of objects do not correlate with spiral arms. They are found in the galactic halo and throughout the galactic plane with maximum concentration toward the galactic nucleus. Among these objects are planetary nebulae, novae, RR Lyrae variables, and type II cepheids. Taking all the evidence together—the distribution of stellar classes and objects, the optical evidence, the distribution of gas and dust clouds, the distribution of neutral hydrogen, and the evidence for galactic rotation—and considering both the evolutionary evidence suggested by the Hertzsprung-Russell diagram and the luminosity-color-index diagram for the clusters, we must conclude without reservation that the Milky Way is a spiral galaxy.

SUMMARY

Not all stellar spectra fall neatly into place on the H-R diagram, nor, for that matter, on the period-luminosity or the mass-luminosity diagrams. Whether all of the unusual or peculiar spectra represent transition stages between well-ordered categories or whether they represent mutations has not yet been positively established.

The peculiar spectra include those of rapidly rotating class O and B stars, stars with extended, even detached atmospheres, called in the respective cases, shell stars and planetary nebulae. Some stars seem to be exchanging atmospheres with companion stars. Still other stars appear to be so nebulous that the light from a normal companion star can shine through them.

Neutron stars represent the ultimate gravitational collapse of stars whose final masses are too large for termination at the white-dwarf stage. The existence of neutron stars seems verified in the discovery of pulsars: small, super dense systems emitting sharp bursts of radio energy of very short duration several times each second. But even so, the *exact* nature of pulsars is not yet known.

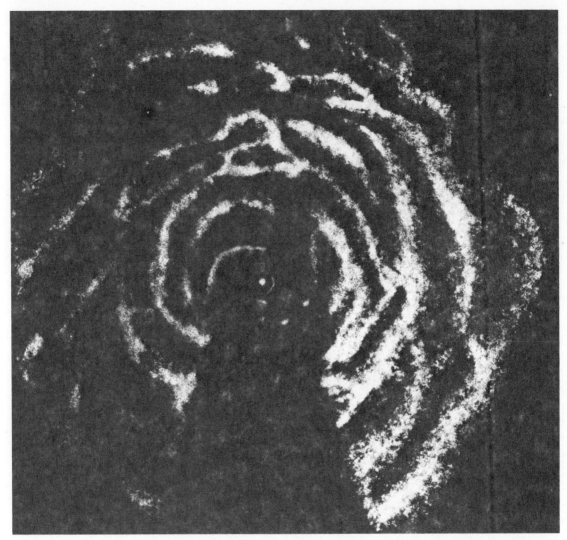

FIGURE 13-28 Map of the location of neutral hydrogen in the Milky Way as disclosed by 21-cm radio noise; the right half was mapped in Australia, the left half in Leiden, The Netherlands (After Westerhout).

Some of the near stars are associated in loose, gravitationally-bonded clusters traveling more or less as a group. Associations even more loosely bonded than galactic clusters may represent young groupings still in the process of forming into one of the stellar subcommunities. Globular clusters, relatively dense and compact aggregations, attend the galaxy rather than exist within it. Most of them are well removed from the galactic plane and are clustered around the galactic nucleus. Apparently globular clusters have had an independent history; they are among the oldest systems known.

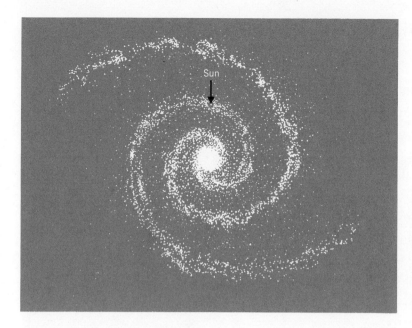

FIGURE 13-29 Schematic plane view of the Milky Way Galaxy as seen from the north galactic pole. Rotation is clockwise. The position of the arms is adjusted to the present evidence for 21-cm radio noise (After McLaughlin, Struve, and others).

Two distinct populations of stars are recognized, although some investigators assert that more than two categories should be considered. Population I stars are the main-sequence stars, found almost exclusively in the spiral arms of galaxies in close proximity to clouds of gas and dust. They are thought to be young stars. Population II stars appear to have completed their sojourn on the main sequence; most are red-giant stars or stars called cluster variables. Most population II stars are found in the vicinity of galaxian nuclei, in elliptical galaxies, and in globular clusters; all are regions that are essentially free of gas and dust.

The broad outlines of the Milky Way Galaxy have been determined from radio exploration of the neutral hydrogen gas distribution. The Milky Way is quite certainly a spiral galaxy with at least several arms, and it is about two-thirds the size of the Andromenda Galaxy. Observation of the residual radial velocities of many stars in the vicinity of the sun discloses a galactic rotation parallel to the plane of the

Milky Way. This apparent rotation is in good agreement with gravitational theory; a stationary assemblage of stars would surely gravitate into a catastrophic collapse.

The presence of interstellar gas and dust is well verified. The dust, mostly in clouds called dark nebulae, obscures the stars behind it and reddens and weakens the light from stars within it. Emission nebulae testify to the existence of gas clouds; the sharp spectral lines that do not partake of the Doppler shift originate from interstellar gas of extremely low density.

The positions of the gas and dust clouds and the association with them of early main-sequence stars, the distribution of the stars in a slab or wheel-like configuration called the Milky Way, the rotation of the assemblage in the plane of the Milky Way, the distribution of the globular clusters, and the certainty that many known nebulous appearing objects are external galaxies—all contribute to our knowledge of the fine structure of the galaxy.

Questions for Review

1. What characteristics of shell stars are deduced from their spectra?

2. What is a planetary nebula? Describe its properties.

3. How are pulsars identified?

4. Distinguish between *galactic cluster* and *association*.

5. How do we deduce that globular clusters are exceedingly old systems compared to galactic clusters?

6. What is meant by *population I* and *population II* stars? Discuss the general differences between these populations.

7. Do we know with absolute certainty that the Milky Way is a spiral galaxy? Explain.

8. How can the rotation of the Milky Way Galaxy be verified?

9. Discuss the evidence for interstellar gas. Does the gas permeate the space between galaxies?

10. Name and give the characteristics of the several types of nebulae. What is wrong with the apellation, *Andromeda nebula*?

Prologue

How did the world begin? How did it become what it is now? These questions kindle our imaginations today, just as they intrigued the speculators of old. Ancient civilizations gave answers to such questions in the form of cosmological legends and myths. These variously describe how the gods created the world out of nothing, or at least out of the primeval chaos that they themselves brought into being.

At present we can go only so far in our scientific analysis of the nature of the universe. Sooner or later we come up against questions such as: What is out beyond that which we presently know? What happened before that?

We do not yet know how the universe originated, if indeed it had an origin within our understanding of the term. We may never know. At present these questions reach beyond the frontiers of our knowledge and are thus more properly in the domain of the philosophers; certainly they are beyond scientific purview.

In any event, it is essential that we separate myth from fact to avoid being ensnared in pure speculation. That is why we should apply the laboratory proven laws of physics toward the understanding of astronomical phenomena. Of course, there is no *harm* in speculation, provided we recognize it for what it is. Indeed, the activity is certain to enrich our lives, and in connection with astronomy, make our whole searching effort eminently worth while.

All Else
14

We noted earlier that the concept of the Milky Way as an organized system of stars pervaded astronomical and philosophical thinking during the 18th century. Among the proponents were the English astronomer, Thomas Wright, the German philosopher, Immanuel Kant, and the Italian astronomer Fr. Secchi. Wright and Kant, together with E. Swedenborg in Sweden and J. H. Lambert in Alsace, discussed the possibility of other stellar systems beyond ours as being consistent with a natural hierarchy. With the aid of telescopes, faint patches of nebulous light, mostly elliptical or circular in appearance and possessing unmistakable symmetry, had been found in large numbers by the middle of the 18th century. Concerning them, Kant wrote in his *Universal Natural History and Theory of the Heavens,* " . . . *analogy between them* [the nebulae] and the system in which we find ourselves . . . is in perfect agreement with the notion that these objects are [island] universes, other Milky Ways."

Nevertheless, the recognition that these particular nebulae were extragalactic was slow in coming, in spite of the vigorous activity in discovery and cataloging during the 100 years from 1750 to 1850. The Magellanic clouds, the Andromeda galaxy (M31), and a bright spiral in Triangulum (M33) are the only galaxies visible to the unaided eye. All four of these were known in the Middle East as early as the 11th century. Subsequent discoveries were paced by developments in telescopic techniques. Charles Messier, a comet hunter, in 1781 com-

piled a list of 103 nebulous objects including some clusters and a few gaseous nebulae that might be mistaken for comets, as an aid in his and other comet hunter's searches. The Messier catalog contains so many objects accessible with binoculars and modest amateur telescopes that it is reproduced in Appendix 8.

The first large efforts at compilation of the nebulae were undertaken by Sir William Herschel in England and by his son, Sir John Herschel, in South Africa. Between them they produced several catalogs that by 1864 jointly contained nearly 5100 entries. These lists were expanded into the *General Catalogue* in 1888 by J. Dreyer, with 7840 clusters and nebulae identified. By 1908 nearly 15,000 objects were listed in the *New General Catalogue* (NGC) and two later supplements. The identification such as NGC 224 refers to the 224th entry in the catalogue. NGC 224 is the Andromeda Galaxy; it is also listed as number 31 in the Messier catalog (M31).

Some of the NGC objects are now known to be star clusters, and a few others are gaseous nebulae. But by far, the nature of these objects was originally a matter of controversy. Some investigators held that the nebulosities were small, symmetrical gas clouds near at hand. Others maintained that the objects were extremely remote, and that as such they were unresolved systems, or galaxies, of stars. These were the objects described by Kant and later popularized by the German naturalist, Baron von Humboldt, as *island universes.*

The controversy was not settled until 1924. During the first two decades of the 20th century, the 60-in. and 100-in. reflectors were placed in operation at Mount Wilson in California. Both instruments resolved individual stars in the nearer nebulae, although not all astronomers, including those taking the photographs, agreed that the stars were indeed real. However, during 1923 and 1924 Edwin Hubble at Mount Wilson discovered cepheid variables in several of the nebulae, including M31. The cepheids were only eighteenth magnitude; clearly, these supergiant objects and the systems containing them were very remote. According to Hubble's calculations, M31 was found to be (in error, it later developed) about 750,000 LY distant. Thus, with the discovery of the cepheids, the resolution of M31 into stars proved to be real; in reference to the previously controversial objects, the word *nebulae* had to be dropped and *galaxies* substituted in its place.

Distribution of Galaxies

Once he had ascertained that the symmetrical nebulae were galaxies, Hubble turned his attention toward mapping their distribution. The field of view of the 100-in. telescope is so small that scrutinizing every bit of the celestial sphere that is visible from Mount Wilson was out of the question. The task would have taken thousands of years. Hubble was forced to compromise with a statistical sampling covering 1283 representative areas.

His results were electrifying. Rather than being scarce, and therefore extraordinary objects, galaxies appeared to dominate the distribution of matter in the visible extent of the universe. Hubble's actual count showed 44,000+ galaxies. Extension of the results of the sampling to the entire celestial sphere led him to estimate that at least 75 million systems were within the range of the 100-in. telescope. Today, with the availability of the 200-in. Hale telescope at Mount Palomar and with modern fast, red-sensitive plates, it is estimated that over 1 billion galaxies are within our grasp. Except for a marked tendency for local clustering (Figures 14-1 and 14-2), the largest counts of galaxies per unit sample area are found near the galactic poles. The counts fall off with decreasing galactic latitudes, until finally galaxies appear to be nonexistent near or within the Milky Way. Originally these latitudes were referred to as the *zone of avoidance;* now we know that interstellar obscuration is solely responsible.

Another feature of the distribution is that for each successive fainter magnitude, the count of galaxies increases by a factor of 3 or 4. This is contrary to the statistics of analogous stellar counts that showed an actual decrease with fainter magnitudes. Presumably, the galaxies statistically are uniformly distributed in space. The tendency toward clustering does exist, but this does not disturb the large-scale averages. Other than this apparent uniformity of distribution, nothing else can be learned about the real space distribution of galaxies (except the relative distances between clusters) without knowledge of intergalaxian distances. Even today this problem can be only partially solved.

Galaxian Associations. More than half of all of the observed galaxies appear to be associated with one or more other galaxies. Associations often occur as galaxian pairs and triplets, but more often they tend to cluster in groups from a dozen or so up to groups numbering in the thousands of galaxies. Thus, the Magellanic clouds form a close pair; their association with the Milky Way forms a loose triplet. Two satellite elliptical galaxies are associated with M31, forming another triplet. All six of these are members of a cluster of 17 galaxies known as the Local Group. There may be other members hidden by the Milky Way, and there may be many dwarf galaxies too faint for detection (p. 376).

Beyond the Local Group, other clusters are seen in all directions; clustering is the rule rather than the exception, at least as far out as we can probe into the universe. Recent evidence developed from new sky surveys suggests that our Local Group is on the outskirts of a cluster of clusters of galaxies, a supergalaxy as it were. The supergalaxy is on

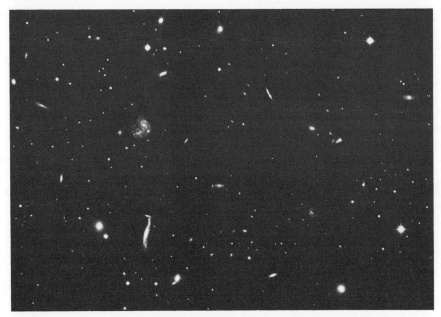

FIGURE 14-1   A cluster of galaxies in the constellation Hercules. Many different types of galaxies are present in this one cluster (Courtesy Hale Observatories).

FIGURE 14-2   A portion of the large cluster of galaxies in the constellation Coma Bernices (Courtesy Hale Observatories).

the order of 120 million light-years across, and is perhaps 30 million light-years in thickness. Membership in this system runs into the tens of thousands of clusters, some of which in turn may contain as many as 10,000 galaxies. Radio observations and radial velocity studies show that the supergalaxy is flattened, and that it is probably rotating around a center marked by the Virgo cluster. Using methods applied to mass determinations of a single galaxy (p. 349) we find that the supergalaxy has a mass of perhaps 1 million billion solar masses. There is some evidence that one or two other supergalaxies exist within the range of our largest telescopes; and more of these distant systems are coming to light each decade or so.

Upon examination of the plates taken in the National Geographic-Mount Palomar Sky Survey, George Abell found that, on the large scale, clusters of galaxies are uniformly distributed out to the extreme limits of visibility. But on the small scale, the clusters tend to clump into second-order clusters whose diameters are on the order of 40 to 50 million parsecs. The most remote clusters found in this survey are estimated to be 2 or 3 billion light-years distant.

One intriguing statistic is worth mentioning here, because it helps to put the sheer immensity of the universe in proper perspective. If we could suspend an infinite number of tennis balls in space, in all directions such that every ball is 50 feet from each of its nearest neighbors, we would have a fairly reasonable scale model of the apparent distribution of the galaxian clusters in the known universe.

Relative Distances to Clusters of Galaxies. Even though the problem of galaxian distances is not yet satisfactorily solved, the *relative* distances of clusters of galaxies can be found with an acceptable degree of confidence. As with stars in star clusters, all of the member galaxies of a cluster of galaxies are at the same mean distance from us. In rich clusters (ones containing many member galaxies of various types) there is a strong probability that some of

the members are highly luminous. The same situation holds true statistically for other rich clusters. One can therefore assume that, statistically, the intrinsic brightness of these most luminous galaxies is the same from cluster to cluster. Hence we can invoke the inverse square law to find the relative distances of the clusters from us. Suppose, for example, the brightest galaxies in cluster *A* appear to be 9 times fainter than similar galaxies in cluster *B*. Then cluster *A* must be 3 times farther away than cluster *B*. In practice, galaxies of middle luminosities are used because statistically there should be less variance in their intrinsic brightnesses than there is in the highly luminous galaxies. Nothing here, however, tells us how far away *any* of the clusters are.

If we could find the actual mileage to *any* nearby rich cluster, say by resolving stars in one of its member galaxies, we would have a zero point on a galaxian scale of distances that would enable us to compute the dimensions of the known universe. But we are unable to establish such a zero point—yet. No rich cluster is near enough. Our estimates of the extent of the universe depend upon a different, more gross statistical yardstick, one that has great cosmological implications. But before examining this final distance measuring technique, we need to clear up a few other details about galaxies in general.

Galaxian Distances. Individual indicators in the form of supergiant class O and B stars, novae, and periodic variables, each type of which has a known average absolute luminosity, can be used to mark the distances to galaxies in much the same manner as was employed for star clusters. But galaxies are so remote that the method works only for the nearer systems. As with the stars, the *distance modulus* (the difference between absolute intrinsic brightness and apparent luminosity) would lead directly to the distances of the galaxies in question, provided they were all of the same brightness. But at the present state of our knowledge, we have no reason to suppose that all galaxies have the same intrinsic luminosity. Indeed, galaxies are typed according to gross differences in appearance; the existence of different types argues against a common absolute luminosity hypothesis.

On the other hand, the distance modulus for marker objects within the galaxies can be used, since they are effectively at the same distance as the systems in which they are located. Among the variable stars which are useful as indicators, novae are less reliable than cepheid variables: there is a greater spread in the absolute luminosities of the novae than there is with cepheids. Hubble was aware of this, and he accordingly appealed to the cepheids and their period-luminosity relationship to arrive at his estimate for the distance of the Andromeda Galaxy of about 231,000 pc, or 750,000 LY.

In the early 1950s, Walter Baade at Mount Palomar ascertained that the brightest population II red giants in M31 were just at the limit

of visibility in the 200-in. telescope; the cluster type (RR Lyrae) variables by which the zero point of the period-luminosity diagram was determined could not be seen. The observed cepheids in M31 had to be population I classical cepheids, brighter as a class by 1.5 magnitudes than the cluster type variables. As a result, the distance to M31, and to all other systems for which cepheids were used as indicators, was revised upwards by at least twice the original estimate. When correction was made for interstellar dimming effects, the Andromeda Galaxy was found to be more nearly 2 million light-years distant.

Other primary indicators must be employed for more distant systems. Out to about 80 million light-years, the novae and the brightest class O and B stars are useful. Best present estimates using the mean of the values obtained with the various markers places M31 at 2.28 million light-years distant, the Magellanic clouds at about 190,000 LY, and the Virgo

FIGURE 14-3 The Whirlpool Galaxy in the constellation Canes Venatici (M51; NGC 5194). The satellite galaxy in the lower center is NGC 5195 (Courtesy Lick Observatory).

FIGURE 14-4 The elliptical galaxy NGC 1201. External globular clusters can be seen in this long exposure photograph (Courtesy Hale Observatories).

FIGURE 14-5 The large Magellanic cloud. With a declination of −70°, this irregular galaxy is not visible in the United States (Courtesy Lick Observatory).

cluster of galaxies at from 40 to 50 million light-years. These distances are presently believed to be reliable within 20%.

But galaxies within 80 million light-years are near neighbors of the Milky Way systems! In actual count they barely dent the statistics; the vast majority of galaxies are much more distant. We must resort to other and, unfortunately, more uncertain statistical methods in the distance determination of remote systems.

Statistical studies show that on the average, the nearer galaxies are all of about the same intrinsic brightness, equivalent to 100 million suns, within a factor of 2. Therefore, *on the average* the apparent brightness of a galaxy is a measure of its distance from us. The reliability of this statistic is only 50%. And even this level of accuracy is limited to groups or clusters of galaxies whose various types can be ascertained. Indeed, among the nearer galaxies, the *range* in intrinsic brightness that yields the

FIGURE 14-6 The barred spiral galaxy NGC 3992 (Courtesy Lick Observatory).

average value given above approaches 1000 to 1.

Within its limitations, this method for estimating galaxian distances is applicable out to about 150–200 million light-years, far short of the suspected limit of visibility of galaxies. Beyond this distance, although galaxies can be counted, it is not possible to identify types, which is essential if we are to arrive at mean absolute luminosity values. Fortunately, we have one further recourse—to be argued presently.

Types of Galaxies

The galaxies' sizes, their masses (estimates) their star counts, their stellar populations, and their absolute luminosities are related among the galactic types. Three main categories of galaxies are recognized: spiral (Figure 14-3), elliptical (Figure 14-4), and irregular (Figure 14-5). There are also subgroups within the first two categories, of which the barred spiral (Figure 14-6) is the most prominent. When seen in, or nearly in, plan view the spirals show a central condensation of unresolved stars, with patterns of spiral arms winding helically inward toward the center. Invariably the spiral arms contain dust and gas; it is in the arms that practically all of the population I stars reside. The central nuclei consist almost exclusively of population II objects. When seen edge-on a spiral galaxy appears more elliptical, attesting to its rapid rotation, with a belt of obscuring matter around its girth (Figure 14-7). An unusual type, a source of radio noise, is shown in Figure 14-8. About 75% of the visible galaxies are spirals.

Galaxies without spiral arms are called *elliptical galaxies*; about 20% of the known galaxies are of this form. In appearance they range from nearly circular (or spherical) to systems definitely flattened in their equatorial planes. The elliptical appearance is generally a perspective effect. In plan view, as with spiral galaxies, most of the elliptical galaxies would appear circular. On the other hand, the very large percentage of circular-appearing elliptical galaxies that we observe cannot result solely from a surfeit of plan views. Many of the galaxies must be truly spherical.

Elliptical galaxies defied resolution of their individual stars until adequate red-sensitive plates and large telescopes became available. In 1944 Walter Baade showed conclusively that the stars in elliptical galaxies are almost pure population II types. With one or two exceptions, no elliptical galaxy shows evidence of neutral hydrogen or of obscuring dust clouds.

Irregular galaxies, exemplified by the Magellanic clouds (Figure 14-5), are genuinely rare. They have no identifiable symmetry and appear only as ragged clouds of stars containing occasional bright patches of nebulosity. The large Magellanic cloud does have a longitudinal concentration of stars, but the significance of this is not yet known.

Numerous attempts have been made to classify the various galactic types into an ordered evolutionary scheme. Indeed, the presence of dust and gas, together with a high percentage of type A stars in the irregular galaxies, suggests that they are young systems. Similarly, the absence of gas and dust together with stars of almost pure population II types argues persuasively that the elliptical galaxies are old systems. There are, however, numerous technical difficulties with such a hierarchy that need not trouble us here. We do not have sufficient evidence to firmly support an evolutionary hypothesis. We had best consider this subject as unfinished business, leaving it to future investigators to prove or disprove the theory.

Spectra and Composition of Galaxies. When spectrograms are taken across the extent of a galaxy, we observe that the spectrum is composite, with the dominant lines at any moment being those of the stars predominating at the spectrograph slit. In general, when the slit is focused on the nucleus of a galaxy, we are

FIGURE 14-7 The galaxy NGC 4594 in the constellation Virgo seen edge-on (Courtesy Hale Observatories).

FIGURE 14-8 A galaxy of unusual type, NGC 5128, in Centaurus. The galaxy is a source of radio noise (Courtesy Hale Observatories).

able to determine the nature of the predominant class of stars, since the bulk of the stars are usually concentrated in the nucleus.

In most cases the spectral lines are broadened, showing that turbulent motions of the stars occur internally. Sometimes the spectral lines are inclined or slanted with respect to the laboratory reference spectrum, as are the spectral lines for Saturn's rings. This is taken to be conclusive evidence for the relatively rapid rotation of a system.

Spectral classes of galaxies range mostly from early G to late K, thus showing the influence on the spectra of the older populations in the galactic nuclei. Some of the open spirals have spectra ranging from class A to G. The early-type spectra are associated with the spiral arms. The letter references here indicate similarities to the corresponding stellar spectral classes; the galaxian spectra show merely strong resemblance to, not identity with, the stellar classes.

Nearly all elliptical galaxies are devoid of population I stars, and they do not show any evidence of emission spectra. Spiral galaxies do have emission spectra in the regions of the arms. These spectra are reminiscent of hydrogen excitation, but there are some emission lines of oxygen that have a low excitation probability. Such spectra are associated with spiral galaxies that have much higher gas concentrations than we find in the Milky Way. The oxygen lines, formerly referred to as forbidden lines because they cannot be produced in laboratory situations, are found only in the spectra associated with interstellar material. In the laboratory the greatest rarefaction of a gas that can be produced is vastly more dense than the tenuous gases of interstellar space. All of the normal emission lines arise from collisions through which the previously excited atoms reach lower energy states. Because of the dearth of collisions, atoms in tenuous gases excited into some higher, but stable energy-configuration through absorption of particular photons may stay thus for long periods before spontaneously emitting the photons. For oxygen, some of these *metastable states*, as they are termed, have periods on the order of 11 million years. That is why the lines associated with transitions from these states are called forbidden; the transitions seldom occur.

As mentioned earlier, irregular galaxies have spectra dominated by class A main-sequence supergiants. In the Magellanic clouds, the 21-cm radio spectrum line is strong, showing large amounts of neutral hydrogen. Gaseous emission nebulae are also present within the confines of these galaxies. While no dust is directly visible, it must be present because the galaxies represent zones of avoidance for more distant galaxies in the same direction.

Mass and Size of Galaxies. Once, if ever, the distance to a galaxy is known, its apparent size is a measure of its true dimensions. The same laws of geometry apply here as with the diameters of solar system bodies, but the statistics are sketchy. For the overwhelming majority of galaxies, we are troubled with uncertainty about distances. But assuming that whatever information we have is essentially reliable, we find that spiral galaxies range from 15 to 20 thousand light-years in diameter, and that they are from one-third to one-half again as large as the elliptical galaxies. Dwarf ellipticals, which are very open, almost nebulous appearing aggregates of stars, generally do not exceed 7000 LY in diameter. The irregular systems are not much larger. We must exercise care in specifying the limits of elliptical galaxies. Successively longer photographic exposures disclose galactic stars at increasing distances from the center. Apparently the stellar concentrations trail off with distance from the galactic nucleus until they imperceptibly merge in view with the background stars of our own galaxy and are thus lost to view.

As far as we know, the Milky Way and the Andromeda Galaxies are giants among galaxies; no evidence of any other galaxy even approximately as large as they are has yet been discovered. We would expect the size of a galaxy to bear some relation to its mass, because the galaxies do consist of stars whose range of mass is comparatively small. The difficulty in resolving individual stars in the nucleus of M31 gives some indication that the mass of the system is large; at least the stellar density is not low, and the system is extensive.

Computation of the mass of a galaxy follows along these lines. Suppose we have determined the distance to a representative galaxy. Then its apparent angular size is a measure of the diameter of the galaxy, and we can express this dimension in terms of convenient distance units. For simplicity, let us further suppose that the subject galaxy is a compact spiral, and that we are successful in determining the mean radial velocity of a compact group of stars, preferably an aggregation closer to the nucleus than to the outskirts of the galaxy. Since we know the size of the galaxy, we can calculate the mean orbital radius of the group of stars, and therefore the orbital path length, which is just 2π times the radius. Also, from the radial velocity we have the speed of the group along the path. The period for one circuit of the stars is just the path length divided by the speed.

Now we invoke Kepler's harmonic law which yields a rough figure for the mass; admittedly Kepler's law is not completely valid, because the mass of the galaxy is not concentrated at the center of this system. But by obtaining radial velocities of member stars at two or three widely separated points, say at the one-third and two-thirds points from the center, perhaps even one on the periphery of the system, the methods of higher mathematics allow us to deduce a reasonable value for the mass. In this way we estimate that M31 has a mass equivalent to about 300 billion suns; a dwarf elliptical galaxy may have a mass as little as 5 billion solar masses. On the average, the mass of galaxies ranges between 60 and 70 billion solar masses.

The Red Shift

One of the most provocative discoveries in modern astronomy, one that has profound cosmological implications, is the red shift observed in the spectra of galaxies. This red shift is indistinguishable from the Doppler shift (Figure 14-9), and it applies to radio as well as to optical spectra. The spectral lines of all distant galaxies are shifted toward longer wavelengths by amounts that are proportional to the distances to the galaxies! Moreover, the shift is proportioned properly among all of the observed wavelengths. Therefore, the presumption is that the red shift indicates progressively higher velocities of recession for galaxies at increasing distances from us. Edwin Hubble found that the recession velocities increase by 20–25 km/sec for every million light-years' increase in distance. The remotest galaxies that we know about are receding the fastest, with speeds in the neighborhood of 125,000 km/sec, or slightly more than 40% of the speed of light! We shall return to this topic presently and examine its implications.

Radio Galaxies and Radio Sources

Galactic radio noise is not limited to the 21-cm wavelength of neutral hydrogen. After Jansky's classic discovery of cosmic radio waves in 1931, succeeding investigators discovered a fairly intense source of 1- to 2-m radio waves in the plane of the Milky Way. The most intense emanations come from the direction of the galactic center. In the early 1940s British radar disclosed periodic radio noise originating from the sun. But they feared that this announcement would tip off the enemy that radar was an accomplished fact, so public disclosure of the solar phenomenon was not made until after World War II.†

Since then, several thousands of discrete

† *Radar,* an acronym for *Radio Detection and Ranging,* was one of the best kept secrets of World War II.

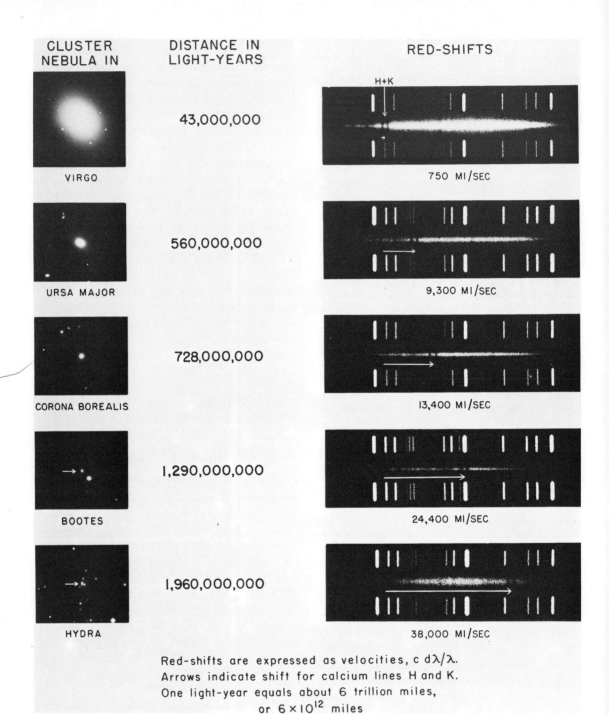

FIGURE 14-9 The relation between red shift (velocity) and distance of galaxies (Courtesy Hale Observatories).

FIGURE 14-10 The Hydra cluster of galaxies, estimated to be more than one billion light-years distant (Courtesy Hale Observatories).

radio sources have been cataloged; the growing evidence is that most are extragalactic. By the early 1970s about 100 of these sources had been identified with individual galaxies, most of them spirals like the Milky Way. The radio emissions are now known to cover the entire radio spectrum.

For some inscrutable reason certain *peculiar* galaxies emit thousands of times more radio energy than does an average galaxy. One such peculiar galaxy is shown in Figure 14-8; another, M87 (NGC 4486 in the Virgo cluster), is seen in Figure 14-11. With ordinary pho-

tographic exposures, this latter galaxy appears to be a giant, but otherwise normal, elliptical galaxy. But with very short exposures that suppress the stellar images in the outer portions of the galaxy, an unusual "jet" is revealed. The radio noise identified with this luminous feature is believed to be emitted by very high speed electrons that are spiralling around lines of force in an exceedingly strong magnetic field. This type of energy release is known as synchrotron emission.

One of the strongest sources of radio energy known originates in association with an optical

FIGURE 14-11 Elliptical galaxy NGC 4486 (M87) showing the nuclear "jet" (see text) (Courtesy Hale Observatories).

system in the Cygnus cluster, estimated to be about 700 million light-years distant. The apparent radio brightness is many times greater than the apparent optical brightness. This object, known as Cygnus A, appears to be two galaxies in collision (Figure 14-12). The astronomical use of the term collision here has a different connotation than we customarily employ. Two galaxies can collide, indeed, *even pass through each other,* with little or no danger or expectation of a single collision between stars! The effect resembles that of two puffs of smoke that intermingle without losing their separate identities. In spite of their compact appearance, galaxies are overwhelmingly empty space. The collision would not be evident to inhabitants of "solar" systems within the galaxies, except through long-term studies of the motions of the stars. Probably, no two stars ever come within 1 or 2 LY of each other. The radio noise apparently arises from collisions between the atoms of gaseous clouds associated with each of the two galaxies that have extraordinarily high velocities relative to each other.

There is one remarkable feature of the radio emissions from some of the peculiar sources. Often *two* centers of radio emission are located on opposite sides of the galaxy, at distances on the order of 100,000 LY from the center!

Many hypotheses have been advanced in attempts to account for the phenomenal amounts of radio energy being released from a given galaxy, amounts that approach in magnitude the visible light output of the galaxy. Among the theories are these: that the radio energy originates in nuclear chain reactions between supernovae; that fission of galactic nuclei is responsible; or that it arises from potential energy release in the gravitational condensation of a galaxy from primeval matter. Perhaps the most acceptable hypothesis concerns the nature of the double sources of radio noise in peculiar galaxies. These radio sources are believed to be associated with huge clouds of gas that have been, or are being, ejected symmetrically from galaxies, and hence are being rapidly dissipated in space. Evidence that some of the extended sources are fading in intensity lends credence to this theory. But as with so many other problems of astronomy on the universe scale, we must classify as unfinished business the subject of radio sources. It should be considered as business that is just getting underway.

Quasi-Stellar Objects. In December, 1960 Allan R. Sandage of the Hale Observatories announced discovery of a starlike object at the precise position of the strong radio source 3C 48 (source number 48 in the *Third Cambridge Catalogue*) (Figure 14-13). If indeed this were a star, it would mark the discovery of the first *stellar* radio source other than the sun. The spectrum of the object was, however, unlike that of any known celestial object.

In 1963 another radio source, 3C 273 (Figure 14-13), was identified with an even brighter starlike object by Maarten Schmidt of the Hale Observatories. Schmidt succeeded in obtaining a spectrum of the optical source, and he identified the spectral lines as having been

FIGURE 14-12 The radio galaxy Cygnus A (Courtesy Lick Observatory).

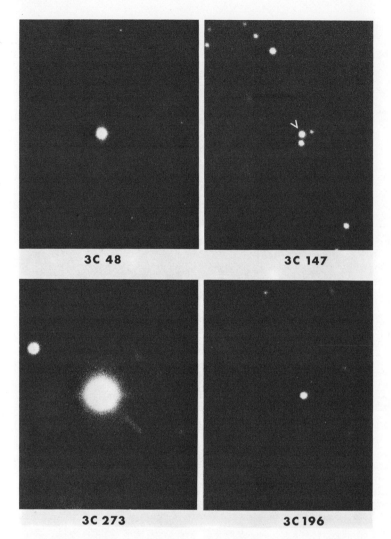

FIGURE 14-13   The quasi-stellar radio sources 3C-48, 3C-147, 3C-196, and
3-273, photographed with the 200-in. telescope (Courtesy Hale Observatories).

red shifted by 16%. Immediately, J. Greenstein and T. Matthews of the Hale Observatories identified the hitherto inscrutable lines of 3C 48 as those of familiar substances that had been shifted toward the red by 37%! The percentage here refers to the speed of light. Between 1963 and 1971, almost 1000 optical objects were identified with radio sources.

All of the spectra obtained from these mysterious objects show enormous red shifts that, as is the case with the galaxies, cannot be distinguished from Doppler shifts. In one case, the red shift corresponds to a velocity of recession of 82% of that of the speed of light. By comparison, the red shift of the remotest galaxy thus far known is only 46%. In every case the optical objects associated with the radio sources appear as fuzzy points of light, defying further resolution. Evidently all of the energy emission, light *and* radio noise, originate from a source of extraordinarily small dimensions, one that is stellar, not galactic, in size. For these reasons such objects have been named *quasi-stellar radio sources,* or *simply quasi-stellar sources* (QSS). Popular writers have referred to them as quasars, a term somewhat frowned upon by many investigators.

The optical appearance of four QSSs is shown in Figure 14-13. Although the quasi-stellar objects differ considerably in brightness from each other, they are nevertheless all extremely luminous objects. At optical frequencies they radiate strongly in the blue and violet region of the visible spectrum, and they are excessive emitters in the ultraviolet. If the red shift is a real measure of their distances, the QSSs are the most luminous objects known, being exceedingly brighter than the brightest elliptical galaxies. Moreover, the QSSs vary in the intensity of both radio and visible light emission, making abrupt changes of several magnitudes in relatively short periods of time, on the order of a few weeks.

Thus, based upon acceptance of the red shifts as true distance indicators, and by following the same rules applicable to distant galaxies, we have a very enigmatic situation. Here are extremely luminous objects of very small size (no more than a few light-months in diameter), emitting 100 times the luminosity of our own galaxy from a volume 10^{17} (100 million billion) times smaller, and varying this brightness sixfold or tenfold in devastatingly short periods.

Interpretation of the Red Shift of Quasi-Stellar Sources. The only known phenomena that can produce red shifts in spectra are exceedingly strong gravitational fields and velocities of recession of the emitting bodies. In the case of QSSs, gravity fields have not been ruled out, but they seem most unlikely. A mass configuration that satisfies both the luminosity and gravity requirements is theoretically self-contradictory. Because of this, the current temporary consensus is that QSSs do in fact have phenomenally large radial velocities of recession. Incidentally, if even one QSS could be discovered that showed a Doppler blue shift, then the question of the authenticity of the spectral shifts as resulting from radial motion would be answered. But the fact that *all* QSSs show red shifts raises the spectre of doubt, or at least of uncertainty, as to the true explanation.

If the QSSs obey the same velocity-distance relation that applies to the galaxies, then most of them are between 5 and 15 billion light-years distant, and they are accordingly phenomenally bright. It is instructive to recall that the most remote cluster of galaxies known is no more than 3 billion light-years away, if that far, and the individual galaxies in the cluster are so faint that they are just at the threshold of visibility. On the other hand, if the velocity-distance relationship for QSSs is different, then the mandate for superluminosity is waived. But they cannot be very close, otherwise they would show proper motions, a circumstance as yet undetected in any QSS. Thus they are certainly extragalactic objects, over 600 and very likely more than 60,000 LY distant.

But now, if we were to accept the notion that the QSSs are outside our galaxy but much closer than galaxies with comparable radial velocities of recession, we would be tormented with the questions of where they came from, and what gave them their speeds. Because of the turbulent flux of ideas concerning the nature and origin of the QSSs, and because most investigators agree that all of the current ideas are probably wrong anyway, we shall not dwell on the various theories here. The reader who is curious about the various hypotheses will find the reference indicated in the footnote of more than passing interest.†

Quasi-Stellar Galaxies. Since the known QSSs are radio emitters—indeed, they were

† Burbridge, G., and Hoyle, F., "The problem of the quasi-stellar objects," *Scientific American,* **215,** December 1966, pp. 40–52.

found through investigations of the radio sources—the question arises as to whether or not there exist similar appearing objects, having similar red shifts, that do not emit radio noise or, if they do, that the emission is too weak for detection. In 1965 Sandage initiated an investigation aimed at answering this question of the very blue, very faint objects considered up to then to be member stars of the galactic halo. His discovery of very large red shifts marked some of them as quasi-stellar objects that had escaped earlier notice because of the absence of radio emissions to serve as beacons. It now begins to appear that these nonradio quasi-stellar galaxies (QSGs), as Sandage has named them, outnumber the QSSs, and that the latter may represent just an earlier, transitory phase in an evolutionary process leading ultimately to QSGs. Studies of the QSSs and QSGs are presently at the very frontiers of modern astronomical investigation.

The Cosmological Ferment

And so we come to the end of the exploratory phase of this book. We have traced astronomical growth from its earliest subtle imagery in the minds of men, through millenniums encompassing the advance of astronomical technologies and beliefs. The perceptive reader will note that the slow beginnings of astronomy gathered momentum equally as slowly, until suddenly, progress in astronomy erupted in a headlong rush. The impetus, of course, was given coincidentally by the invention of the telescope and by the freedom of thought that man began to enjoy after the slumber of the Dark Ages.

From this point astronomy becomes highly speculative. And, as is necessarily the case, the arguments pro and con anent the origin and evolution of the universe, and its organization, reflect the convictions and biases of those who do the arguing. In this usage, the term *argument* refers to the reason or reasons offered in a proof of a statement or belief. There

are those who will insist on a mechanistic view of creation, one that refers to a natural process that is mechanically determined and that can be explained according to the laws of physics and chemistry. Then there are others who cannot surrender a teleological philosophy, where ends are achieved or processes are determined in accordance with a design or purpose. In fact, the term *cosmology* itself denotes a branch of metaphysics that treats of the character of the universe as an orderly system, or *cosmos*. Thus by definition cosmology is philosophical. And lastly, we must admit that religious convictions, ranging all the way from atheism to theology, are bound to color individual cosmological thought.

But that is not to say cosmologists are casting about in a void of unprovable assertions in search of the ultimate explanation for the universe as we find it. Indeed, a scientific cosmology necessarily includes the philosophy of *cosmogony*, which deals with theories concerning the origin and evolution of the universe. We, mankind, do not have all of the tools or all of the data with which to finally resolve the issue. But based upon what we have learned, we can avoid the ghostly pitfalls of emotion, of mythology, and of astrology, and thereby intelligently explore the various cosmological hypotheses in search of one that is personally satisfying. As it should be, the choice is a matter for the individual conscience.

An anecdote, of more than casual importance because it focuses attention on the emotional aspects of cosmological thinking is appropriate here. It concerns a certain archbishop, a biblical chronologist, who called together his bishops more than 300 years ago to inform them that through his intensive studies of the Scriptures he had discovered the precise date and time that the Almighty created the universe. It was accomplished, the archbishop said, on Sunday, October 23, at 2:30 in the afternoon, in the year 4004 B.C. After an agonizing period of silence one bishop is reported to have asked, "And pray, Holy Father, *what* was God doing *before* he created the universe?" The archbishop drew himself up to his full height of 4 feet 10 inches, and thundered, "He was creating Hell for those who ask questions such as that!"

Most teachers of astronomy have, at one time or another, been challenged during discussions on cosmology by students with strong fundamentalist beliefs concerning the biblical account of the Creation. To which the astute instructor can respond by pointing out that *two different and independent accounts* of the Creation appear in the Book of Genesis, and that even here the individual must choose between the alternatives to the satisfaction of his conscience.

The Cosmological Dilemma. A dilemma is an argument presenting an antagonist with two or more alternatives, each equally conclusive against him. Or saying it another way, a dilemma is a choice between equally unsatisfactory alternatives. Whether the choices available to a modern cosmologist are *equally* unsatisfactory remains to be seen. But certainly all choices presently open to him are unsatisfactory. Let us see why and how:

Our instantaneous snapshot view of the universe is, on the large scale, one of galaxies uniformly distributed in direction. Moreover, the red shift of galaxies increases in the same manner, uniformly in all directions. The galaxies are composed of stars all behaving in the same way, that is, doing the thing that stars *do*. The detailed differences among individual stars are found to be the same uniformly in all directions. (This is not to be confused with local, small scale situations.) The physicist would call the universe *isotropic* (having the same properties in every direction).

But what, we may ask, was the snapshot view yesterday, or a million yesterdays ago, or a million million yesterdays before that? And what will the instantaneous picture be tomorrow, or a million tomorrows after tomorrow?

The task of the cosmologist is to sift the evidence available only in today's view, and then attempt to reconstruct the past and predict the future on the basis of the universe as he can see it at this instant. The evidence does not lead to a single conclusion in either direction.

The task is complicated by a situation that does not confound our every day behavior. We are aware that the speed of light is constant *and* finite. In the end, except for information provided by all of our senses excluding sight, which senses by their very nature are woefully circumscribed, information that is transferred to and from us rides along on a beam of electromagnetic radiation — light, radiant heat, or radio waves. There exists a lag between the occurrence of an event and the arrival of an electromagnetic signal about it, of 1 millisecond for every 300 km distance. This is equivalent to a lag of one ten-millionth of a second over a distance of 30 m. This lag offers no inconvenience; our lives are unaffected by such minute discrepancies between the time of an occurrence and our apprisal of it.

But the cosmologist is faced with a different situation. Information about an event on the sun, say a solar flare, arrives at the earth some 8 minutes after its onset. The annihilation of a star by a supernova explosion, say in M31, will not be known on earth until nearly 2 million years later. Indeed, the apparent distribution of galaxies is immersed in the complication provided by the time of travel of light waves over a variety of cosmic distances. What we observe about a galaxy at a distance of 1 billion light-years is *where it was and what it was doing* 1 billion years ago! But in the instantaneous snapshot, we are often comparing information about the space position, the luminosity, the motion, and the physical properties of this galaxy with information received from another galaxy, that may have taken only 10 million years to get here. The time lag between the two information signals is 100 millionfold!

Once (if ever) the cosmologist can accommodate the complications of the time element into his viewpoint, even more serious problems will plague his efforts. Suppose that all galaxies were intrinsically the same brightness, a naive supposition, but one that illustrates a serious problem in a relatively simple way. Then from the inverse square law we can infer the *relative* distances to all galaxies. If the distribution of galaxies is homogeneous as well as isotropic, then doubling the distance should increase the galaxy count eightfold; tripling it should produce a galaxy count 27 times as large. Actual counts of galaxies show a rate of increase with distance substantially less than this. If allowed to stand without correction, this feature of the galaxy counts implies a thinning out with distance in all directions, and that we are therefore at the very center of the highest concentration of matter in the universe, a highly improbable situation. Also, the velocity of recession of galaxies appears to increase with distance in every direction. This further supports the contention that we are at the original center of creation.

But now let us incorporate dimming effects into our argument and see where this leads. With respect to the light emitted, the rapid recession of a galaxy results in the dimming of the light from an energy effect and an attenuation effect. According to the Doppler principle, light waves are stretched out, or *attenuated*, because of the recessional motion; the initial and final positions of a wave leave from different points. Longer wavelength means lower frequency; lower frequency implies less energetic photons. There is nothing new here; this is a fundamental principle of radiation theory. In addition, the attenuation means that fewer photons will arrive at the photographic plate each second. The attenuation has no impact on the constancy of the speed of light. The photons that do arrive will still be traveling at this limiting speed. Both effects contribute to the need for longer exposures than would otherwise be necessary. In other words, the photographic record shows galaxies to be dimmer than they would be if they were at rest.

When the galaxy counts are adjusted for these dimming effects, it appears that the number of galaxies per unit volume of space *increases* with distance! Cosmologically, we are no better off than we were; we still appear to be at the center of the universe, but now it coincides with the point of *least concentration* of matter! The fact that we appear to be at the center at all impales us on one of the horns of a dilemma. There is no reason to believe that we should enjoy such a preferred position. Nay, the whole history of astronomical thought from the homocentric, or geocentric, universe, to the heliocentric universe, to a universe centered on our galaxy, and finally to a grander scheme of countless numbers of galaxies, argues against such an assumption.

We become painfully aware of the second horn of this dilemma when we attempt to account for the existence of nearby young galaxies, the irregulars and the spirals, when the evidence cited above points to a universe of galaxies all of the same age. A third antagonism arises from the observed increase in magnitude of the red shift with distance. Two interpretations are open to us. Either the velocity of recession of remote galaxies is unchanging, being the same in the past and in the future as we now observe for each individual galaxy, or the velocity of recession increases as galaxies reach ever greater distances. The speed of recession in this case must be such as to preserve a value characteristic of, or consistent with, the corresponding distance.

The first choice leads to a uniformly expanding universe, one that contains a point in space and time corresponding to its very origin. The other alternative also leads to an expanding universe, but now it is an accelerating one. The principal effect of acceleration is simply to shorten the time span required between the origin of the universe and arrival at the presently observed state. Such a time scale can be shown to lead to an age for the universe that is *less* than that which we currently accept for our own galaxy! This is disturbing, to say the least.

The Cosmologies

The existence of problems such as these, by no means a complete list, provides the cosmologist with a justification for his existence. His task is to find a set of principles, which may even necessitate the discovery of new physical laws, that can fully account for the present (and incomplete) space-time view of the universe. In practice, and amid many variations, two basic cosmological schemes stand out. Of all of the schemes and their variations that have been aired, each does a better job in some areas than its competitors; none produces a model that matches the observed universe with absolute fidelity. As the observational data mount, one of the theories may well prevail over all of the others. Or all may fall by the wayside, to be replaced by newer theories that may or may not suffer the same fate.

The Cosmological Principle.

If the universe appears isotropic from our present vantage point, then it should appear that way from any other point. There can be no preferred vantage point; that is to say, the portion of the cosmos we observe and of which our vantage point is the center is no different from any or all other parts. Whereas there are local variations in the details, all observers everywhere in the universe would see the same large-scale picture. This is known as the *cosmological principle,* and it is an essential postulate for any theory of the universe.

The Expanding Universe.

For a starting point in the examination of the general cosmological models, let us accept the evidence at hand: that isotropy prevails and that the red shifts are truly Doppler shifts. Then we must conclude that the universe is expanding, and as it does so it thins out; the concentration of matter becomes more tenuous with passing time. Provided no new matter is created to fill the void left by departing galaxies, the total mass of the universe remains constant. All of it must have originally been together in one location, in the form of a primeval atom composed of subatomic particles, or even just a blob of energy at extremely high temperature. This would constitute an *initial state,* the "original" creation, in the sense that whatever may have gone before would leave no trace, no record of having been in existence.

For some reason, the primeval atom exploded, sending matter in all directions, with the fastest traveling material being farthest from the center at any given time. Out of this matter condensed all of the cosmic features that we see. Extrapolation from these observed features back to the original cataclysm yields an age for the universe in the vicinity of 15 billion years, with a 50% uncertainty in either direction. The uncertainty arises from lack of precise knowledge of the distances to the remotest galaxies.

Supporting Evidence — the 3°K Background Radiation.

Radio astronomy observations by Arno Penzias and Robert Wilson of the Bell Telephone Laboratories led to the discovery in 1965 of a weak, isotropic, continuous background radiation in the wavelength range from 0.2 cm to 21 cm. The wavelength distribution and intensity of this radiation corresponds to the energy spectrum of a blackbody at 3°K.

The existence of such radiation was theoretically anticipated by G. Gamow several years earlier. Gamow argued that the expansion of the primeval fireball would result in a decrease in temperature with time. Not all of the energy would be converted to matter. Most of it would continue to be thermal in character, degrading ultimately to a temperature of about 3°K in from 10 to 20 billion years. All of the modern estimates for the age of the universe lie within this range!

Its significance is momentous for the "big-bang" theory. The steady expansion and degradation of the initial density and energy levels of the primeval fireball would result in the observed radio noise arriving uniformly from all directions in space. It represents the residual energy of the initial cataclysm. Thus its presence is in theoretical agreement with the big-bang hypothesis, and pending future, contrary evidence, the 3°K radiation must tentatively lend support to it.

Arguments Against the Expanding Universe Theory. Some of the difficulties inherent in the big-bang theory are these. It is predicated on a Doppler shift interpretation of the red shift of the galaxies. Some strong objections have been voiced against this interpretation because of the magnitudes of the velocities of recession that it leads to. One alternative has been suggested: that because of the extremely great distances involved, photons would eventually become tired, that is to say, lose energy along the way. Radiation theory, we recall, relates the energy of photons to the frequency, and thus the wavelength, of the light with which they are associated. Tired light appears redder because it is of lower frequency, and hence of longer wavelength. Thus the apparent red shifts can be accounted for without *necessitating* any recessional motion at all. Except for the violation done to other presumed universal laws such as the law of gravity, a *static* universe could produce the observed phenomena, and therefore it is distinctly possible. It is extremely significant, however, that there is no supporting evidence for such a hypothesis concerning the red shifts, either in observation or in radiation theory.

Another difficulty with the expanding universe cosmology is that there is no evidence that galaxies are slowing down in their headlong flight, nor that they have done so in the past, as they surely should because of the mutual gravitation between the galaxies. Also, the laws of nature, believed to be universally applicable, would have to be severely bent in order to allow a static (motionless) universe. Both of these problems can be reconciled with observations if we abrogate the evidence of the Doppler effect and substitute for it the concept of the reduction in photon energy with distance of travel. On the other hand, an original big bang could have propelled the matter initially at such speeds that the escape velocity dictated by the quantity of mass was exceeded. But even so some slowing down should have taken place before the entities got out of gravitational range from each other, and there ought to be some evidence for it. We cannot test this hypothesis without knowing precisely the density of matter in the universe. We do not as yet have this information.

If the matter that ultimately formed into galaxies had an initial velocity less than the critical escape value, the expansion will slow and eventually stop, to be replaced with a contraction that subsequently will bring all of the matter back together again, perhaps initiating another big bang. Thus an oscillating universe is conceivable. Each cycle, however, would be viewed by any of the universe's inhabitants as a separate Creation. They too would have no record of prior occurrences, or knowledge of the inhabitants of such periods.

A third difficulty plagues the big-bang theory in the form of late comers into the family circle of galaxies. We see galaxies in what appear to be a range of ages from very young to very old. But the big-bang hypothesis requires all galaxies to be of the same approximate age. Attempts to resolve this difficulty led to an alternative, much simpler cosmological model, the steady-state universe.

The Steady-State Theory. The notion that the universe had a finite beginning, and faces certain extinction when all of the remaining gaseous material ultimately condenses and all of the stars cool down from the white-dwarf stage and become nonradiating black dwarfs, is unacceptable to many. Even the idea that the universe is oscillating and only becomes discontinuous periodically is repugnant to those for whom it poses a threat to teleological, even theological, beliefs. Several astronomers, among them H. Bondi, T. Gold, and F. Hoyle, devised a new, perfect cosmological principle, that included *time*. Thus, their cosmological model extended the principle of isotropy so that not only does the universe appear the same from any vantage point; it appears the same at all times: past, present, and future. The notion of the expansion of the universe is retained, but as galaxies move apart, matter is spontaneously created to fill the void; from the new matter, new galaxies are formed. Thus the process of creation is continuous; the universe is infinitely large and infinitely old. There was no beginning and there will be no end.

The continuous-creation theory, as it is sometimes called, accounts for the existence of young and old galaxies appearing side-by-side. Moreover, the new creation will proceed at just the precise rate necessary to preserve a constant density of matter in space and the existing proportions of young to old galaxies.

The Argument Against the Steady-State Theory. The quasi-stellar sources provide the most cogent current argument against the steady-state theory, although it is not the only one. All QSSs have red shifts which, if the velocity-distance relation of the galaxies is applicable, place the overwhelming majority of them at the remote outskirts of the known universe. In other words, QSSs represent objects that were created, or events that occurred, long, long ago in the earliest period of the universe of which there is any record. *These events have not repeated since!* But the steady-state theory requires a more homogeneous distribution of QSSs than the current evidence supports. Of course, if the red shift of the spectra of QSSs is not a velocity-distance relationship, then perhaps they are not all so remote, in which case the steady-state requirements are not violated. But on the basis of the existing observational evidence, we are not justified in pushing such a hypothesis.

Tests for Cosmological Models. There are a number of tests to which the various cosmological models can theoretically be subjected. All of them require accuracy and precision in observations not yet attainable. The steady-state theory requires a constant proportion between old and young galaxies in all directions, however far we look into space. The

evolutionary theory, in contrast, asserts that because we are looking back further in time when viewing more remote galaxies, the most remote system should appear to be the youngest. On the one hand, we should find young galaxies at all distances from us. On the other hand, the oldest galaxies should be the nearest to us. Observations cannot yet clear up this dilemma.

Another test involves the determination of the density of matter in the universe. According to the big-bang hypothesis matter is becoming more tenuous with time everywhere because of the continuing expansion. Hence the nearer regions should contain less matter per unit volume than those more remote regions. The steady-state theory argues that the density of matter on the large scale is isotropic and remains constant. But we cannot as yet determine the density distribution with anything resembling the required precision.

In spite of appearances, that we are at the center of an expanding universe does not necessarily follow from the isotropic recession of galaxies. The following analogy is not difficult to comprehend, and it is easy (in part) to demonstrate. Let us inflate an ordinary spherical balloon to some predetermined size: well below the breaking point, but not flabby. Next, we mark a series of spots all over the surface in such a way that each spot is the same distance, say 1 cm, from each of its neighbors. Finally we resume the inflation of the balloon at a slow, constant rate. We observe that (depending on the uniformity of the balloon material) that each spot moves away from every other spot at the same rate. Moreover, the expansion is in three dimensions; any spot on one side of the balloon moves away from the corresponding opposite spot in a direction perpendicular to the surface. By practicing a little imagination, we can visualize spots uniformly spaced within the volume of the balloon, so that when inflation resumes, they too move away at a uniform rate from each of their neighbors. Thus the apparent isotropy of the universe becomes more understandable. The cosmological principle appears sound. All observers everywhere in the universe will see the same expansion effect, just as we would with respect to any spot on the balloon if we could ride along on one of them. Hence, we are not necessarily at the center of the universe. Nor is there any clue as to where the center *is*.

Relativity and Cosmology

Inevitably any cosmological theory has to take into consideration the possibility that geometries, other than the familiar Euclidean geometry of plane surface or its extension to three-dimensional solids, may prevail in the universe. Other geometric concepts that may more accurately portray the universe arose in connection with certain mathematical investigations by several theoreticians, among them Albert Einstein, around the turn of the 20th century.

In light of these mathematical advances, Euclid's geometry is absurdly rigorous in its requirements for standards of length and shape. It necessitates transformations in the positions of plane figures in order to achieve congruencies that are physically impossible in practice. One of the fundamental Euclidean axioms holds that a straight line is self-evident, a concept that defies proof. Implied in the definition of the term *straight* is an invariance in direction, such as the path of propagation of a ray of light in empty space or in a medium of constant density. That light rays on occasion might *not* follow straight paths was unknown until comparatively recent times. Hence, Euclid did not consider this fact in the development of his geometric proofs. Albert Einstein was among the first to direct attention to this potential weakness in the foundations of existing physical science. This came with the publication of his theories of relativity.

Einstein's *special relativity,* a restricted innovation, rests upon two bold hypotheses. First, Einstein held that absolute measurement is impossible. In the universe, only relative measurements have meaning; that is to say,

there exists no absolute frame of reference against which comparisons in length, mass, and time can be made. Secondly, any observer, anywhere in the universe, will always obtain the same relative value for the speed of light, regardless of his motion with respect to the source, provided the motion is unaccelerated.

Thus, special relativity applies to systems and observers moving uniformly with respect to each other in straight lines and at constant speed. We cannot pursue the details here but some of the important consequences, all of which have been verified in the laboratory, are these:

1. The measured mass of a body in motion differs from the value for the mass obtained when the body is at rest. The relationship is,

$$m = \frac{m_0}{\sqrt{1 - v^2/c^2}} \qquad (14.1)$$

where m_0 is the rest mass, v is the speed of the body relative to an observer, c is the speed of light, and m is the effective or *relativistic mass* of the body in motion.

2. Similarly, the measured dimension of a moving body parallel to the direction of motion has a lesser value than the same dimension measured when the body is at rest. That is, the body contracts in the direction of its motion.

$$L = L_0\sqrt{1 - v^2/c^2} \qquad (14.2)$$

Here, L is the relativistic length, L_0 the at-rest length, and v and c are the same quantities as before.

3. Because it is impossible to synchronize clocks in absolute space (that is, without reference to some kind of coordinate system), the best we can do is use, say, light flashes to indicate starting and ending times for any synchronization method. But the velocity of light is finite. The effect of this is to make the *advance* of time, measured relative to a moving system by a stationary observer, proceed more slowly than his own.

$$t_2 = t_1 \sqrt{1 - v^2/c^2} \qquad (14.3)$$

Time measured with respect to the moving system is t_2; t_1 is the time measured with respect to the stationary observer.

But wait! Nothing in the theory of special relativity specifies *which* observer in *which* system must be considered as the stationary reference. Either one can be selected; the relative motion is equivalent in either case. This gives rise to a seeming paradox. Between any two systems moving relative to each other, *which one* really experiences a mass increase, or a length contraction, or a time dilation? Special relativity does not provide an answer. Since by specification the systems with their observers are moving uniformly (not accelerating) with respect to each other, their respective paths can cross, or their position coincide, only once; neither of the observers in the two systems can come back to check whether one or the other, if either, becomes more massive, contracts in the direction of motion, *or ages more rapidly!*

In its efforts to fathom the nature of the universe, astronomy attempts to describe the distribution of matter and energy in space and the physical processes occurring therein. But the answer to a more fundamental question still eludes us: What *is* space? To the classical physicists and astronomers space was inscrutable except for a few properties. Space was characterized by a total absence of matter but yet was three-dimensional. Space permitted the unimpeded propagation of electromagnetic energy; indeed, light was propagated in a straight line.

Einstein's *theory of general relativity* represented an attempt to describe space in relationship to time; time became an indispensable fourth dimension. The theory also attempted to deduce the properties of the space-time continuum in the presence of matter and energy. Among other features, general relativity was not restricted by the notion of uniform, straight-line motion; accelerations were permitted.

How does general relativity square with our common-sense view of space? We see space as some sort of unlimited, immovable container which is the three-dimensional stage for every visible happening. This sort of boundless box, in the sense that we cannot locate the sides, top or bottom, is uniform throughout. This is the common-sense view of space that was mathematized by Euclid, and accepted *in total* by the Newtonian classicists.

If we assume the classical view of space to be the true situation, where do the consequences lead us? If we were to embark on a journey in a straight line in any direction, only three possible outcomes exist. We may eventually run out of stars and galaxies, which means that matter as we know it is but a small oasis in a vast ocean of nothingness. Or, we may never reach the boundaries of matter-populated space, meaning there is an infinite amount of matter in an equally infinite volume of space. Or lastly, we may find that we ultimately return to our starting point, completing an unanticipated closed circuit, proving that space after all is limited and finite.

If this last conjecture seems too fantastic for acceptance, there are almost equally as grave difficulties with the other two possibilities. If the stellar population is finite, gravitation would eventually produce a condensation of all matter into one clump, or else the whole system, stars and energy, would dissipate into infinite space. If the stellar population is infinite, then the force of gravity should everywhere be infinite. Every particle, every body would be subject to an infinite pull in every direction, hardly a situation in which survival could be anticipated.

Moreover, as the German astronomer H. W. M. Olbers pointed out in 1826, if there were an infinite number of stars (and galaxies), then the night sky would be as blindingly bright as the surface of the sun. This is because no line of sight anywhere would fail to intersect the surface of a star. There would be no dark patches between stars.

General relativity provides a major break with

these ideas by asserting that the path of a ray of light will be slightly *curved* in the vicinity of a large mass, such as the sun. Thus if the propagation path of light describes the geometry of space, then *we must conclude space is curved.*

Many theoretical assaults have been mounted against the problem of just how space is curved, all to no avail at present. Einstein's attempt to apply the principles of general relativity to the universe as a whole was marked by a significant lack of success. In part, Einstein's model of the universe was static: unchanging and permanent. The distribution of matter in it was homogeneous and isotropic. Thus, matter in space could be said to have a certain average density. Also, the space-time path joining two nearby points in the universe resembled that of Euclidean geometry applied to spherical surfaces.

The defects in the Einstein model proved fatal. The curvature of space was a function of the density of matter in it. Einstein's universe was therefore finite and closed. Only an empty universe, one with zero density, could have a curvature of infinite radius, that is, be characterized by a Euclidean space of three dimensions in which light would travel in straight lines.

An even more serious defect concerned the distribution of matter in such a static universe. Observations showed then, as now, that the average density is extremely low; space is truly *almost* empty. On the average the density of matter in the universe is equivalent to about the mass of one hydrogen atom in a cubic meter of space. The density of matter in the vicinity of the solar system is more than 1000 times greater than this. As we shall see in a moment such local concentrations of matter proved troublesome.

The pressure exerted by matter (and energy) of such low concentration would be negligible. Hence, the inexorable tug of gravity would eventually cause a collapse. Einstein's static universe was inescapably unstable. To overcome this difficulty Einstein proposed a

counter force called *cosmic repulsion* that had the strange property of being negligible at small distances and most effective on the universal scale. The reason for postulating this property is that no repulsive force is detectable on a local scale such as solar-system distances.

Thus, the Einstein universe, poised between two unbalancing forces, was in unstable equilibrium. Sooner or later, the random motions of matter in such random concentrations as are known to exist would disturb the balance. Either matter would accelerate into a headlong rush toward collapse and oblivion, or else it would fly apart, expanding indefinitely ever faster and faster into an equally certain oblivion.

Einstein soon repudiated his own postulate of cosmic repulsion. With it went the motion of a static universe. General relativity is quite successful on the small scale, say for systems of solar-system dimensions or less. But it has proven inadequate when applied to the universe as a whole. Nevertheless the theory of general relativity made a great contribution to astronomy when it was published in 1916, in that it ushered in the era of modern cosmology. The ferment is still waxing strong.

The Fate of the Universe

If the steady-state universe proves to be the true situation, there are objectionable, although not crucial, problems concerning the inviolability of the conservation of the mass-energy principle. The irreversible aging of the stars into ultimately invisible, cold black dwarfs does not destroy mass, it just changes its form. Therefore, the creation of new matter that eventually forms into new stars and new galaxies, according to the steady-state hypothesis, inevitably increases the mass of the universe. But acceptance of this requirement is mandatory if a key feature of the steady-state universe, that the density of matter in space remains constant, is to be preserved.

As to the violation of the principle of conservation of mass-energy, Hoyle argues that the creation of just 35 atoms of hydrogen per cubic meter *every billion years* is sufficient to maintain the density of the universe constant. Now our acceptance of the preceding conservation principles is based upon experience on a relatively local scale. Certainly no one has tested them to the degree of precision necessary in order to disclose the magnitude of the violation of the conservation of mass argued by Hoyle. Thus, the violation does not seriously jeopardize the conservation of mass and energy principle in our *operational* universe, any more than does the existence of special relativity according to Einstein mitigate against Newton's laws under ordinary circumstances.

If the oscillating big-bang theory eventually prevails, then recognizable forms of matter will periodically be annihilated with each successive cataclysm, only to reform again in each new cycle. Certainly, the oscillating universe is teleologically acceptable in spite of the antagonism it generates among those who hold that the currently conceived universe is infinite and everlasting in its present form.

We cannot overlook the possibility that new physical laws and principles may someday be discovered that will radically change many of our current notions. All we have to do is to look back a few hundred years to see what changes were wrought in man's thinking by the then new discoveries of Copernicus, Galileo, Kepler and Newton in celestial mechanics; of Faraday, and Gauss, and Maxwell in electromagnetic theory; of Einstein, Stefan, Wien, Boltzmann, and Planck in radiation theory and quantum physics, to name but a few. We would be naive in the extreme to deny the possibility of future discoveries of momentous consequence merely because they cannot be seen on the immediate horizon. The case for the future in astronomy cannot be better expressed than in the words of Edwin Hubble in his last paper published in 1936:

"For I can end as I began. From our home on earth, we look out into the distances and strive to imagine the sort of world into which we are born. Today we have reached out into space; the exploration of space must always end on a note of uncertainty. And necessarily so. We are by definition at the very center of our observable region. Our immediate neighborhood we know rather intimately. But with increasing distance our knowledge fades, and fades rapidly. Eventually we reach the dim boundary — the utmost limits of our telescopes. There, we measure shadows, and we search among the ghostly errors of measurement for landmarks that are scarcely more substantial.

The search will continue. Not until the empirical resources are exhausted need we pass on to the dreamy realms of speculation. The urge is older than history, and it will not be suppressed."

SUMMARY

The question of whether or not many external nebulae were really galaxies was answered in 1924 at Mount Wilson Observatory by Edwin Hubble who, using the new 60-in. and 100-in. telescopes, was able to resolve individual stars in the nearer nebulae. Additional verification came with the discovery of cepheid variables in the same nearby galaxies; this discovery placed the distances of these galaxies vastly beyond the domain of the Milky Way Galaxy.

Even though galaxies have a definite tendency toward clustering, they are found in every direction in space, and as far out in space as we are able to probe. The distances to the remote galaxies are found statistically from estimates of their intrinsic brightness.

There are several types of galaxies, ranging from ellipticals to spirals. Apparently, the series represents different stages in galaxian evolution, with the elliptical galaxies being the oldest. Galaxian spectra show the features present in the spectra of the Milky Way. Rotation is commonly observed. The overall spectra of most galaxies are similar to those of the early G to late K-type stars; this demonstrates the dominant influence of the older stars in the galaxian nuclei. Elliptical galaxies are almost totally devoid of population I stars.

Galaxian masses, found by application of Kepler's and Newton's laws to the observed rotation of galaxies, range from 60 to 70 billion solar masses, although some of the dwarf elliptical galaxies have masses as small as 5 billion solar masses. The Milky Way and Andromeda appear to be giants among galaxies.

All galaxies external to the Milky Way show red shifts of their spectral lines relative to the center of our galaxy. This is taken to be an indication that each galaxy is receding from every other galaxy, with speeds that increase in proportion to increasing distance. Many galaxies emit radio energy far in excess of predicted values. The explanation for this anomaly has yet to be discovered. The quasistellar sources, some of them galaxies, are all regions of unprecedented absolute luminosity, and they show extraordinarily large red shifts. The nature of these objects is presently the enigma of the universe.

Cosmology is presently in a state of ferment, with the search for answers to questions concerning the form and structure of the universe, its extend, and its origin and destiny being (from the point of view the apparently stately pace of astronomy) feverishly pursued. At present, two main theories concerning the origin of the universe are being argued; the expanding universe, or big-bang theory, with its extension to an oscillating, or repeating, model, and the steady-state theory. Both (or all) of these theories have their strengths and weaknesses. At present, the weight of observational evidence in the form of galaxian red shifts, the existence of the quasi-stellar objects, and the existence of the 3°K background radio noise weigh heavily in favor of the expanding universe theory. The extension of the expanding universe concept to an oscillating universe enjoys much less support. Both, however, seem in a stronger position than the steady-state theory.

Questions for Review

1. How was the controversy settled concerning whether or not certain nebulae were extragalactic?

2. What is a galaxian association? A galaxian cluster?

3. How is the distance to remote galaxies determined?

4. What is the primary persuasive evidence that elliptical galaxies are older systems than spiral galaxies?

5. What are the steps in estimating the distance to distant galaxies?

6. What are the arguments for and against the expanding universe hypothesis?

7. What are the arguments for and against the steady-state universe hypothesis?

8. Now that you have examined the available evidence as presented by the contents of this text, what is your position on the origin and destiny of the universe? Write a short paper logically stating *and defending* your position.

Appendixes

APPENDIX 1
Physical Constants and Useful Tables

Length

1 meter (m): 100 cm; 1000 mm; 1.09 yd; 3.28 ft; 39.4 in.
1 kilometer (km): 1000 m; 0.62 mi
1 centimeter (cm): 0.01 m; 0.39 in.
1 inch (in): 2.54 cm
1 foot (ft): 0.305 m
1 mile (mi): 1.61 km; 1609 m
1 micron (μ): 0.0001 cm
1 angstrom unit (Å): 1/100,000,000 cm (10^{-8} cm)
1 astronomical unit (AU): 1.49598×10^8 km
1 light-year (LY): 9.460×10^{12} km
1 parsec (pc): 206,265 AU; 3.26 LY; 3.086×10^{13} km

Mass

1 gram (g): the basic unit of mass
1 kilogram (kg): 1000 g
1 solar mass ($M_\odot$): 1.989×10^{30} kg

Time

1 second (sec): basic unit of time; $1/3.155693 \times 10^7$ of a tropical year (1900)
1 hour (hr): 3600 sec
1 day (mean solar day): 86,400 sec
1 sidereal second: 0.99727 solar sec
1 sidereal day: $23^h56^m4.099^s$ of mean solar time

Velocity (Speed)

1 meter per second (m/sec): 0.37 mi/min; 3.28 ft/sec; 3.6 km/hr
1 kilometer per second (km/sec): 1000 m/sec
1 mile per minute (mi/min): 60 mph; 96.6 km/hr; 88 ft/sec; 26.8 m/sec

Acceleration

1 centimeter per second per second (cm/sec^2): 0.03 ft/sec^2
1 meter per second per second (m/sec^2): 3.28 ft/sec^2
gravitational acceleration (standard) at earth's surface: 980.665 cm/sec^2; 9.8 m/sec^2; 32.174 ft/sec^2

Force

1 dyne (dyn): the force that will impart an acceleration of 1 cm/sec^2 to a mass of 1 g
1 newton (N): 100,000 dyn; 0.202 lb
1 pound (lb): 445,000 dyn
1 gram (g) (weight): approximately 1/28 oz
1 kilogram (kg) (weight): 2.21 lb; 980,000 dyn
1 metric ton: very nearly 2200 lb

Physical and Astronomical Constants

speed of light (c): 2.99793×10^{10} cm/sec; 2.99793×10^5 km/sec; approximately 186,000 mi/sec; acceptable approximation, 3×10^{10} cm/sec

constant of gravitation (G): 6.668×10^{-8} dyn/cm^2/g^2

solar mass ($M_\odot$): 1.989×10^{30} kg

earth mass ($m_\oplus$): 5.997×10^{24} kg

earth mean radius: 6371.23 km

solar equatorial parallax: 8″.794 (8.794 seconds)

mass ratio, sun/earth ($M_\odot/m_\oplus$): 332,700

Orbital Data for the Planets

Planet	Semimajor axis (AU)	Sidereal period (days)	Sidereal period (years)	Orbital eccentricity	Orbital inclination	Mean orbital speed (km/sec)
Mercury	0.387	87.98		0.2056	7°.004	47.8
Venus	0.723	224.7		0.0068	3°.394	35.0
Earth	1.000	365.26		0.0167		29.8
Mars	1.524	686.98	1.8809	0.0934	1°.85	24.2
Jupiter	5.203		11.862	0.0765	1°.305	13.1
Saturn	9.534		29.458	0.00557	2°.49	9.7
Uranus	19.18		84.013	0.0472	0°.773	6.8
Neptune	30.06		164.793	0.0086	1°.774	5.4
Pluto			247.686	0.250	17°.17	4.7

Source: Supplement to *The American Ephemeris and Nautical Almanac* (mean equator and equinox of 1960).

Planetary Motion and Position Data†

	Mercury	Venus	Earth	Mars	Jupiter	Saturn	Uranus	Neptune	Pluto
Motion									
Sidereal period	88^d	225^d	365^d	687^d	11.8^y	29.5^y	84^y	164.8^y	248^y
Synodic period	115.9^d	583.9^d	—	779.9^d	398.9^d	378.1^d	369.7^d	367.5^d	366.7^d
Mean distance from sun (AU)	0.387	0.72	1.0	1.52	5.2	9.54	19.2	30.1	39.4
Mean distance from sun (km $\times 10^6$)	57.9	108.	149.6	228	778	1426	2868	4490	5900
Precession (advance of perihelion)	43″	8″.6	3″.8	1″.35					
Orbital elements									
a— semimajor axis	(Equal to mean distance from the sun)								
e— eccentricity	0.21	0.007	0.017	0.09	0.077	0.056	0.045	0.0086	0.25
i— orbital inclination	7°	3°24′	—	1°51′	1°18′	2°29′	0°46′	1°46′	17°12′
Ω— heliocentric longitude of ascending node††	48°	76°.42		49°.34	100°.16	113°.43	73°.87	131°.5	110°
ω— longitude of perihelion point††	77°.4	131°.17		335°.54	13°.77	93°.30	172°.95	21°.16	223°.66

† Data from the Supplement to *The American Ephemeris and Nautical Almanac*, Epoch 1060.
†† These elements vary with time: values given are for the epoch of September 6, 1971.
(Notation such as 8″.6 and 77°.4 means seconds and tenths, and degrees and tenths, respectively.)

APPENDIX 2
Mathematical Review

Symbols and Notation

$=$	equals
$\approx$	approximately
$\cong$	approximately equals
$\neq$	not equal to
$\equiv$	identical to, is defined as
$>$	greater than
$<$	less than
$\geq$	equal to or greater than
$\leq$	equal to or less than
$\pm$	plus or minus
$\propto$	proportional to
Δ	small increment of, or small interval in
— or /	fraction, ratio, divided by

Exponents

An exponent is a superscript which indicates successive multiplications of a number by itself. For most purposes 10 is the number used. Hence the superscript is called a power of ten. Thus

$$10^2 = 100 = 10 \times 10$$
$$10^3 = 1000 = 10 \times 10 \times 10$$
$$10^4 = 10,000 = 10 \times 10 \times 10 \times 10$$
$$10^5 = 100,000 = 10 \times 10 \times 10 \times 10 \times 10$$
$$10^6 = 1,000,000 = 10 \times 10 \times 10 \times 10 \times 10 \times 10$$
$$10^9 = 1 \text{ billion.}$$

We define negative exponents as

$$10^{-1} = 1/10 = 0.1 = 1/10^1$$
$$10^{-2} = 1/100 = 0.01 = 1/10^2$$
$$10^{-3} = 1/1000 = 0.001 = 1/10^3 \text{ etc.}$$

There is a great advantage in using powers of ten notation when dealing with very large and very small numbers. For example the speed of light is approximately 30 billion centimeters per second. This can be written 30,000,000,000 cm/sec or more conveninetly in powers of 10 as 3×10^{10} cm/sec. Similarly a number such as 3 one-hundred-millionths can be written as 0.00000003, 3×10^{-8}.

Geometry

A few theorems from elementary plane geometry are presented here for their utility, without proof. The proofs can be found in any high school geometry text.

1. There are 360° (degrees) in a full circle.
2. An acute triangle contains three internal angles, each less than 90°. An obtuse triangle contains three internal angles, one of which is greater than 90° but less than 180°. A right triangle has three internal angles, one of which is a right angle (90°).
3. In a right triangle, the square of the hypotenuse (side opposite the right angle) is just the sum of the squares of the other two sides. Thus in Figure A2-1, $r^2 = x^2 + y^2$. Said another way, if $x^2 + y^2 = r^2$ then the triangle is a right triangle (Figure A2-1).
4. In any triangle the area is equal to one-half the base times the altitude. Thus in Figure A2-2, the area $A = 1/2\,(ba)$ where a is the altitude and b is the base.
5. In any triangle, a line connecting the midpoints of any two sides is parallel to the third side and equal to one-half of its length. This relationship is evident in Figure A2-3.

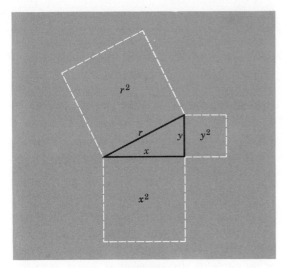

FIGURE A2-1

Algebra

Algebra is a branch of mathematics which deals with the general properties of and relations between numbers by means of signs and symbols. The utility of algebra is immeasurable; many physical relations cannot be solved without its use.

Consider the following relations between numbers: $2 + 3 = 5$; $1 + 5 = 6$; $3 + 4 = 7$. In general we can write $a + b = c$. This expression is always valid; if any two numbers are known there is one and only one other number which satisfies the relationship.

We now invent an extension to our familiar number system such that -2 is the same magnitude less than zero that 2 is greater than zero. In general,

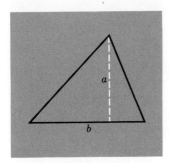

FIGURE A2-2

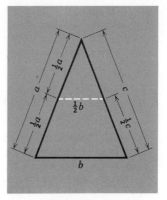

FIGURE A2-3

any negative number is defined in the same way with respect to a "normal" number. The general expression $a + b + c$ still holds. Thus $2 - 3$ means $2 + (-3)$. If we draw a line graph of the real number system centered on zero, with equal units or length in each direction, the significance of $2 - 3$ becomes apparent; it is *equivalent* to moving two units to the right of zero, then 3 units in the opposite direction. The terminous of this motion is -1. Hence $2 - 3 = -1$ which is the same as $a + b = c$ where a equals 2, b equals -3, and c equals -1.

Four axioms from elementary arithmetic are useful here.

1. *If equals are added to equals, the sums are equal.* Or, since $2 + 3 = 5$, $(2 + 3) + 4 = 5 + 4$. That is, if $a + b = c$, $(a + b) + d = c + d$.
2. *If equals are subtracted from equals, the remainders are equal.* Again, since $2 + 3 = 5$, $(2 + 3) - 4 = 5 - 4$, or $(a + b) - d = c - d$.
3. *If equals are muliplied by equals the products are equal.* Thus: $(2 + 3) \times 4 = 5 \times 4$ or $(a + b) \times d = c \times d$.
4. *If equals are divided by equals, the quotients are equal.* Thus: $(2 + 3) \div 4 = 5 \div 4$ which can also be written as $(2 + 3)/4 = 5/4$. Then also, $(a + b)/d = c/d$.

A study of the above indicates the value of symbolic notation and also illustrates a need for signs of grouping. In Axiom 3 if the parentheses are omitted we might read "2 added to 3 times 4 is equal to 5×4," which is not true; $2 + 3 \times 4 = 14$, not 20.

But the quantity $2 + 3$ multiplied by 4 is 5×4 which *is* 20.

We can now generalize as follows:

$$a + b = c$$
$$a - b = d$$
$$a \times b = e \text{ (more simply } ab = e)$$
$$a/b = f$$

In the following

$$\frac{(a + b)c}{d} - e = f$$

means "a is added to b, the sum is multiplied by c; the resulting product is next divided by d, and e is subtracted from the quotient. The whole operation is then equal to f.

Generally, in algebraic notation lower case letters a, b, c . . . denote constant or unchanging quantities, and letters x, y, z . . . denote variable quantities. Observe that all letters may take on different values in different problems. For particular kinds of quantities such as density, pressure, force, etc., standard notation has been adopted.

Finally, some quantities are known to depend upon the way in which other quantities vary. We say that $A \propto B$ when we wish to indicate that any change in A is proportional to the change in B. If we can determine by experiment how the proportional changes are related, we can replace the proportional symbol by a *constant of proportionality*. Thus

$$A = kB$$

provided the change always occurs in the same proportion. When we are more interested in the *way* the change takes place rather than the value of the change at any given time we generally write

$$\Delta A = k\Delta B.$$

This expression means "the difference between two values of A (which of course is the change in A) is equal to some constant proportion of the difference between two values of B (that is, the change in B)."

Angular Measure

An angle is a measure of the amount of turning required to bring into coincidence two straight lines that meet in a point. Thus, two lines that are perpendicular to each other form a right angle; the angular measure of a right angle is 90 degrees (90°). A turn through a full circle generates an angle of 360°.

There are 60 minutes (60') in 1°, 60 seconds (60") in 1', and 3600" in 1°.

Another system of angular measure, based upon the *time* it takes the celestial sphere to make one complete apparent revolution about the earth, is of great convenience in astronomy. A rotation through a full circle generates an angle of 24 hours (24^h). As in the degree system, each 1^h is subdivided into 60 minutes (60^m), and each 1^m is further subdivided into 60 seconds (60^s).

The relation between the two systems is clearly evident. $15° = 1^h$; $1° = 4'$; $1' = 4''$.

A third form of angular measure that is often used in mathematical treatment is known as *radian measure*. A radian is the angle subtended by an arc of a circle equal to the circle's radius (Figure A2-4).

The ratio of the length of the circumference of a circle to its diameter, C/D, is constant: 3.14159 . . . (a never-ending decimal). Then we have $C = 2\pi r$. (A diameter is twice the radius.) From the preceding, a radius turning through a full circle generates 360°, and in radian measure it generates an angle of 2π radians (rad). Hence 1 rad is $360°/2\pi = 57°.3$; 1 rad contains 206,265" of arc.

Properties of Circles and Spheres

Consider a circle or a sphere with circumference C, diameter D, and radius r. The circumference of a sphere is defined as that *great circle* generated at the surface by a plane cutting through the sphere that also contains its center. Thus for both geometrical figures, $C = \pi D$; $C = 2\pi r$. The area of a circle is $A = \pi r^2$; also $A = \pi D^2/4$. The area of the surface of a sphere is 4 times the area enclosed by one of its great circles: $A = 4\pi r^2$; $A = \pi D^2$.

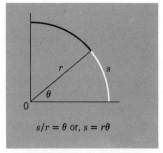

$$s/r = \theta \text{ or, } s = r\theta$$

FIGURE A2-4 Radian measure. All units are length units.

The volume of a sphere is

$$V = \frac{4}{3}\pi r^3; \quad \text{also,} \quad V = \pi\frac{D^3}{6}.$$

The ratio between the areas of two circles is just the ratio between the squares of their radii,

$$\frac{A_1}{A_2} = \frac{\pi r_1^2}{\pi r_2^2} = \frac{r_1^2}{r_2^2}.$$

Similarly, the ratio between the volumes of two spheres is just the ratio between the cubes of their respective radii,

$$\frac{V_1}{V_2} = \frac{\frac{4}{3}\pi r_1^3}{\frac{4}{3}\pi r_2^3} = \frac{r_1^3}{r_2^3}.$$

APPENDIX 3
Instruments and Methods

The human senses, unaided, are almost totally inadequate to the task of probing the secrets of the universe and its celestial inhabitants. Indeed, except for the earth and a few isolated contacts with the lunar surface (so far), the senses of touch, taste, hearing, and smell are to no avail in our investigation of the nature of the universe. We must of necessity interpret the evidence and unravel the mysteries of the universe from information secured by the sense of sight alone. And even vision is inadequate to the task without the aid of instruments that *gather* electromagnetic energy (primarily light and radio waves), *resolve* detail, *analyze* complex evidence, *record* the stimuli received, and *preserve* the information for leisurely future study.

Astronomical instruments collectively must accomplish the tasks imposed by these requirements. We shall not here consider all of the categories of instruments, or the many variations within the categories. We shall consider briefly the basic instruments in broad, general terms. These are: the telescope and its first cousin, the camera; the spectrograph; the photoelectric photometer; the radio telescope; and the deep-space probes. All of the instruments require an initial collection of electromagnetic impulses which must be brought to a focus for further processing. The primary machines for performing the collection and focusing tasks are the *mirror* and the *lens*.

Image Formation by a Mirror

Electromagnetic energy is propagated from each point of a source in the form of an advancing, ever expanding envelope of waves. At a sufficient distance from the source advancing envelopes for all practical purposes consist of plane parallel waves. The required distance is surprisingly short. An ordinary household light bulb behaves essentially as a point source emitting plane parallel waves at a distance of about 3 m.

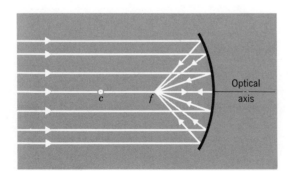

FIGURE A3-1 Focus of monochromatic light from a point source by a concave mirror.

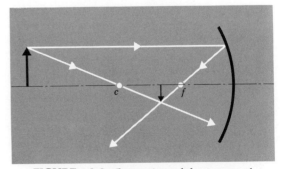

FIGURE A3-2 Formation of the image of an extended source beyond the center of curvature of a concave mirror.

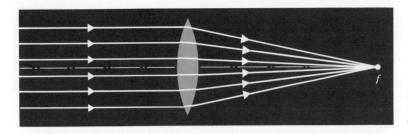

FIGURE A3-3 Focus of monochromatic light from a point source by a convex lens.

We can conveniently think of the successive space positions of a point on one of the wave fronts as generating a straight line, and the energy pulse that follows such a path can be considered a ray. Under these conditions, adjacent rays are parallel. A bundle of closely adjacent rays is called a *pencil* of rays.

When a pencil of parallel rays impinges on a flat (plane) mirror, the reflected rays are still parallel. If on the other hand the surface is randomly irregular, the rays are reflected in random directions; the pencil becomes *diffused*. If the surface is a smooth, curved mirror, the rays *converge* toward or *diverge* from each other depending on whether the mirror surface is *concave* (hollowed) or *convex* (bulged) toward the source (Figure A3-1).

If the curvature of a concave mirror is mathematically correct, all of the rays from a point source impinging on it will converge to a point called the *focus*. The distance from the center of the mirror to the focus is the *focal length* of the mirror. Pencils of parallel rays from different points of an extended source, or from different point sources, converge to different focal points that are all in the same *focal plane*. The collection of such focal points constitutes a picture of the source or sources (Figure A3-2).

Image Formation by a Lens

Figure A3-3 illustrates the formation of an image of a point by a converging lens. Light rays are *refracted* (bent) as they pass through the interface between media of different densities. Excluding the special case of rays perpendicular to the interface, light entering a denser medium is refracted *toward* the perpendicular to the interface at the point of entrance. On leaving a more dense medium the refraction is away from the perpendicular (Figure A3-4). If the two opposite interfaces bounding a more dense medium, such as a rectangular glass solid in air, are parallel, the emergent rays travels parallel to their original direction, but they are displaced from the original ray path toward the denser material. If the interfaces are not parallel (and provided the angle between them is not too great) a ray is successively refracted in the same direction (Figure A3-5). The figure clearly indicates that the rules of refraction are satisfied. The *prism* is a common example of non-

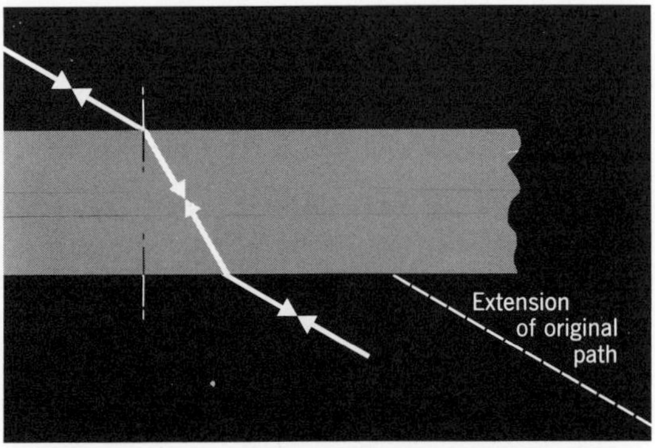

FIGURE A3-4 Refraction of monochromatic light in passing through a transparent medium of higher density than air.

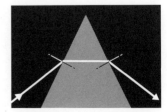

FIGURE A3-5 Refraction of monochromatic light by a prism.

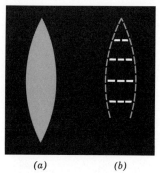

(a)   (b)

FIGURE A3-6   A con-
verging lens (a); sketch
showing that a con-
verging lens is essentially
composed of a series of
prisms (b).

parallel interfaces.

The opposite faces of a lens can be thought of as an infinite array of prism segments arranged as in Figure A3-6. The amount of refraction of any given ray depends upon the angle between the surfaces at the points of entrance and exit. Light rays of a single wavelength from a point source all converge to a single point, the *focus*. Lenses with two convex surfaces, or one convex and one plane surface, are *converging* lenses. Lenses with one (or both) concave surface(s) *diverge* the light rays; the emergent paths of a bundle of rays appear as if the rays originated on the *other side* of the lens from an imaginary or *virtual* focus. Some lenses have combinations including one concave and one convex surface. Whether such a lens is a converging or diverging lens depends upon the relative curvature of the two surfaces. The details need not concern us here.

Telescopes

A telescope is basically nothing more than an *objective* mirror or an *objective* lens conveniently supported on a suitable mounting that permits tracking of the rotating celestial sphere, together with auxiliary appurtenances for studying the images formed by the objectives. The term objective refers to the mirror or lens nearest the source. Lens-type telescopes are called *refractors;* mirror-type telescopes are called *reflectors.* Invariably the elements comprising a refracting telescope are mounted in a closed tube in order to control or eliminate transient air currents and undesirable stray light, both of which degrade the image formed (Figure A3-7). Small-apperture reflecting telescopes usually have the mirror mounted at one end of an open tube. In most large reflectors, the tube is replaced by a framework, or cage, in order to save bulk and weight (Figure A3-8). The telescope proper is mounted on one of many variations of the equatorial mount as described in Appendix 5.

Functions of a Telescope.   The primary function of a telescope is to collect light rays from a source and bring the bundle of rays to focus in a clear image. The overwhelming difficulty is in securing an image of acceptable quality. Some of the problems and the (partial) solutions to them are considered in the section on *aberrations.*

Light Gathering Power.   Just as more rain per unit time can be caught in an open rain barrel than in a teacup, so can more light from a source be collected by a telescope of larger aperture than one of smaller aperture. With respect to a star, essentially a *point source*, the light-gathering power of a telescope is proportional to the square of the aperture (diameter of the objective). This is because the area A of a circle is related to its radius r, or diameter $r/2$ according to the formula,

$$A = \pi r^2 = \pi (r/2)^2 = \pi d^2/4.$$

Clearly if the radius, and hence the diameter of an objective, is doubled, its area is increased fourfold. If the radius is 10 times larger, the area is 100 times greater, and so on.

Now the quantity of light gathered per unit time is a function of (depends on) the *area* of the objective; for example, the image of a point source is 4 times brighter in a telescope that is twice the diameter of its counterpart. This theoretical relationship is only approximately true for stars, as we shall see shortly. For extended sources, the focal length of the objective is involved. In brief, the light rays originating from the extreme edges of an extended source, such as the moon, a planet, or a galaxy converge toward the objective. The *size* of the image formed at the focus is the product of the angle (expressed in radians) between the rays from the extremities and the focal length of the objective. Radian measure is defined in Appendix 2. For example, the full moon subtends an angle of about 1/2°, or nearly 0.008 radian. The usual 6-in. reflecting telescope has a focal length of 48 in. The size of the moon's image in such a telescope is 0.008 × 48 = 0.384 in. or very nearly 1 cm. For the 200-in. Hale telescope with its 55 ft (660 in.) focal length, the image size of the moon is 0.008 × 55 = 0.44 ft or a little over 5 1/4 in.

Image Brightness. The image of an extended source (a source with an apparent size) is proportionally less bright than the image of a point source. Clearly, the light brought to a focus from an extended source must be distributed over its extended image. The intensity of the illumination depends on the image area. There is no way to optically increase the brightness of an image except to increase the size of the objective. It is not particularly difficult to see qualitatively that the brightness of an image of an extended source is both increased in proportion to the square of the aperture and decreased in proportion to the square of the image diameter. With telescopes, as with cameras, the ratio of the focal length to the diameter of the objective is called the "f-number"; the brightness of the image of an extended source is proportional to $1/(f\text{-number})^2$. The reader should satisfy himself that any given *point* on the extended image formed by a 6-in. telescope of 48 in. focal length, is only 1/64 as bright as a point on the source viewed directly.

The human eye may not recognize this fact, because it in itself is a form of telescope with a variable focal length and a variable aperture. The usual sensation of brightness depends upon the concentration of the light on the retina. A camera, on the other hand, recognizes the true situation. A camera with an aperture of f4 will secure a photographic image of equal brightness to that secured by an aperture of f8 in just one-fourth the time.

Resolving Power. Even if a lens or mirror is perfectly shaped and is of excellent optical quality, it cannot produce a perfectly sharp, detailed image. This is the case because of the wave nature of light propagation. Light waves are *diffracted* when passing close to a sharp edge, much as water waves are in passing through an opening in a breakwater (Figure A3-9). One can observe the diffraction effects when sunlight forms a shadow of a stick or a pole. The shadow appears sharp and distinct close to the ground; the quality of the shadow rapidly degrades

FIGURE A3-7 The 36-in. refractor of Lick Observatory (Courtesy Lick Observatory).

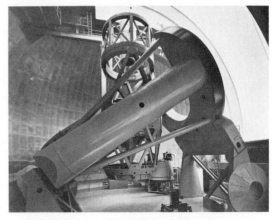

FIGURE A3-8 The 200-in. Hale reflector telescope at Mount Palomar (Courtesy Hale Observatories).

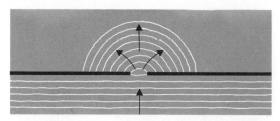

FIGURE A3-9 Diffraction of water waves through an opening in a breakwater.

at increasing distances from the object. The light diffracts around the edge of the object and is more pronounced the farther it travels into the diffraction zone, which in this case is the shadow of the object.

The perimeter of a lens or mirror forms such a sharp edge. Light from a point source is diffracted in the vicinity of the edge, but it is not diffracted a little further in toward the center of the lens or mirror. The image thus formed is not a true point image; it appears as a *diffraction disk* surrounded by alternate, concentric, light and dark rings (Figure A3-10). The pattern is caused by the interference of the diffracted portion of the light beam with the undiffracted portion.

The *resolving power* of a telescope is a measure of its ability to form recognizable, separate images of closely adjacent point sources. It is expressed as the smallest angle of separation of point sources that can be resolved into separate images. The property is a function of (depends on) the aperture of the objective. The theoretical resolving power of a telescope is expressed as $\alpha = 4.56/d''$ of arc when d is the diameter of the objective and the wavelength of the light is near the middle of the visible spectrum. A 1-in. telescope will not resolve two stars that are

FIGURE A3-10 Diffraction pattern of a star (see text).

closer than 4.56″ of arc, whereas a 10-in. telescope will resolve a pair that are separated by only 0.456″ of arc.

Magnification. Magnification involves nothing more or less than examining the image formed by an objective with a magnifier. The latter, known as the *eyepiece,* is in reality an auxiliary telescope of short focal length. The term *magnifying power,* or simply *power,* of a telescope is generically meaningless since the magnification achieved depends upon the ratio of the focal lengths of the objective and the eyepiece (Figure A3-11). The eyepiece extracts information from the focal point, or focal plane, of the objective, and first diverges, then collimates (renders parallel) the emergent light rays. Magnification is physiologically related to the image size appearing on the retina of the eye. For an objective of 50-in. focal length, a magnification of 50 times is produced by a 1-in. focal length eyepiece; a 1/2-in. eyepiece magnifies 100 times, and a 1/10-in. focal length eyepiece produces magnification of 500 times. It is clear, then, that the resolving power of a telescope depends upon the focal length of the eyepiece used at the moment.

No amount of magnification will improve the clarity of an image. Any defects present will surely be magnified equally along with the desired detail. Nor will magnification increase the brightness of an image. Indeed, it does just the reverse since magnification enlarges the image area over which the incident light is spread.

Aberrations. Imperfections in image quality are collectively called *aberrations*. Some aberrations can be eliminated, or at least reduced below objectionable minimums, by good engineering and craftsmanship. *Spherical aberration* is the term given to the image defect that arises from the inability of a spherical lens or mirror to bring all of the light rays from a point source to a common focus (Figure A3-12). Spherical aberration can be eliminated in a mirror by modifying the shape of its surface from spherical to *parabolic* (Figure A3-13). This latter is the shape of the familiar reflecting surface in an automobile headlight. This solution is not applicable to lenses; for these, the image quality can be greatly improved by "stopping" down the effective aperture with an adjustable diaphragm, but this is done at the sacrifice of much of the light grasp. Spherical

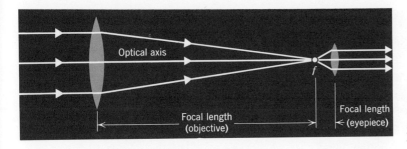

FIGURE A3-11 Diagram showing magnification of an image formed at the focal point by an eyepiece lens.

aberration in lenses is usually corrected at the same time *chromatic aberration* is eliminated.

Chromatic aberration, a defect not suffered by mirrors, is a term applied to the inability of lenses to bring light of *all* colors to a common focus (Figure A3-14). In passing through a lens or prism, light of decreasing wavelengths is refracted by increasing amounts as Figure A3-14 illustrates. Depending on which focal point or focal plane is used, the image shows either red or blue color fringing. Chromatic aberration, and with it, spherical aberration, is reduced to acceptable limits by means of a compound objective. Two pieces of optical material of different densities are cemented together; the surfaces of each component are shaped as shown in Figure A3-15. The end result of such a combination is to increase the focal length relative to blue light and to decrease it relative to red light. The focal lengths are adjusted so as to produce a common focal point, not only for red and blue light, but for all colors in between. Such an objective is termed *achromatic*, or color corrected.

There are a number of other aberrations that more or less plague optical systems, but space does not permit detailed discussions of them here. Most of

them can be eliminated or controlled by properly shaping the optical surfaces and by keeping the object being viewed on, or very close to, the optical axis. Perhaps the most interesting of these is *coma*, especially troublesome in large parabolic mirrors. axis, are shaped like little comets, with the tails pointing outward.

Perhaps the most successful in eliminating this, and other aberrations as well, are complex telescopes of the *Schmidt* configuration (Figure A3-16) and various spin-off systems such as the Maksutov, the Gregory-Maksutov, and other such *catadioptric* systems. All of the above use combinations of spherical mirrors and either lenses or, in the case of the Schmidt telescopes, a thin correcting plate that corrects for spherical aberration. All of these systems are almost coma free, and they give outstanding image quality clear out to the edge of the viewing field. Unfortunately, because of mechanical instability in the thin correcting plates, Schmidt telescopes

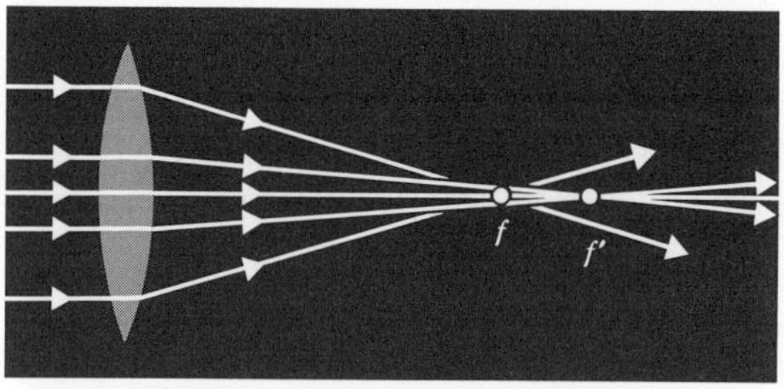

FIGURE A3-12 Spherical aberration in a lens. The central rays come to a focus farther from the lens than the peripheral rays.

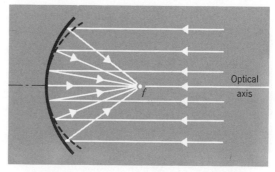

FIGURE A3-13 Correction for spherical aberration in a mirror; the corrected surface is a paraboloid of revolution (in cross section, a parabola).

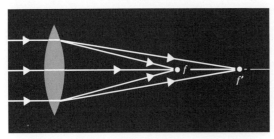

FIGURE A3-14 Chromatic aberration. Blue light f focuses closer to the lens than red light f'.

generally cannot have apertures exceeding 48–60 inches.

<u>Accessibility of the Image in a Telescope.</u> Except as used by amateurs for whom direct viewing is a prime consideration, most telescopes are used as cameras. Usually, the plate—or film—holder is placed at the prime focus of the telescope. Under these circumstances the magnification that may eventually be desired is most easily accomplished in the photographic darkroom by an enlarger.

Reflecting telescopes form their images in front of the objective surface. This poses no serious problem with large-aperture telescopes such as the Lick 120-in. and the Hale 200-in., even when they are used for direct viewing or photography at the prime focus. The observer merely sits in a cell or cage at the prime focus. The cage does block off some of the incoming light rays, but the light loss is surpris-

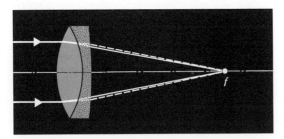

FIGURE A3-15 An achromatic doublet lens. The densities of the two glass elements are so selected that light of all colors comes to a focus essentially in the same plane.

ingly modest, no more than 10–15% at most. Other techniques employ secondary mirrors to divert the image-forming rays either at right angles off the optical axis and out of the way of the objective, or else to reflect them through a hole in the objective to a focus behind it. Various modern configurations of reflecting telescopes are diagrammed in Figure A3-17. In some large telescopes, when used in conjunction with a spectrograph, the optical train is elongated with a diverging mirror and then diverted down the polar axis into a special room or spectrographic laboratory. This system is known as the Coudé focus (Figure A3-18).

The Spectrograph

As the name implies, the spectrograph permits analysis of light from celestial objects by spreading it out into a visible spectrum. If analysis is made by direct viewing, the instrument is termed a *spectroscope;* when used in conjunction with a photographic plate, it becomes a *spectrograph.*

A spectrum is produced in one of two ways. By far the most commonly used one is an optical grating. It consists of a glass plate ruled with many thousands of fine, parallel scratches per inch. The phenomena of diffraction and interference of light waves produces the spectrum. Each scratch behaves as a small slit, each edge of which diffracts the light that hits it. Different wavelengths of light are diffracted through different angles. Thus the spectrum is produced by interference effects between the light diffracted from adjacent slits.

A *prism spectrograph* operates on the principle of dispersion, as we discussed earlier in connection with lenses. The schematic arrangement of the optical train of a prism spectrograph is illustrated in Figure A3-19. The light from a star is brought to a focus on a small slit. The beam, which is diverging beyond the focal point, is collimated, or rendered parallel, by a collimating telescope. The parallel rays are then dispersed by the prism, or train of prisms, and they are finally brought to focus by a camera lens on a photographic plate. The spectrum of dispersed light is again a line of images of the slit so close to each other that they form a continuum. If for any reason the light at any wavelength is missing, or is substantially reduced in intensity, a dark

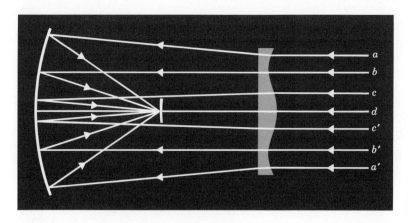

FIGURE A3-16 Optical system of a Schmidt reflector.

image of the slit, called an *absorption line* appears in the spectrum at the precise wavelength position. If the light consists of only very few wavelengths, or is somehow enhanced at very few wavelengths, bright-line images of the slit appear at the respective wavelength positions; this is known as an *emission spectrum*. Most celestial objects, except those shining by reflected light, produce a *continuous spectrum* upon which is superimposed either or both an absorption and an emission spectrum.

The optical train of a spectrograph is enclosed in a light-tight box that has openings only for the admission of the light beam and for exposing a photographic plate. The photographic exposure time for

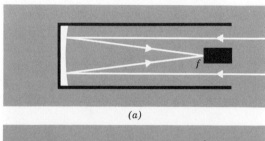

(a)

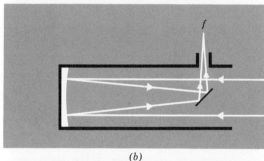

(b)

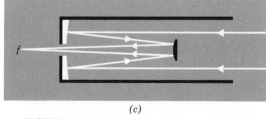

(c)

FIGURE A3-17 Several reflector telescope configurations; (a) *prime* focus; (b) *Newtonian* focus; (c) *Cassegrain* focus.

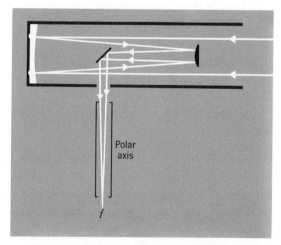

FIGURE A3-18 The Coudé focus for a reflector telescope.

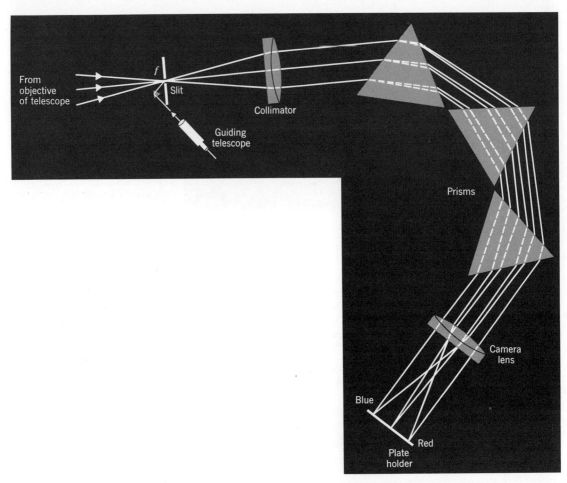

FIGURE A3-19 The optical train for a three-prism spectrograph.

a very faint star or galaxy can run into hours, even days. The spectrograph is, of course, mounted at the focal plane of the telescope, which is in an observatory open to the outside atmosphere. Thus during the course of an exposure the temperature may vary considerably, a situation that will seriously degrade the quality of the spectrum obtained. Therefore, the spectrograph is internally heated and maintained very precisely at constant temperature with thermostatic controls.

Radio Telescopes

Radio telescopes operate on the same general principle as optical reflectors, except that unlike optical telescopes, they are limited to seeing only one frequency at a time. Compared to a visible light, radio waves have extraordinarily long wavelengths. For example, the wavelength of the radio signal emitted by neutral hydrogen, an exceedingly important emission in astronomy, is 21.0 cm. The wavelength of light near the middle of the visible spectrum is only 5×10^{-5} cm, *400,000 times shorter!*

A typical parabolic radio telescope is shown in Figure A3-20. Radio telescope objectives need not have smooth surfaces, and the surface need not be continuous. All that is required is that the irregularities, or the openings, be very much smaller than the radio wavelengths to be received. Wire mesh, similar to that used in fencing, makes an excellent reflecting surface. The telescope objective focuses the radio waves on the end of a *waveguide*

FIGURE A3-20 The 210-ft Goldstone radio telescope (Courtesy NASA).

which conducts the signal to the electronic decoding center. The spectrum is traced on a moving strip of paper by a pen recorder.

Types of radio telescopes other than parabolic, in many cases resembling connected arrays of TV antennas, are very effective for the limited purposes they serve. Since the arrays can be strung over many miles, great improvement in resolution can be secured over that of the parabolic reflectors. Even so, an array that can equal the resolving power of the human eye would have to be more than 155 km long.

The resolving power of radio telescopes is quite low. The resolution of the 250-ft Jodrell Bank radio telescope in Manchester, England for the 21.0 cm wavelength of neutral hydrogen is about 0.19°, roughly a third of the diameter of the full moon. Radio viewing is thus quite fuzzy. But just as the center of the diffraction disk of a star can be used to locate its position with accuracy that exceeds the resolving power of the telescope, so can radio telescopes pinpoint the location of radio sources that are smaller than the resolving capability of the telescope.

APPENDIX 4
The Magnitude System and Its Use

As early as the 2nd century B.C. a magnitude system of sorts was in use. Hipparchus compiled a catalogue of 1022 visible stars that he placed in six categories of apparent brightness. The most luminous stars he called first-magnitude objects. Those just perceptibly fainter Hipparchus called second-magnitude stars, and so on, down to magnitude six. Stars in this last category were just at the threshhold of visibility.

During the late 1840s or early 1850s William Herschel devised a simple, direct method, albeit an approximate one, for measuring the relative apparent intensities of the light from various stars. His method, essentially the same as that discussed in connection with the limiting stellar magnitudes of telescopes, when applied to the older empirical magnitude system yielded a hundredfold difference in brightness between first and sixth magnitude stars.

Under the assumption that Herschel's law of photometric intensities was exact, Norman R. Pogson in 1859 proposed a quantitative scale of (apparent) stellar magnitudes. Contemporary physiological experiments had strongly suggested that equal intervals of brightness perceived by the eye (the basis for Hipparchus' magnitude system) were in reality equal *ratios* of intensity. The ratio of intensity of first to sixth magnitudes was taken by Pogson to be exactly 100:1. Moreover, each magnitude differed from the next by the fifth root of 100, or approxi-

mately 2.512. Thus a fifth-magnitude star is just 2.512 times brighter than a sixth-magnitude star; a fourth.magnitude star was just 2.512 times brighter than a fifth-magnitude object, and therefore it was 2.512×2.512 $(2.512)^2$ times brighter than a sixth-magnitude star, and so on. A first-magnitude star is thus $(2.512)^5 = 100$ times brighter than a sixth-magnitude star.

In practice, fractional magnitudes are necessary simply because the stars do not cooperate in emitting light intensities in exact multiples of 2.512. Fractional magnitudes are determined by precise photoelectric photometry methods. Incidentally, these methods showed that some stars were brighter than first magnitude. Sirius, for example, is magnitude -1.4. The limiting magnitude for the average unaided eye is about magnitude 6.5.

Limiting Stellar Magnitudes for Telescopes

The average effective diameter of a dark-adapted pupil of the human eye is about 7 mm, or roughly 1/3 of an inch. A 1-in. telescope objective has 9 times the area of the pupil and therefore gathers 9 times as much light from a given star. When the eye is placed at the focus of a 1-in. telescope, it is able to distinguish stars 9 times fainter than it can when it is unaided. This is equivalent to a gain of 2.4 magnitudes. But the unaided eye is capable of seeing stars down to magnitude 6.5, hence the limiting magnitude of the 1-in. telescope is roughly magnitude 9.

Now a 10-in. telescope gathers 100 times as much light as the 1-in. instrument, hence it can detect stars 5 magnitudes fainter. Thus its limiting magnitude is magnitude 14. Table A4.1 gives various magnitude data.

In establishing his photometric scale Herschel reduced the effective apperture of a moderate sized telescope by diaphragms, or stops, in steps, and noted the different magnitudes that became invisible at each step. Suppose a ninth-magnitude star was just barely visible when the objective was stopped down to 1 in., and a fourteenth-magnitude star was at the limit of visibility when the objective was stopped down to 10 in. Clearly, a difference of 100 times in the effective area of the objective was equivalent to 5 magnitudes (Q.E.D.).

Absolute Magnitude. By definition *absolute magnitude* is the magnitude a star would have if placed at a standard distance of 10 pc from the observer. Absolute magnitude determinations are essential in the determination of differences in intrinsic brightness of stars. For example, consider the sun's measured apparent magnitude of −26.5. A parsec is equivalent in distance to 206,265 AU; a distance of 10 pc is roughly 2 million AU. Thus if the sun were removed to a distance of 10 pc, according to the inverse square law (Chapter 10) its apparent brightness would be decreased by a factor of 4×10^{12}. This corresponds to about 31.5 magnitudes. At 10 pc the sun's absolute magnitude would therefore be −26.5 + 31.5 = magnitude +5.

Visual and Photographic Magnitudes

The color response of an optical device, including the human eye, greatly influences the magnitude assigned to a star. This is because the stars vary greatly in color. Hence it is necessary to be succinct in specifying the kind of magnitude measured. *Visual* and *absolute visual* magnitudes are measured in the yellow-green portion of the visible spectrum. This is the range of wavelengths to which the human eye is most sensitive. *Photographic and absolute photographic* magnitudes are those recorded on standard blue-violet sensitive photographic emulsions. Emulsions that are treated with dyes so as to have a peak response in yellow light yield *photovisual* and *absolute photovisual* magnitudes.

Color Index. The difference between photographic and photovisual magnitudes is defined as the *color index, CI*:

$$CI = m_{\mathrm{phg}} - m_{\mathrm{phv}}.$$

Color index is independent of stellar distances since the inverse square law applies equally to all wavelengths of light. Hence a color index is the same no matter whether apparent magnitudes or absolute magnitudes are considered.

A red star will appear brighter on the photovisual emulsion, and with respect to it will have a smaller magnitude. (Remember the magnitudes go down as the brightness increases.) Hence the CI of a red star will be positive. On the other hand a blue star will appear brighter on the blue-sensitive plate, thus its CI will be negative. How *much* negative or positive the color index is depends upon an arbitrary zero point. The photographic emulsions are so adjusted that the star Sirius, a spectral class A star with a temperature of about 10,000°K will appear equally bright on both. Stars hotter than Sirius will have negative CI's; cooler stars will have positive CI's. The range in color indices is from −0.6 for the bluest stars to +2.0 for the reddest.

TABLE A4.1 Magnitude Data

Object	Magnitude
Sun	−26.5
Moon (full)	−12.5
Venus (when brightest)	−4.0
Jupiter	−2.0
Mars (at closest opposition)	−2.0
Sirius	−1.4
Unaided eye limit	6.5
50 mm binoculars	10.0
4-in. telescope	12.0
6-in. telescope	13.0
10-in. telescope	14.0
100 in. telescope	19.0
200-in. telescope (visual limit)	20.5
200-in. telescope (photographic limit)	23.5

<u>Use of the Magnitude System.</u> When the star field is photographed with both blue-sensitive and yellow-sensitive plate, the resulting color index for any observed star is readily obtained from the comparative intensities of the star images. The color index is calibrated for a wide range of stellar colors, hence it is an exceedingly powerful tool in the study of stellar temperatures. (See Chapter 10 for the relation between stellar temperature and color.) The techniques employed thus far exceed in precision the use of color photography.

APPENDIX 5
Radiation Theory

As is so often the case in scientific investigations, digressions must be made in order to develop an understanding of the terminology employed and the frame of reference. So it is here; a qualitative appreciation for the meanings of radiation, frequency, wavelength, and spectrum is essential to our progress in understanding the universe. Physicists employ the term radiation when speaking of an energy-transfer process that requires no physical medium for its propagation. Radiant energy is that form of energy which is transferred by radiation. These are abstract concepts. In our ordinary Newtonian frame of reference, radiant energy has no mass. It is not visible. In fact, radiant energy is undetectable except by specially designed receivers or detectors. These detectors are usually quite selective in the fraction of the total radiant energy to which they respond. We shall find it most helpful to turn to physics for the preliminary information with which to attack our problem.

Physicists discovered a remarkable relationship between the various manifestations of radiant energy. Electromagnetic radiation, for this is what radiant energy proves to be, has fundamental characteristics. We tend to categorize these characteristics in terms of the sorting devices which selectively respond to them. As we shall see, this is an artificial classification made for our convenience.

Physics provides us with fundamental assumptions or concepts about electromagnetic energy which, like time and mass and space, cannot be proved. This does not detract from their reality in any way.

1. There exists a property of matter known as electric charge. There are two kinds of charge: positive and negative.

2. Charge is usually associated with charge carriers; charge is not described by its carriers.

3. Charges exert force on each other. The presence or absence of matter is of no consequence.

4. The region of space in which a charge experiences force is called an electric (or magnetic) field. Strangely, this mental construct is a vector quanity; it has magnitude and direction. The magnitude of a field is determined by the acceleration a charge would experience when released in the field. The direction of a field is that in which a positive charge moves.

5. Electric charges can oscillate. This postulate is connected with the idea of a uniformly varying electric field, one that reverses direction in a cyclic manner. An oscillating electric charge generates such a field by transmitting energy into the surrounding region. The energy is propagated with the speed of light, which experimentally is approximately 30 billion centimeters per second or, in more familiar units, about 186,000 mi/sec.

6. Associated with a varying electric field is a varying magnetic field. The magnetic field oscillates in a plane perpendicular to that of the electric field. Both field disturbances are perpendicular to the direction of energy propagation. Taken together they comprise an electromagnetic field.

7. The existence of an electromagnetic field is deduced by observing its effect on test charges; if the field is a varying one, the inference regarding oscillators follows.†

† The discussion on electric and magnetic fields is intentionally brief and incomplete. Our purpose here is to establish a notion of oscillators whose existence can be postulated from observed effects on their environment.

These are difficult notions because they are not abstracted directly out of experience. They form, in part, the foundation of the work of James Clerk Maxwell, one of the great theoretical physicists of all time. Maxwell's equations, first presented in 1873 in his famous treatise on electricity and magnetism, are still fundamental in electromagnetic theory.

Our attention to this much detail is justified. It is difficult to appreciate how enormous quantities of energy can traverse essentially empty space and then react with matter in the variety of ways we observe. The concepts of charge, fields, action-at-a-distance, and speed of light are beyond our direct experience. How then can we rationalize them into meaningful concepts? We shall have to do so in terms of observed effects. These, at least, are real.

A mechanical analogy is helpful to us here. Let us suppose that we are holding the end of a partly slack rope, the other end of which is secured to a post or similar stationary object. If we abruptly raise our hand a short distance and then return it to its starting position, a kink or wave appears in the rope. The wave travels at uniform speed down the rope to its tied end (Figure A5-1).

Let us analyze what we have seen. The wave must have transmitted a pulse of energy, since we put energy into the system with the flick of our hand. Now if energy is conserved, the vertical displacement of the rope associated with the wave must have been a consequence of this energy input. The kink, or wave, of energy did not involve a displacement of the rope along the direction of propagation, even though the energy pulse traveled down the rope.

We can reason in a similar fashion with respect to electromagnetic energy propagation in a region of space. In terms of the oscillation of the electric field, we can visualize a wave motion traveling at 30 billion centimeters per second. A continuous succession of these waves transmits energy. How energy can be propagated in the absence of a supporting medium even today remains a mysterious, inexplicable phenomenon.

Most readers will subconsciously envision the wave propagation of energy undulating, as it races along through space, like a scurrying caterpillar. This is not the case. Waves really expand uniformly in all directions in free space. The undulations are change in amplitude (intensity) that likewise occur in three dimensions. If the amplitude changes take place in a *single plane* but always at right angles to the direction of propagation, we say the wave is *plane polarized*. It is possible for the amplitude changes to rotate through different successive planes. These waves are said to be *circularly polarized*. For simplicity let us consider a plane-polarized wave disturbance as it passes successive points lying in a straight line in space (Figure A5-2). Points 1 through 14 in the figure are equally spaced. We consider what these points see as an electromagnetic wave disturbance passes each in succession. The electromagnetic disturbance completes a wave cycle 83.3 times each second. Now the propagation velocity is the speed of light, 3×10^{10} cm/sec. If the points are 300 km apart the time interval illustrated as the disturbance travels from point to point is 0.001 second. Each line of points in the figure, then, represents what the array sees every

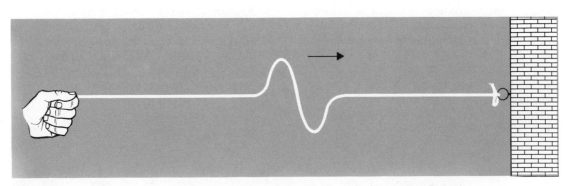

FIGURE A5-1 Energy pulse sent down a rope in the form of a traveling wave. Rope is not displaced longitudinally (From Cole, F. W., *Introduction to Meteorology*, John Wiley and Sons, New York, 1970).

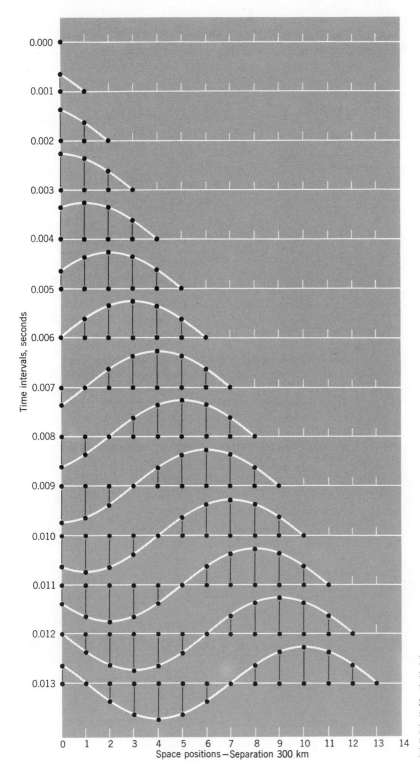

FIGURE A5-2 Graphical representation of electromagnetic wave propagation. Each line represents the same space points at intervals of 0.001 second. Each point on the line is 300 km apart.

0.001 second. For example, in line 2 the disturbance has reached point 1; it is imminent at point 2.
In line 3 the disturbance has progressed to point 2
and is now imminent at point 3, and so on. In fact,
when the cycle is completed at point 1 it is just imminent at point 13. The succession of intensity variations at each point, called changes in amplitude,
has the graphical appearance of a particle being
displaced vertically, such as a cork in water waves,
or the fixed particles of the rope in Figure A5-1. But
each point downstream experiences a given change
in amplitude exactly 0.001 second later than the
preceding point. This gives the amplitude change
the appearance of undulating as the disturbance is
propagated through space. This is what we mean by
wave motion.

Wavelength, Frequency, and Speed. Three quantities used to describe the propagation of electromagnetic energy are easily related by way of an
analogy. Let us visualize a train of cars each 50 ft
long, passing a particular reference point such as a
crossing sign (Figure A5-3a). Suppose we observe
that one car passes the sign each second. We can
describe the motion in the following manner.

1. The *frequency* with which cars pass the reference
    point is one per second (written 1/sec).

2. The *length* of each discrete unit (one car) is 50 ft.
3. The *speed* of the train is equivalent to the
    frequency (1/sec) with which each unit of length
    (50 ft) passes the point. In the example the speed
    is 50 ft/sec.

We can generalize as follows. Let ν represent
frequency, λ be the unit of length, and c be the particular speed. Then

$$c = \nu\lambda \qquad (A5.1)$$

$$\nu = \frac{c}{\lambda} \qquad (A5.2)$$

$$\lambda = \frac{c}{\nu}. \qquad (A5.3)$$

For electromagnetic radiation in a vacuum, c has
a constant value of approximately 3×10^{10} cm/sec†.
For all pratical purposes the speed of propagation in
the atmosphere is the same. The lower case Greek
letter *lambda* (λ) refers to the distance between corresponding points on two successive waves (Figure

† In the sciences and engineering, it is customary to write
very large and very small numbers as powers of 10. A brief
discussion of this notation is contained in Appendix 2.

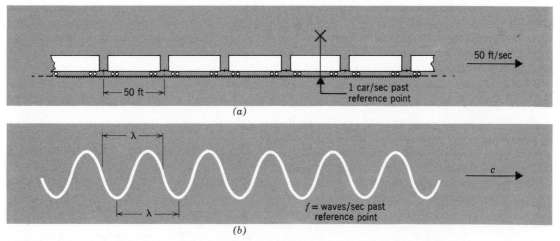

FIGURE A5-3 Wavelength, frequency, and speed analogies. (*a*) *Speed* of train carries
cars of a given *length* past a reference point with a *frequency* determined by the speed
and car length. (*b*) Graphical analogy of traveling *waves* (from Cole, F. W., *Introduction
to Meteorology,* John Wiley and Sons, New York, 1970).

A5-3*b*). The lower case Greek letter *nu* (*ν*) is customarily used to represent frequency. From Equation A5.3 we see an important relationship between frequency and wavelength. At constant speed, as the wavelength decreases, the frequency must increase to preserve the equality.

Atoms, Ions, and Molecules. Atoms, ions, and molecules are the electromagnetic oscillators that respond to and generate electric and magnetic fields. In particular it is the charged particle portion of atoms and molecules that is responsible for the electromagnetic effects. Atoms are the smallest elemental subdivisions of matter that still retain the physical characteristics of a given substance. They consist of a nucleus containing relatively massive positively charged particles called *protons;* most nuclei also contain similarly massive uncharged particles called *neutrons.* Each normal atom has associated with it negatively charged, lightweight particles called *electrons* on a one-to-one correspondence with the protons. Atoms are therefore electrically neutral.

An ion is an atom that has lost or gained one or more electrons. It has a net charge associated with it in contrast to the normal atom. Molecules consist of two or more atoms, either of the same kind or in certain complex combinations of different kinds. In general, the physical characteristics of a given molecule are entirely different from any of the characteristics of its constituent atoms. The atoms in a molecule are in general bound together through sharing one or more electrons between them. Molecules, like atoms can be ionized.

Atomic and molecular electrons reside in preferred energy levels, loosely referred to as orbits, or orbital shells. The analogy to planetary orbits is very crude; electron orbits are complex, and it is more accurate to say that such an orbit represents the zone or region where the probability of finding a given electron is greatest. We shall presently observe the role of these electromagnetic oscillators in the production of spectra.

Blackbody Radiation. One of the most important of all classical physical theories concerns the *manner* in which energy is absorbed or given off by an object. In 1859 a German physicist, Gustav Kirchhoff, discovered that the emitting power of a given substance was directly proportional to its ab-

sorbing power, although in general the rates of emission and absorption were not equal. Kirchhoff's discovery marked the first step toward the solution of radiation rate problems. His discovery suggested that there may exist an *ideal substance* that could absorb completely all radiation falling upon it. Such a substance would also be the best possible emitter of radiation. Such a substance, when cool enough that radiation from it was not in the visible range, would be totally jet black since it would *reflect* no light; a body made of this substance was termed a blackbody. Such a body would emit all energy wavelengths (or frequencies) at maximum efficiency, even though the intensity might be different for different wave lengths. The *total radiation and its wavelength distribution would depend solely upon the temperature.* No such material has been found, but blackbody radiation characteristics have been synthesized within 99% accuracy in the laboratory.

We can gain some insight into the concept of radiation (or absorption) efficiency from our own experience. Recall that if an object, say a heating element of an electric range, is continuously heated to higher and higher temperatures, we can feel or sense at a distance the increase in heat given off. Light is often also emitted and it ranges from darkest red to perhaps straw color as the temperature increases. We associate the highest temperature with the lighter color; this is consistent with the physical description of the electromagnetic spectrum. Qualitatively at least, we are aware of a greater *rate* of radiated heat energy with increased temperature.

Let us suppose we could heat an object in a furnace through a series of discrete steps of higher temperature. At each temperature step we employ a little black box to measure and record the intensity and the frequency of the emitted radiation from the far-infrared to the far-ultraviolet†. At the conclusion we plot a graph of the data as in Figure A5-4.

It appears that the intensity of the total energy output is greater for each higher temperature. The area between a temperature curve and the reference

---

† Little black boxes are indispensable to physicists. They signify that measurement is accomplished, or phenomena are displayed, in which the instrument itself is of no consequence. Presumably the design and construction of an instrument that performs black-box functions are merely engineering problems.

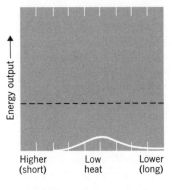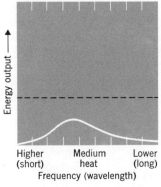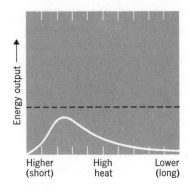

| Higher (short) | Low heat | Lower (long) | Higher (short) | Medium heat | Lower (long) | Higher (short) | High heat | Lower (long) |

Frequency (wavelength)

FIGURE A5-4 Representative energy emission curves for an object heated to successively higher temperatures (from Cole, F. W., *Introduction to Meteorology,* John Wiley and Sons, New York, 1970).

abscissa (the bottom of the graph) is greater for higher temperature. Indeed, the area under a given curve is proportional to the intensity of the total energy emitted at that temperature.

We also learn that the frequency (and wavelength) of the intensity maximum shifts toward higher frequencies with succeeding higher temperatures. Finally, we observe similarities in the shape of each temperature curve; there is a very narrow range of frequencies where most of the energy is emitted, and the energy emission nearly vanishes at *both* higher and lower frequencies.

Physicists have learned that when different materials are heated through the same series of experimental temperatures, they have different appearances at each temperature. Black-box measurements also show wide variations in the intensity and frequency distribution of the emitted energy. On the other hand, measurements show that energy emitted through a tiny hole connected to an interior cavity (Figure A5-5) of an object differs in intensity and frequency from that emitted by the exterior surface (Figure A5-6). Furthermore, the *cavity radiation* characteristics depend only on the temperature, not

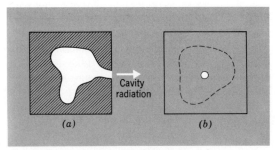

FIGURE A5-5 Cavity radiator: (*a*) cross-section view and (*b*) front view. Cavity radiation emerges from the tiny hole leading to exterior (from Cole, F. W., *Introduction to Meteorology,* John Wiley and Sons, New York, 1970).

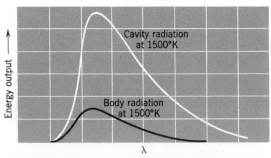

FIGURE A5-6 Curves showing energy radiated by body compared to energy radiated by cavity within it. Cavity radiation is called "blackbody" radiation (from Cole, F. W., *Introduction to Meteorology,* John Wiley and Sons, New York, 1970).

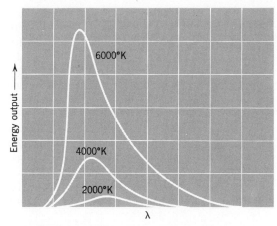

FIGURE A5-7 Characteristic radiation curves for a blackbody at different temperatures (from Cole, F. W., *Introduction to Meteorology*, John Wiley and Sons, New York, 1970).

on the substance or even the shape of the cavity. That is, for a given temperature all cavities, whatever the substance, radiate at the same efficiency. Such cavity radiation approaches the theoretical ideal and is therefore termed *blackbody radiation*.

Characteristic black body radiation curves for different temperatures are illustrated in Figure A5-7. The distribution of radiation frequencies for each temperature is well known. Also, the rate at which energy is emitted (e.g., the intensity of the radiation), the energy distribution over the range of frequencies, and the total radiation intensity are all well established theoretically.

Radiation Laws. The discovery that a cavity emits blackbody radiation resulted in the experimental verification of two radiation laws that had followed on the heels of Kirchhoff's work. The *Stefan-Boltzmann* law asserts that the total emissive power per unit time per unit area (radiation intensity) is proportional to the fourth power of the absolute temperature of the emitting body.

$$E \propto T_{(\text{°K})}^{4} \qquad (A5.4)$$

Absolute temperature, expressed in degrees Kelvin (°K), is derived from the thermodynamic properties of real substances. The Kelvin temperature scale is numerically equal to °C plus 273.15 (for practical purposes, this number is rounded off to 273).

A second radiation law derived in 1893 by Wilhelm Wien asserts that as the temperature of an emitting body increases, the peak intensity of the emission shifts toward shorter wavelengths.

$$\lambda_{\text{peak}} = \frac{k}{T_{(\text{°K})}} \qquad (A5.5)$$

where k is an experimentally determined constant. Figure A5-7 shows graphically both of these relationships. The total radiation intensity is proportional to the area bounded by a given temperature curve and the base of the graph. The high point on the curve is the intensity peak, and the height of the curve at any point on it is a measure of the radiant intensity at that wavelength.

As impressive as these discoveries are, the 19th century physicists were unable to predict the shapes of a series of temperature curves for a cavity radiator *unless* a curve for some particular temperature was experimentally determined. This deficiency did not hinder laboratory progress particularly. But if general radiation theory was to be extended to the study of the stars, theoretical methods were essential; the stars could not be brought into the laboratory.

In a paper presented in 1900 to the Deutsche Physikalische Gesellschaft a German physicist, Max Planck, broke the theoretical bottleneck and propelled physical science along the road toward

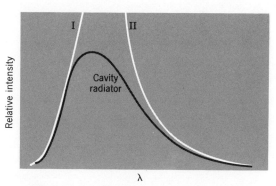

FIGURE A5-8 Relative deviation from the intensity-wavelength cavity radiation curve by I, Wien's early work, and II, the Rayleigh-Jeans formula.

quantum theory. The latter is a way of looking at reality in which energy is not emitted or absorbed continuously by atoms or molecules, but in discrete steps. Each step involves the emission or absorption of a discrete amount or packet of energy called a *quantum*. Planck's success was not just the fruit of blind trial-and-error methods. He began with the earlier work of Rayleigh and Jeans, and of Wien, none of whom could produce a mathematical formula that agreed in all respects with the observed cavity radiation (Figure A5-8).

Planck struck forward with the bold quantization hypothesis, and with it modified Wien's formula for cavity radiation. The result is history; the theory of ideal blackbody radiation was now complete. And with it astronomy was free to move forward out of the relative stone age into the future. The energy distribution curve for cavity radiators could now be predicted for any temperature independently of prior experimentation. The mystery of the sun's surface temperature was a mystery no longer. By assuming it radiates as a blackbody and by utilizing radiation theory, we can arrive at an independent estimate of the sun's temperature and its radiant intensity. An energy distribution curve for the sun is plotted, and the graph is compared to blackbody radiation curves; a good fit is found with the 5770°K curve. This, then, is the sun's effective surface temperature.

APPENDIX 6
Observational Astronomy for the Beginner

There is no reason why one cannot acquire a telescope of his or her choice, set it up on the first clear, reasonably dark night, and begin to enjoy active involvement in observational astronomy. It takes about a year to become thoroughly familiar with the celestial sphere and astronomical time as well as proficient in the use of the telescope. But this does not detract from the thrill of zeroing in on the first *selected* target. These brief guidelines are designed to help you do just that.

Telescope Selection. The telescope is an excellent case in point where the biggest is best. The larger the telescope, the brighter the image, and the more the detail, including separation of close stars. From a cost standpoint, a *reflector* telescope (one that employs a mirror as the primary optical component) is the best buy. However, smaller sized refractors (telescopes employing lenses) may be cheaper than the smallest *practical* reflector. Each type has its advantages and disadvantages, but for carefully made and assembled instruments, both types perform well. A 6-in. reflector, one whose mirror is nominally 6 in. in diameter, is about the smallest that should be considered, although a 4-in. instrument is suitable for limited observing. A 2-in. refractor is as small as one should consider. In my experience, observers who settle for telescopes smaller than these will soon be trading up to larger ones, at added expense, of course.

The Mount. Unless a telescope is to be used exclusively for observing distant scenes or objects on the earth, there is only one type of mount that is acceptable: the *equatorial mount*. Figure A6-1 shows one of several versions; all have in common a *polar axis* and a *declination axis*. The polar axis is aligned *parallel* to the earth's axis, and thus allows the telescope to be rotated exactly in an east-west direction. Once an object is brought into the center of the field of view, it can be followed as it apparently travels from east to west, by rotation of the telescope about this one axis.

The declination axis is permanently fixed at right angles to the polar axis. It enables the telescope to be pointed north and south. In operation, the tele-

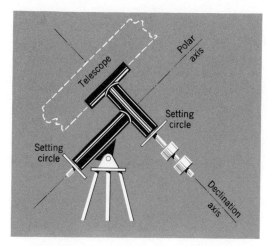

FIGURE A6-1 An equatorial telescope mount.

scope is set to the declination of the object to be viewed, and the declination axis is then locked. Then the telescope is rotated about the polar axis until the desired object is in the field of view; thereafter slow rotation about this axis toward the west, at a rate corresponding to the rotational rate of the earth toward the east, will hold the object in view until it sets.

Telescope Auxiliaries. The main telescope should be equipped with a wide field (6° or more) *finder* telescope, mounted exactly parallel with the optical axis of the main instrument. Practically all commercial telescopes are furnished with a finder telescope. Three *eyepieces* will do for a start. Their respective focal lengths (specified by the supplier) should be about 1-in., 1/2 in., and 1/4 in. In metric measure the focal lengths are approximately 25 mm, 12.5 mm, and 6 mm, respectively. If desired, a wide-angle eyepiece of about 45 mm focal length is a valuable accessory for viewing the moon.

Setting Up the Telescope. For a casual evening's observations, absolute accuracy is not necessary in orienting the telescope mount with respect to the celestial sphere. As nearly as you can judge, align the telescope tube parallel to the polar axis, and lock the declination axis. Place the mount on a level surface, with the least obstructed view toward the south. Aim the whole instrument toward the north star, *Polaris,* by moving the mount feet or the tripod as the case may be. Polaris is a moderately bright star, due north; it can be located by the *pointers,* the two stars in the end of the bowl of the Big Dipper. This constellation traces a circle about Polaris, when viewed from latitudes from about 35° N, once every 24 hours. It can be found anywhere on this circle at a given hour, depending on the date (Figure A6-2).

Look for Polaris in the finder telescope. If the polar axis is aimed too low or too high for your latitude, loosen the inclination lock nut and adjust the angle of the polar axis until Polaris is in the center of the field of view of the finder. Retighten the polar axis inclination nut; it is now permanently adjusted for your location.†

Shift the mount slightly to improve the alignment; then mark the position of the mounting feet on the surface for future reference. The telescope is now ready for use.

† Most vendors will set the inclination of the Polar axis to the local latitude prior to, or upon, sale.

Celestial Coordinates. Celestial coordinates, which specify particular locations on the celestial sphere, are discussed fully in Chapter 2. Briefly, the sky appears to rotate once in 24 hours about the celestial poles, which are directly above the earth's poles. It is the earth, of course, that is rotating toward the east; hence the sky appears to rotate toward the west, 15°/hr (360° ÷ 24 hr = 15°/hr).

Midway between the celestial poles is an imaginary circle called the celestial equator; it is everywhere directly above the earth's equator. Celestial latitude, corresponding to terrestrial latitude, is termed *declination.* It varies from 0° at the celestial equator, to +90° at the north celestial pole, and to −90° at the south celestial pole. Declination, like latitude, is fixed relative to the earth. Once a selected object is centered in the field of view of the telescope, the declination axis is locked; *the only motion of the telescope required in following the object is rotation about the polar axis.*

Celestial longitude, corresponding to longitude on the earth, is called right ascension. The *circles of right ascension* each terminate at the celestial poles; they cross the celestial equator at right angles. They are numbered in *hours* toward the east, from 0^h to 23^h. Each *hour circle* is 15° from its neighbors. The name *hour circle* indicates that the right ascension circles successively cross the *local meridian* every hour.†† The local meridian is an imaginary line containing the earth's poles and the point directly above the observer. Thus the local meridian runs due north and south.

Table A6.1 shows which hour circle is approximately on the local meridian for selected dates at 8 P.M. local mean solar time. Local time differs from *standard* time unless the observer is exactly on the meridian of longitude associated with the standard time of the given time zone. At most, local time will not differ from standard time as much as an hour anywhere in the continent. This is still within the limits of accuracy required for the following observing methods.

Locating Celestial Objects. Sky charts, Appendix 7, and lists giving the right ascension and declination of the celestial objects should be at hand. Suggested sources for additional materials are given

†† Actually they cross the local meridian every $59^m 50^s$. The reason for this is detailed in Chapter 4. For the present purposes, the difference can be ignored.

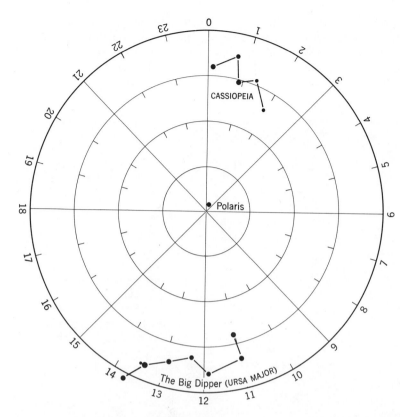

FIGURE A6-2 Polaris and the Big Dipper and Cassiopeia. Use this chart to find sidereal time by facing north and turning the chart so that the dipper in the figure is oriented the same as the Big Dipper in the sky. The *number* at the extreme top is the sidereal time.

at the end of this appendix. Table A6.2 lists a few selected objects which will get you started. Check Table A6.1 to determine whether or not the desired object is above the horizon at the observing time, for the observing date. In general, a span of nine hour circles, centered on the local meridian, are suf-

TABLE A6.1   Approximate Sidereal Time at 8 P.M. Local Mean Time

Sidereal time	Date	Sidereal time	Date	Sidereal time	Date	Sidereal time	Date
0^h	Nov. 21	6^h00^m	Feb. 19	12^h00^m	May 22	18^h00^m	Aug. 21
0^h30^m	Nov. 28	6^h30^m	Feb. 27	12^h30^m	May 29	18^h30^m	Aug. 29
1^h00^m	Dec. 5	7^h00^m	Mar. 7	13^h00^m	June 6	19^h00^m	Sept. 5
1^h30^m	Dec. 13	7^h30^m	Mar. 14	13^h30^m	June 14	19^h30^m	Sept. 13
2^h00^m	Dec. 21	8^h00^m	Mar. 22	14^h00^m	June 21	20^h00^m	Sept. 21
2^h30^m	Dec. 28	8^h30^m	Mar. 29	14^h30^m	June 29	20^h30^m	Sept. 28
3^h00^m	Jan. 5	9^h00^m	Apr. 6	15^h00^m	July 6	21^h00^m	Oct. 6
3^h30^m	Jan. 12	9^h30^m	Apr. 13	15^h30^m	July 14	21^h30^m	Oct. 13
4^h00^m	Jan. 20	10^h00^m	Apr. 22	16^h00^m	July 22	22^h00^m	Oct. 21
4^h30^m	Jan. 28	10^h30^m	Apr. 29	16^h30^m	July 29	22^h30^m	Oct. 28
5^h00^m	Feb. 4	11^h00^m	May 7	17^h00^m	Aug. 6	23^h00^m	Nov. 5
5^h30^m	Feb. 12	11^h30^m	May 14	17^h30^m	Aug. 13	23^h30^m	Nov. 13

Sidereal time is numerically equal to the number of the right ascension circle on the local meridian. Remember, local mean time will in general differ by up to an hour from local standard time.

ficiently above the eastern and western horizons for good viewing. For example, on December 21, at 8 P.M. 2^h right ascension is on the local meridian. Objects whose right ascension range between 22^h and 6^h are favorably situated above the horizons. Remember, the circles of right ascension advance toward the west at the rate of 1/hr. Thus, at midnight on the above date, 6^h00^mRA is on the meridian; good viewing can now be had for objects whose right ascension range between 2^h (now 60° west of

TABLE A6.2 Selected Celestial Objects

	Right Ascension	Declination	Remarks
Double stars			
γ Arietis	1^h51^m	+19°03′	Beautiful fixed pair.
ι, 6 Triangulii	2^h09^m	+30°04′	Fine yellow and blue pair.
β Orionis	5^h12^m	−8°15′	*Rigel.* Attendant star is bluish.
α Geminorum	7^h31^m	+32°00′	*Castor.* Very fine object.
11 Monocerotis	6^h26^m	−7°00′	Beautiful fixed triple star.
88 Leonis	11^h29^m	+14°39′	Yellow and lilac pair.
ζ Ursae Majoris	13^h22^m	+55°11′	*Mizar.* Second star from the end of the handle of the Big Dipper. Forms a naked eye pair with *Alcor*, often called the *Horse and Rider.* With low power Alcor remains in the field with the double star Mizar.
ζ Coronae Borealis	15^h37^m	+36°48′	Beautiful object.
α Herculis	17^h12^m	+14°27′	Orange and green pair.
ν Scorpii	16^h09^m	−19°21′	Both components are close doubles.
α Scorpii	16^h26^m	−25°29′	*Antares.* Red and green pair.
β Lyrae	19^h29^m	+27°51′	*Albireo.* Grand yellow and blue pair.
ε Lyrae	18^h43^m	39°37′ 39°34′	The "double-double" star. Separation 208″; each component is a binary star.
Nebulae and clusters			
NGC 663	1^h42^m	+61°00′	Fine open cluster, visible in the finder.
NGC 457	1^h16^m	+58°03′	Condensed cluster about 18′ in diameter.
NGC 7662	23^h23^m	+42°12′	Remarkably bright, oval planetary nebula, bluish.
NGC 869	2^h17^m	56°55′	The famous double cluster in Perseus (see text, Figure 13-8).
NGC 2244	6^h30^m	+4°54′	Beautiful open cluster visible to the naked eye.
NGC 2440	7^h39^m	−18°05′	Bright, bluish planetary nebula; requires moderately high power.
NGC 3242	10^h22^m	−18°23′	Planetary nebula with brighter inner ring.

Note: the Messier objects, listed separately in Appendix 7, make excellent observational targets. Those marked (●) are especially fine.

the local meridian) and 10^h (60° to the east of the local meridian).

Our principal concern, here, is locating a bright reference star, one that we can identify with certainty, whose right ascension and declination are known. The telescope is pointed at the star, and it is centered in the field of view. Within tolerable limits of accuracy, the telescope is now pointed at the specific right ascension and declination as given by the star's address. We use such a reference star as a home base, and scan into any other desired position according to the following procedures.

<u>How to Find Celestial Objects.</u> The first order of business is to ascertain the aspect of the sky on the date and at the time of observing. Table A6.1 is useful here. Suppose, for example that it is October 17, and that observing will start about 9 P.M. From Table A6.1 we note that the date is about halfway between the tabular entries of October 13 and October 21. Then at 8 P.M. the hour circle on the meridian is the one that is about halfway between 21^{h}30^m and 22^{h}00^m, or 21^{h}45^m. At 8:15 P.M. the 22^h circle will be on the meridian; at 9:15 the 23^h circle will be on it. Thus, objects with right ascensions ranging from 19^h to 3^h (possibly even from 18^h to 4^h) are in favorable aspect.

From this point on we will use a definition for *star time,* called *sidereal time.* Thus, if the 16^h circle of right ascension is exactly on the meridian, the sidereal time is 16^h.

Now, we are ready to select objects from the sky charts, or object lists, in the viciity of 23^h RA. See Table A6.2 for an abbreviated list.

<u>Naked-Eye Sighting.</u> Sometimes an object is so prominent that simply sighting along the top or side of the telescope tube will enable one to pinpoint the object in the finder telescope. Then by careful manipulation the object is *centered* in the field of view of the finder, and presto! it is in the field of the main telescope. Lock the declination axis and track the object in right ascension by rotation about the polar axis. Here is an example: date, October 17; time, 9 P.M. Depending on where you are located in your standard time zone, the 23^h circle is near, or on, the local meridian. The sidereal time is nearly, or exactly, 23^h. The aspect of the sky is shown in the star chart in this appendix covering 18^h RA to 22^h RA. For observers in the 48 contiguous United States, say at latitude 40° N, the celestial equator will be 50° above the due south horizon; Polaris will be 40°

above the northern horizon. An object of declination +40° will pass through the point directly overhead, called the *zenith.* The prominent constellation, *Cygnus* the swan, often called the Northern Cross, is about halfway between the local meridian and the western horizon, high overhead. The star *Albireo,* α Cygni, apparent magnitude 3.0, marks the beak or head of the swan, or, if you prefer, the foot of the cross. It is sufficiently prominent to the unaided eye that one should be able to pick it up in the finder by sighting along the telescope tube. Center the star in the field of view, use a low to medium power eyepiece for the main telescope, and discover that Albireo is really a beautiful optical double (the two stars are not gravitationally connected): one component is bright golden in color; the other is a brilliant pale blue.

<u>Star Hopping.</u> The general plan in *star hopping* is to pick a route from a selected reference star to the desired object, so that at each step, at least two identifiable stars are in the field of view. The reference objects should be as widely separated as possible. As the telescope is swept to the next position step, one star leaves the field; a new one appears. Each step is made preferably in either declination or right ascension. After sufficient practice, you will be able to use a combination sweep in both coordinate directions simultaneously.

In Figure A6-3 the numbered cirlces represent the successive positions of the finder telescope field of view, superimposed on the star field. The circles are drawn to the scale of the star chart. In the last step, the field of the main telescope is represented by the smaller circle. Generally with a lower power eyepiece (40–45× magnification), this field of view is about 1°.

Here is an example showing the use of this method. The date and time are the same as before: October 17; 9 P.M. The same star chart applies as for the previous example. This time we will search for M57, the Ring nebula (Figure 13-5), not visible to the naked eye.

Figure A6-3 is a section of the star chart showing the more prominent stars in the constellation Lyra. The dominant star is *Vega,* the second brightest star visible in northern latitudes. Sight in on Vega; place it to the east edge of the field of the finder†. Two other stars, *Epsilon Lyrae* and *Zeta Lyrae,* will form a

† If the finder is an *inverting* telescope as is usually the case, turn the star chart upside down and reverse the directions.

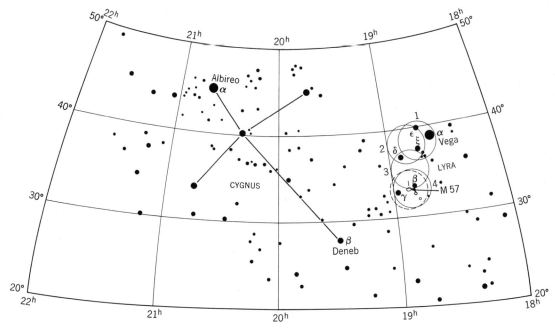

FIGURE A6-3 Star hopping from Vega to M57 (see text).

triangle with Vega (circle 1). Place Epsilon Lyrae at one edge of the field; rotate the telescope in right ascension toward the *east* very slowly. As Vega leaves the field, *Delta Lyrae* will enter from the opposite side. Circle 2 shows the appearance of the stars in the field of the finder telescope. Next, turn the telescope in *declination* toward the south, until both Epsilon Lyrae and Zeta Lyrae leave the field: only Delta Lyrae remains. Continue the southward motion very slowly, almost immediately *Beta Lyrae* enters the field. Circle 3 shows the appearance of this step. Continue moving the telescope south, until Beta Lyrae is near the center of the field. *Gamma Lyrae* will appear at one edge. *M57 is about one-third of the distance from Beta Lyrae toward Gamma Lyrae.* Adjust both declination and right ascension until the finder cross hairs mark this position. Look in the main telescope eyepiece (properly focused, of couse) and you should see the Ring nebula. Change eyepieces to obtain the optimum image.

Right Angle Sweep. A very useful searching technique makes use of the fact that there are only two possible motions in an equatorial mount: right ascension and declination. In principle, the *right-angle sweep* technique involves rotating the telescope in

declination (celestial latitude) to the desired position, locking the declination axis, and then sweeping the skies in right ascension until the desired object is located.

If the telescope mount is equipped with graduated *setting circles* on each of the two axes, the procedure is simplified as shown in the concluding section. We don't need setting circles here.

Our preliminary step is essential, however, for each eyepiece used. Aim the telescope at any convenient point in the sky, preferably where the star counts are reasonably high. The telescope is then held motionless, and the *time* it takes any given star to apparently drift across the diameter of the telescope field is recorded. The apparent motion, of course, is produced by the rotation of the earth at a rate of 15°/hr, or 1° every 4 minutes. Hence, dividing the drift time found above by 4 gives the angular size of the field of view in degrees. For an average reflector with about 40× magnification, the apparent field is about 1°.

This is how the information is put to use. The telescope is sighted in on a previously identified reference star. Many other faint stars will be seen in the field of view, most of which will not appear on the star charts. Select one of these fainter stars at the edge of the field in the direction you wish to step off. Slowly move the telescope in declination until the selected star is at the *opposite side* of the field of view. You have just stepped off an angular distance in degrees equal to the apparent angular field of the telescope. Continue stepping for the required number of degrees, then lock the declination axis and begin sweeping east and west in right ascension. The following example should clarify the procedure.

Assume that the date is July 28, the time is 8 P.M. local standard time. The sidereal time is about 16^h30^h (see Table 6.1) depending on your location in the time zone. The constellation Bootes with its bright star *Arcturus* (RA 14^h14^m, decl. $+19°\ 22'$) is high and to the west of the meridian (Figure A6-4). Center the image of Arcturus in the telescope field of view. Step off $17°\ 18'$ in declination toward the north, arriving at decl. $+36°\ 32'$. Lock the declination axis, and sweep toward the east from RA 14^h14^m to 16^h40^m. The great

globular cluster M13 in Hercules should now be in your field of view. Figure A6-4 shows the general detail of the star field for this problem, together with the search route.

Setting Circle Methods. Some telescopes come equipped with setting circles; most of the others can be equipped with them when desired. The polar axis setting circle is engraved in hours and minutes of right ascension. The declination axis circle has degrees and minutes in four series of 90° each. Fixed (and sometimes floating) indices or pointers are attached to the mount as reference marks. The use of the hour circle is somewhat complicated, but the declination circle is easy to use and is a big help in observation.

In use, a known reference star such as *Antares*, in the constellation Scorpio (a summertime constellation), is centered in the telescope field of view (Figure A6-5). The declination axis is locked in position and the declination setting circle is slip-turned to

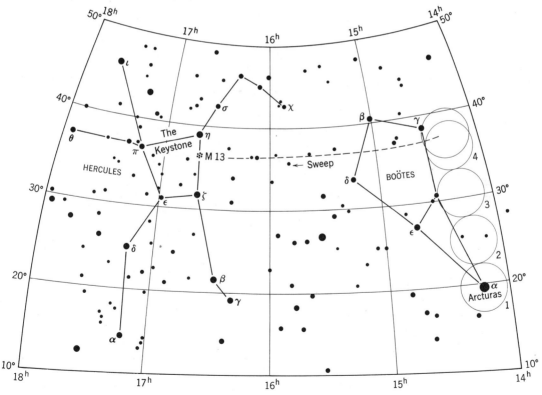

FIGURE A6-4 Right-angle sweep. Star hop from Arcturus to the declination of M13, then sweep in right ascension until object is located.

place −26°19′ under the pointer. It too is then locked securely to the declination axis. Unless subsequently disturbed, the circle will not need resetting for future observing sessions.

The declination circle is now indicating the direction in celestial latitude that the telescope is pointing. To locate any other object whose aspect is favorable for the season, the telescope is rotated about the declination axis, until the object's declination appears under the pointer. The axis is then locked, and the telescope is swept in right ascension until the object is located. Figure A6-5 shows the aspect of the sky around Antares, with several search routes indicated for good objects for viewing. The region around Scorpio is especially rich in Messier objects. Good hunting!

Table A6.2 lists a few worthwhile objects for beginning observing; the Messier objects are not included. The complete Messier catalog appears in Appendix 8.

<u>Supplementary Materials.</u> A good star chart is indispensable in observing (as are simplified guides for the use and care of telescopes) and in astronomical timekeeping. Probably the best, but by no means the only, star atlas for beginning work is *Norton's Star Atlas,* Sky Publishing Company, Cambridge, Mass. Two excellent guides that will immeasurably help you in your first efforts are *How to Use Your Telescope* and *Time in Astronomy,* both published by Edmund Scientific Company, Barrington, New Jersey. While you are about it, ask Edmund for its free catalog. It contains a wealth of interesting information.

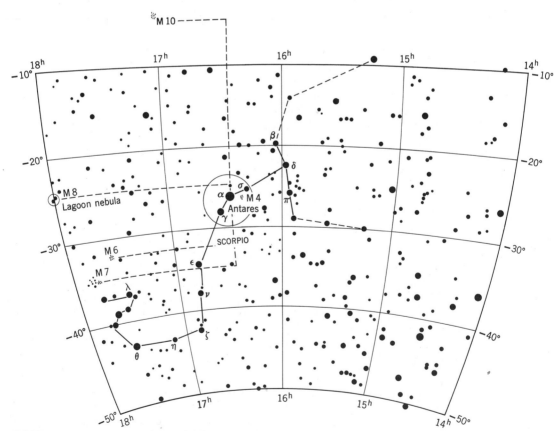

FIGURE A6-5 Setting circle methods (see text).

The Messier Catalogue of Nebulae and Star Clusters

Messier no. (M)	NGC or (IC)††	Right ascension (1950)		Decli-nation (1950)		Apparent visual magnitude	Description†
		h	m	°	′		
1	1952	5	31.5	+22	00	11.3	● Crab nebula in Taurus; remains of SN 1054
2	7089	21	30.9	− 1	02	6.4	Globular cluster in Aquarius
3	5272	13	39.8	+28	38	6.3	● Globular cluster in Canes Venatici
4	6121	16	20.6	−26	24	6.5	Globular cluster in Scorpio
5	5904	15	16.0	+ 2	17	6.1	Globular cluster in Serpens
6	6405	17	36.8	−32	10		Open cluster in Scorpio
7	6475	17	50.7	−34	48		Open cluster in Scorpio
8	6523	18	00.6	−24	23		● Lagoon nebula in Sagittarius
9	6333	17	16.3	−18	28	8.0	Globular cluster in Ophiuchus
10	6254	16	54.5	− 4	02	6.7	Globular cluster in Ophiuchus
11	6705	18	48.4	− 6	20		Open cluster in Scutum Sobieskii
12	6218	16	44.7	− 1	52	7.1	Globular cluster in Ophiuchus
13	6205	16	39.9	+36	33	5.9	● Globular cluster in Hercules
14	6402	17	35.0	− 3	13	8.5	Globular cluster in Ophiuchus
15	7078	21	27.5	+11	57	6.4	Globular cluster in Pegasus
16	6611	18	16.1	−13	48		● Open cluster with nebulosity in Serpens
17	6618	18	17.9	−16	12		Swan or Omega nebula in Sagittarius
18	6613	18	17.0	−17	09		Open cluster in Sagittarius
19	6273	16	59.5	−26	11	7.4	Globular cluster in Ophiuchus
20	6514	17	59.4	−23	02		Trifid nebula in Sagittarius

† A supplemental listing with descriptions, of the more interesting Messier objects follows the catalog entries.
†† NGC is the New General Catalogue of star clusters, galaxies and nebulae compiled by T. L. E. Dreyer in 1888. IC is the Index Catalogue, an extension of the NGC, compiled in the Harvard observatory annals, 1904, and revised in 1908.
● Especially fine observational targets.

Messier no. (M)	NGC or (IC)††	Right ascension (1950)		Declination (1950)		Apparent visual magnitude	Description†
		h	m	°	'		
21	6531	18	01.6	−22	30		Open cluster in Sagittarius
22	6656	18	33.4	−23	57	5.6	Globular cluster in Sagittarius
23	6494	17	54.0	−19	00		Open cluster in Sagittarius
24	6603	18	15.5	−18	27		Open cluster in Sagittarius
25	(4725)	18	28.7	−19	17		Open cluster in Sagittarius
26	6694	18	42.5	− 9	27		Open cluster in Scutum Sobieskii
27	6853	19	57.5	+22	35	8.2	● Dumbbell planetary nebula in Vulpecula
28	6626	18	21.4	−24	53	7.6	Globular cluster in Sagittarius
29	6913	20	22.2	+38	21		Open cluster in Cygnus
30	7099	21	37.5	−23	24	7.7	Globular cluster in Capricornus
31	224	0	40.0	+41	00	3.5	● Andromeda galaxy
32	221	0	40.0	+40	36	8.2	● Elliptical galaxy; companion to M31
33	598	1	31.0	+30	24	5.8	Spiral galaxy in Triangulum
34	1039	2	38.8	+42	35		● Open cluster in Perseus
35	2168	6	05.7	+24	21		Open cluster in Gemini
36	1960	5	33.0	+34	04		Open cluster in Auriga
37	2099	5	49.1	+32	33		Open cluster in Auriga
38	1912	5	25.3	+35	47		Open cluster in Auriga
39	7092	21	30.4	+48	13		Open cluster in Cygnus
40		12	20	+59			● Close double star in Ursa Major
41	2287	6	44.9	−20	41		Loose open cluster in Canis Major
42	1976	5	32.9	− 5	25		● Orion nebula
43	1982	5	33.1	− 5	19		Northeast portion of Orion nebula
44	2632	8	37	+20	10		Praesepe; open cluster in Cancer
45		3	44.5	+23	57		● The Pleiades; open cluster in Taurus
46	2437	7	39.5	−14	42		Open cluster in Puppis
47	2478	7	52.4	−15	17		Loose group of stars in Puppis
48		8	11	− 1	40		Cluster of very small stars; not identifiable
49	4472	12	27.3	+ 8	16	8.5	Elliptical galaxy in Virgo
50	2323	7	00.6	− 8	16		Loose open cluster in Monoceros
51	5194	13	27.8	+47	27	8.4	Whirlpool spiral galaxy in Canes Venatici
52	7654	23	22.0	+61	20		Loose open cluster in Cassiopeia
53	5024	13	10.5	+18	26	7.8	Globular cluster in Coma Berenices
54	6715	18	51.9	−30	32	7.8	Globular cluster in Sagittarius
55	6809	19	36.8	−31	03	6.2	Globular cluster in Sagittarius

† A supplemental listing with descriptions, of the more interesting Messier objects follows the catalog entries.
†† NGC is the New General Catalogue of star clusters, galaxies and nebulae compiled by T. L. E. Dreyer in 1888. IC is the Index Catalogue, an extension of the NGC, compiled in the Harvard observatory annals, 1904, and revised in 1908.
● Especially fine observational targets.

Messier no. (M)	NGC or (IC)††	Right ascension (1950)			Decli-nation (1950)		Apparent visual magnitude	Description†
		h	m	°		′		
56	6779	19	14.6	+30		05	8.7	Globular cluster in Lyra
57	6720	18	51.7	+32		58	9.0	● Ring nebula; planetary nebula in Lyra
58	4579	12	35.2	+12		05	9.6	Spiral galaxy in Virgo
59	4621	12	39.5	+11		56	10.0	Spiral galaxy in Virgo
60	4649	12	41.1	+11		50	9.0	Elliptical galaxy in Virgo
61	4303	12	19.3	+ 4		45	9.6	Spiral galaxy in Virgo
62	6266	16	58.0	−30		02	7.3	Globular cluster in Scorpio
63	5055	13	13.5	+42		17	8.6	● Spiral galaxy in Canes Venatici
64	4826	12	54.2	+21		57	8.5	Spiral galaxy in Coma Berenices
65	3623	11	16.3	+13		22	9.4	Spiral galaxy in Leo
66	3627	11	17.6	+13		16	9.0	Spiral galaxy in Leo; companion to M65
67	2682	8	48.4	+12		00		Open cluster in Cancer
68	4590	12	36.8	−26		29	8.2	Globular cluster in Hydra
69	6637	18	28.1	−32		24	8.0	Globular cluster in Sagittarius
70	6681	18	40.0	−32		20	8.1	Globular cluster in Sagittarius
71	6838	19	51.5	+18		39		Globular cluster in Sagitta
72	6981	20	50.7	−12		45	9.3	Globular cluster in Aquarius
73	6994	20	56.2	−12		50		Open cluster in Aquarius
74	628	1	34.0	+15		32	9.3	Spiral galaxy in Pisces
75	6864	20	03.1	−22		04	8.6	Globular cluster in Sagittarius
76	650	1	39.1	+51		19	11.4	● Planetary nebula in Perseus
77	1068	2	40.1	− 0		12	8.9	Spiral galaxy in Cetus
78	2068	5	44.2	+ 0		02		Small emission nebula in Orion
79	1904	5	22.1	−24		34	7.5	Globular cluster in Lepus
80	6093	16	14.0	−22		52	7.5	Globular cluster in Scorpio
81	3031	9	51.7	+69		18	7.0	● Spiral galaxy in Ursa Major
82	3034	9	51.9	+69		56	8.4	Irregular galaxy in Ursa Major
83	5236	13	34.2	−29		37	8.3	Spiral galaxy in Hydra
84	4374	12	22.6	+13		10	9.4	Elliptical galaxy in Virgo
85	4382	12	22.8	+18		28	9.3	Elliptical galaxy in Coma Berenices
86	4406	12	23.6	+13		13	9.2	Elliptical galaxy in Virgo
87	4486	12	28.2	+12		40	8.7	Elliptical galaxy in Virgo
88	4501	12	29.4	+14		42	9.5	Spiral galaxy in Coma Berenices
89	4552	12	33.1	+12		50	10.3	Elliptical galaxy in Virgo
90	4569	12	34.3	+13		26	9.6	Spiral galaxy in Virgo

† A supplemental listing with descriptions, of the more interesting Messier objects follows the catalog entries.
†† NGC is the New General Catalogue of star clusters, galaxies and nebulae compiled by T. L. E. Dreyer in 1888. IC is the Index Catalogue, an extension of the NGC, compiled in the Harvard observatory annals, 1904, and revised in 1908.
● Especially fine observational targets.

Messier no. (M)	NGC or (IC)††	Right ascension (1950)		Decli- nation (1950)		Apparent visual magnitude	Description†
		h	m	°	′		
91		omitted					
92	6341	17	15.6	+43	12	6.4	Globular cluster in Hercules
93	2447	7	42.4	−23	45		Open cluster in Puppis
94	4736	12	48.6	+41	24	8.3	Spiral galaxy in Canes Venatici
95	3351	10	41.3	+11	58	9.8	Barred spiral galaxy in Leo
96	3368	10	44.1	+12	05	9.3	Spiral galaxy in Leo
97	3587	11	12.0	+55	17	11.1	Owl nebula; planetary nebula in Ursa Major
98	4192	12	11.2	+15	11	10.2	Spiral galaxy in Coma Berenices
99	4254	12	16.3	+14	42	9.9	Spiral galaxy in Coma Berenices
100	4321	12	20.4	+16	06	9.4	Spiral galaxy in Coma Berenices
101	5457	14	01.4	+54	36	7.9	Spiral galaxy in Ursa Major
102		omitted					
103	581	1	29.9	+60	26		Open cluster in Cassiopeia
104	4594	12	37.4	−11	21	8.3	Spiral galaxy in Virgo
105	3379	10	45.2	+13	01	9.7	Elliptical galaxy in Leo
106	4258	12	16.5	+47	35	8.4	Spiral galaxy in Canes Venatici
107	6171	16	29.7	−12	57	9.2	Globular cluster in Ophiuchus

† A supplemental listing with descriptions, of the more interesting Messier objects follows the catalog entries.
†† NGC is the New General Catalogue of star clusters, galaxies and nebulae compiled by T. L. E. Dreyer in 1888. IC is the Index Catalogue, an extension of the NGC, compiled in the Harvard observatory annals, 1904, and revised in 1908.
● Especially fine observational targets.

Selected Messier Objects

M Number	Constellation	Right Ascension		Declination		Description
		h	m	°	'	
M103	Cassiopeia	1	29.8	+60	26	A fine open cluster, 1° field; contains a red star.
M31	Andromeda	0	40.0	+41.0		The Great nebula in Andromeda, a spiral galaxy visible as a hazy spot to the unaided eye (Figure 14-1).
M33	Triangulum	1	31	+30	24	Very large, faint, ill-defined spiral galaxy. Use low power on a clear night. Spiral arms visible only in photographs (Figure 13-26).
M42	Orion	5	32.9	− 5	25	The great Orion nebula visible as a hazy star in Orion's sword Best seen with low power. Probably the most beautiful of the gaseous nebulae accessible to small telescopes (Figure 13–24).
M1	Taurus	5	31.5	+21	59	The Crab nebula (Figure 12–16). Faintly visible as an oval patch; the serrated edge is visible only in large telescopes.
M44	Cancer	8	37.2	+20	10	Praesepe, The Beehive; a large, scattered open cluster (Figure 13–8) almost visible with the unaided eye. Best seen with low power.
M46	Puppis	7	39.5	−14	42	A beautiful cluster about the size of the full moon. The irregular planetary nebula NGC 2438 is on northern edge.
M3	Canes Venatici	13	39.9	+28	38	Beautiful, bright, condensed globular cluster. Best seen with a 6-in. or larger telescope at high power.
M13	Hercules	16	39.9	+36	33	Probably the grandest globular cluster of all (Figure 13–11). Visible with binoculars.
M57	Lyra	18	52.0	+32	58	The Ring nebula; oval in shape resembling a smoky doughnut. Magnifies well.
M27	Vulpecula	19	57.4	+22	35	The Dumbbell nebula; elliptical in appearance, with fairly luminous notches.

Note: instructions for finding several of these objects are included in Appendix 6.

APPENDIX 8
Constellations and Star Charts

Constellations

Name	Genitive	Abbreviation	Meaning	Reference RA and Decl.	
Andromeda	-dae	And	Chained Lady	1^h	$+40°$
Aquarius	-rii	Aqr	Water Bearer	23^h	$-10°$
Aquila	-lae	Aql	Eagle	19^h30^m	$+ 5°$
Aries	Arietis	Ari	Ram	2^h	$+20°$
Auriga	-gae	Aur	Charioteer	5^h30^m	$+40°$
Bootes	-tis	Boo	Herdsman	14^h30^m	$+30°$
Cancer	Cancri	Cnc	Crab	8^h30^m	$+20°$
Canes Venatici	Canum Venaticorum	CVn	Hunting Dogs	12^h30^m	$+40°$
Canis Major	-ris	CMa	Great Dog	7^h	$-20°$
Capricornus	-ni	Cap	Goat	21^h	$-20°$
Cassiopeia	-peiae	Cas	The Queen; also, Lady in the Chair	1^h	$+60°$
Cepheus	-phei	Cep	King (of Ethiopia)	22^h	$+65°$
Cetus	-ti	Cet	Whale	2^h	$-10°$
Coma Bernices	Comae-	Com	Bernice's Hair	13^h	$+25°$
Corona Borealis	-nae	CrB	Northern Crown	15^h30^m	$+30°$
Corvus	-vi	Crv	Crow	12^h30^m	$-20°$
Cygnus	-ni	Cyg	Swan; also, the Northern Cross	20^h	$+40°$
Draco	-conis	Dra	Dragon	18^h	$+70°$
Eridanus	-ni	Eri	The River Eridanus	3^h30^m	$-20°$
Gemini	-orum	Gem	Twins	7^h	$+25°$
Hercules	-lis	Her	Hercules	17^h	$+35°$
Hydra	-drae	Hyd	Sea-Serpent	11^h	$-20°$
Leo	Leonis	Leo	Lion	10^h30^m	$+20°$
Lepus	Leporis	Lep	Hare	5^h30^m	$-20°$
Libra	-ae	Lib	Scales	15^h	$-15°$
Lyra	-ae	Lyr	Harp	18^h30^m	$+35°$
Monoceros	-rotis	Mon	Unicorn	7^h	$- 5°$

Constellations (*Continued*)

Name	Genitive	Abbreviation	Meaning	Reference RA and Decl.	
Ophiuchus	-chi	Oph	Serpent-Bearer	17^h	$0°$
Orion	-nis	Ori	Hunter	5^h30^m	$0°$
Pegasus	-si	Peg	Winged Horse	23^h30^m	$+20°$
Perseus	-sei	Per	Perseus	3^h30^m	$+45°$
Pisces	-cium	Pic	Fishes	23^h30^m	$+5°$
Piscis Austrinis	-ini	PsA	Southern Fish	23^h	$-30°$
Sagittarius	-rii	Sgr	Archer	18^h30^m	$-30°$
Scorpio	-pionis	Sco	Scorpion	17^h	$-30°$
Serpens	-pentis	Ser	Serpent	16^h	$0°$
Taurus	-ri	Tau	Bull	4^h30^m	$+15°$
Triangulum	-li	Tri	Triangle	2^h	$+30°$
Ursa Major	-sae, -ris	UMa	Great Bear, also Big Dipper	11^h	$+60°$
Ursa Minor	-sae, -ris	UMi	Little Bear, also Little Dipper	15^h	$+70°$
Virgo	-ginis	Vir	Virgin	13^h	$-10°$
Vulpecula	-lae	Vul	Fox	19^h30^m	$+25°$

Constellation Charts (Equatorial Zone)

FIGURE A8-1. 0^h–6^h RA

FIGURE A8-2. 6^h–12^h RA

FIGURE A8-3. 12^h–18^h RA

FIGURE A8-4. 18^h–24^h RA

Note: Some distortion in this projection occurs North of decl $+40°$ and South of decl. $-40°$.

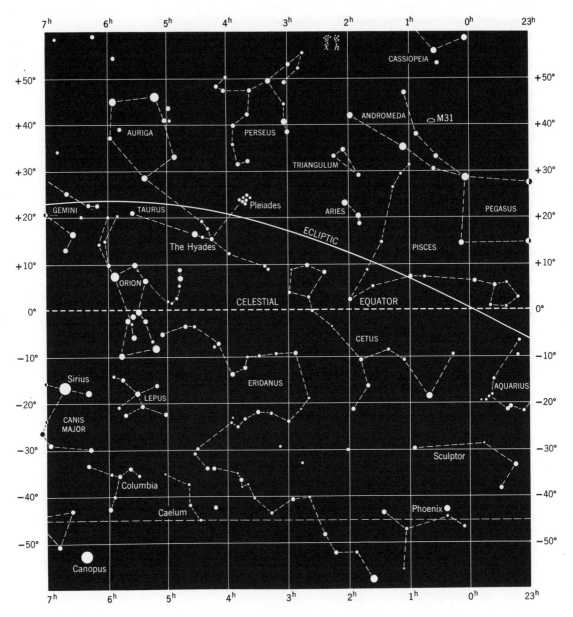

FIGURE A8-1

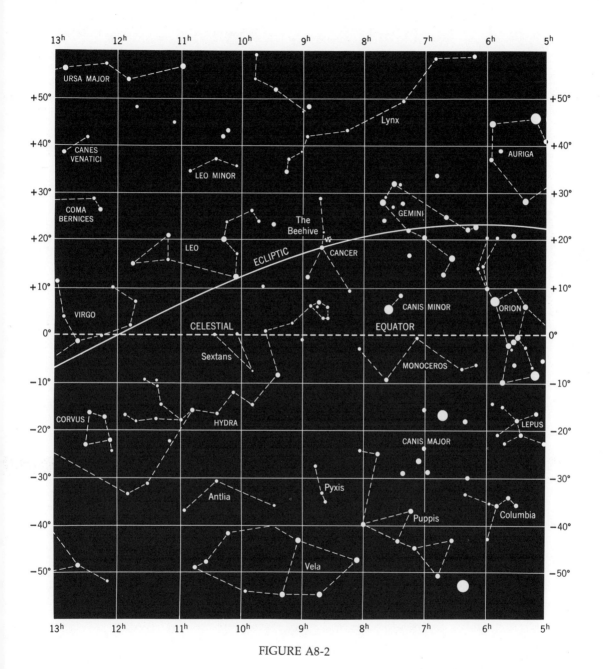

FIGURE A8-2

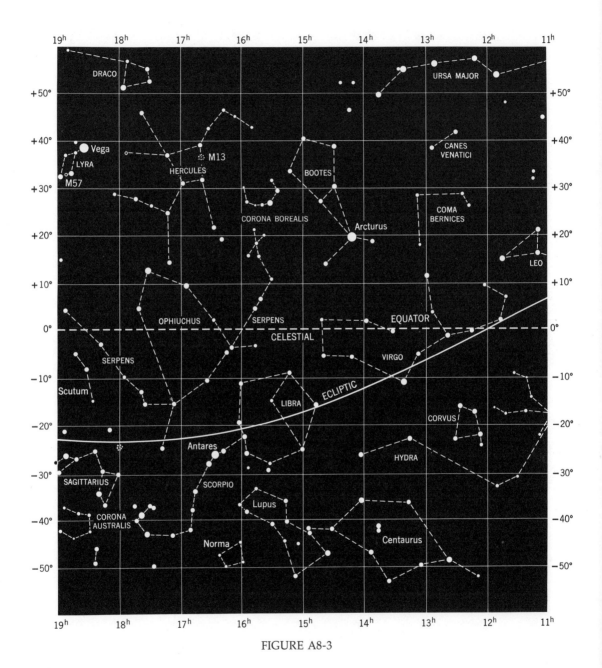

FIGURE A8-3

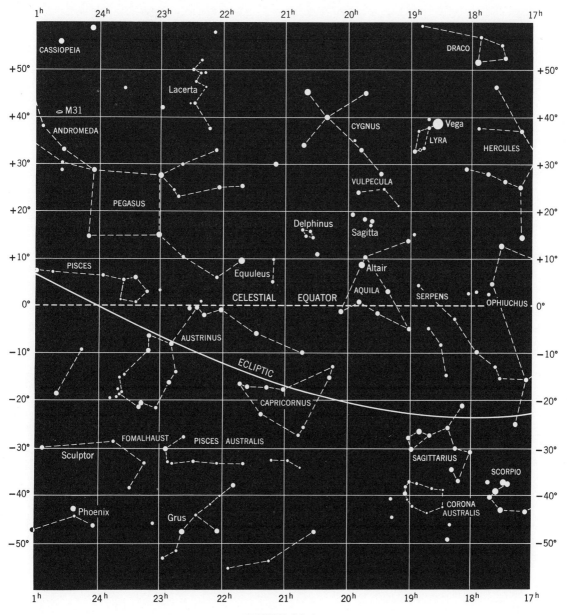

FIGURE A8-4

Glossary

A

Aberration (of starlight) Apparent displacement in the direction of a star due to the earth's orbital motion.

Absolute magnitude Apparent magnitude an object would have at a distance of 10 pc.

Absolute zero Temperature of $-273.16°C$ (or $0°K$) where all molecular motion stops.

Absorption spectrum Spectrum of dark lines upon a continuous background spectrum.

Accelerate Change velocity: to speed up, slow down, or change direction.

Acceleration of gravity Numerical value of the acceleration produced by the gravitational attraction on an object at the surface of a planet or star.

Albedo Fraction of incident sunlight that a planet or minor planet reflects.

Almanac Book or table listing astronomical events.

Altitude Angular distance above or below the horizon, measured along a vertical circle, to a celestial object.

Amplitude Range in variability, as in the light from a variable star.

Angstrom Unit of length equal to 10^{-8} cm.

Angular diameter Angle subtended by the linear diameter of an object.

Angular distance Difference in direction between two points, measured in degrees.

Angular momentum Measure of the momentum associated with motion about an axis or fixed point.

Annular eclipse Eclipse in which the entire perimeter of the eclipsed body remains visible.

Annulus cone That portion of a shadow cone from which an annular eclipse would be visible.

Antapex of solar motion Point on the celestial sphere from which the sun appears to be receding.

Apastron Point in the orbit of a binary star at which it is farthest from its companion.

Aperture Diameter of a telescope objective.

Apex of solar motion Point on the celestial sphere toward which the sun appears to be moving.

Aphelion Point in the orbit of a body revolving about the sun at which it is farthest from the sun.

Apogee Point at which the orbit of a body revolving about the earth is farthest from the earth.

Apparent magnitude Measure of the observed light flux received from a star or other object at the earth.

Apparent noon Instant of time at which the real sun reaches the celestial meridian.

Apparent relative orbit Projection onto a plane perpendicular to the line of sight of the relative orbit of the fainter of the two components of a visual binary star about the brighter.

Apparent solar day Interval between two successive transits of the sun's center across the meridian.

Apparent solar time Hour angle of the sun's center plus 12 hours.

Apsis Point of greatest or least distance from a center of attraction (pl. apsides). Thus the major axis of an elliptical orbit is also referred to as the line of apsides.

Ascending node Point on the celestial sphere at which an orbit crosses the ecliptic from south to north.

Association Loose cluster of stars whose spectral types, motions, or positions in the sky indicate that they have probably had a common origin.

Asterism A named cluster or grouping of stars within a constellation of different name. Thus, the Big Dipper is an asterism within the constellation Ursa Major.

Asteroid Synonym for minor planet or planetoid.

Astrology Pseudoscience that treats with supposed influences on human destiny, of the configurations and locations in the sky of the sun, moon, and planets; a primitive religion having its origin in ancient Babylonia.

Astrometric binary Binary star in which one component is not observed, but whose presence is deduced from the orbital motion of the visible component.

Astronomical horizon Imaginary circle everywhere 90° from the zenith.

Astronomical latitude Angle between a line perpendicular to the surface at a given locality and the plane of the earth's equator.

Astronomical unit (AU) Originally meant to be the semimajor axis of the orbit of the earth; now defined as the semimajor axis of the orbit of a hypothetical body with the mass and period that Gauss assumed for the earth. The semimajor axis of the orbit of the earth is 1.000000230 AU.

Astronomy Branch of science that treats of the physics and morphology of the universe.

Astrophysics That part of astronomy which deals principally with the physics of stars, stellar systems, and interstellar material. Astrophysics also deals, however, with the structures and atmospheres of the sun and planets.

Atmospheric refraction Bending, or refraction, by the earth's atmosphere of light rays from celestial objects.

Atom Smallest particle of an element that retains the properties which characterize that element.

Atomic transition Change in the state of energy of an atom; the atom may gain or lose energy by collision with another particle or by the emission or absorption of a photon.

Attenuation effect Reduction in brightness of a receding source because of the arrival of fewer photons per second from it than from a similar stationary source.

Autumnal equinox Point on the celestial sphere and point in time at which the sun crosses the celestial equator southward.

Azimuth Angle along the celestial horizon, measured eastward from the north point to the intersection of the horizon with the vertical circle passing through an object.

B

Bands (in spectra) Emission or absorption lines, usually in the spectra of chemical compounds or radicals, so numerous and closely spaced that they coalesce into broad emission or absorption bands.

Barred spiral galaxy Spiral galaxy in which the spiral arms begin from the ends of a bar running through the nucleus rather than from the nucleus itself.

Barycenter Center of mass of two mutually revolving bodies.

Base line That side of a triangle used in triangulation or surveying whose length is known (or can be measured) and which is included between two angles that are known (or can be measured).

Big-bang theory Theory of cosmology in which the expansion of the universe is presumed to have begun with a primeval explosion.

Binary star Pair of mutually revolving stars.

Blackbody Hypothetical perfect radiator, which absorbs and reemits all radiation incident upon it.

Black dwarf Presumed final state of evolution for a star, in which all of its energy sources are exhausted and radiation is no longer emitted.

Black holes Stars that have collapsed beyond the theoretical limits for a neutron star; the gravitational field of these stars is so powerful that light waves cannot escape from them.

Blink microscope (or blink comparator) Microscope in which the user's view is shifted rapidly back and forth between the corresponding portions of two different photographs of the same region of the sky.

Bode's law Expression of the approximate regularity in the spacing of the planets.

Bolide Spectacularly bright meteor.

Brightness Intensity of radiation on a scale in which the sun's apparent luminosity at 1 pc would represent one unit of intensity.

Burst Sudden increase in the intensity of an isolated wavelength of solar radio radiation, lasting only a few seconds.

C

Canal (on Mars) Approximately linear marking of unknown significance.

Capture (in celestial mechanics) Modification by a primary of an orbit of a previously independent body so as to bring it under gravitational control or subject to major influence.

Cassegrain telescope Type of reflecting telescope in which the eyepiece is placed behind the objective mirror.

Cassini division Vacancy which separates the two bright portions of Saturn's rings.

Celestial equator Imaginary circle on the celestial sphere midway between the celestial poles.

Celestial latitude Angular distance from the ecliptic.

Celestial longitude Angular distance from the vernal equinox, measured eastward and parallel to the ecliptic.

Celestial mechanics That branch of astronomy which deals with the motions and gravitational influence of the members of the solar system.

Celestial meridian Imaginary circle on the celestial sphere passing through the celestial poles and the zenith.

Celestial poles Points about which the celestial sphere appears to rotate; also, intersections on the celestial sphere with the extensions of the earth's polar axis.

Celestial sphere Apparent sphere of the sky.

Center of gravity Weighted mean position of all the mass elements of a system.

Centrifugal force (or acceleration) Imaginary force (or acceleration) that is often introduced to account for the illusion that a body moving on a curved path tends to accelerate radially from the center of curvature. The actual force present is the one that diverts the body's motion from a straight line and is directed toward the *center* of curvature. It is, however, legitimate to introduce a fictitious centrifugal force field in a rotating (and hence noninertial) coordinate system.

Centripetal force (or acceleration) Force required to divert a body from a straight path into a curved path. It is directed toward the center of curvature.

Cepheid variable Class of pulsating intrinsic variable stars.

Chromatic aberration Defect of lenses due to their inherent inability to focus all colors in the same focal plane.

Chromosphere That part of the solar atmosphere that lies immediately above the photospheric layers.

Circumpolar regions Portions of the celestial sphere near the celestial poles that are either always above or always below the horizon.

Circumpolar star Star which appears never to go below the horizon.

Cluster of galaxies System of galaxies containing from several to thousands of member galaxies.

Cluster variable (RR Lyrae variable) Member of a certain large class of pulsating variable stars, all with periods less than 1 day. These stars are often present in globular star clusters.

Color excess Amount by which the color index of a star is increased when its light is reddened in passing through interstellar absorbing material.

Color index (CI) Difference between the photographic and visual magnitudes of stars or other objects measured in light of two different spectral regions.

Color-magnitude diagram Plot of the magnitudes (apparent or absolute) of the stars in a cluster against their color indices.

Color temperature Temperature of a body determined from the intensity of one or several wavelengths of its radiations.

Coma Diffuse head of a comet; also an aberration of reflecting telescopes which gives off-center star images a cometary appearance.

Comet group Set of comets with near-identical orbits.

Comparison spectrum Spectrum of a vaporized element (such as iron) photographed beside the image of a stellar spectrum, and with the same camera, for purposes of comparison of wavelengths.

Configuration Any one of several particular orientations on the celestial sphere of the moon or a planet with respect to the sun.

Conic section Curve of intersection between plane and right circular cone: a circle, ellipse, parabola, or hyperbola.

Conjunction Configuration in which two celestial bodies have the same celestial longitude or the same right ascension.

Conservation of angular momentum Law that angular momentum is conserved in the absence of any force not directed toward or away from the point or axis about which the angular momentum is referred—that is, in the absence of a torque.

Constellation Configuration of stars or the area of sky which it occupies.

Continuous spectrum Spectrum of light comprised of radiation of a continuous range of wavelengths or colors rather than only certain discrete wavelengths.

Convection Transfer of energy by moving currents of a fluid containing that energy.

Convective zone Layer of rising and falling currents just beneath the photosphere of the sun.

Coordinate Number used to specify location.

Copernican theory Doctrine that the sun is the center of the universe and that the planets revolve around it.

Core (of earth) Central part of the earth, believed to be a liquid of high density.

Coriolis effect Deflection (with respect to the ground) of projectiles or winds or ocean currents moving across the surface of the rotating earth.

Corona Faint, extended outer portion of the sun's atmosphere.

Corona (or halo) of galaxy Outer portions of the Galaxy, especially on either side of the plane of the Milky Way.

Coronagraph Instrument for photographing the chromosphere and corona of the sun outside of eclipse.

Cosmic rays Atomic nuclei (mostly protons) that are observed to strike the earth's atmosphere with exceedingly high energies.

Cosmogony Study of the ultimate origins of physical systems, especially of the universe.

Cosmological constant Term that arises in the development of the field equations of general relativity, which represents a repulsive force in the universe. The cosmological constant is often assumed to be zero.

Cosmological model Specific model, or theory, of the organization and evolution of the universe.

Cosmological principle Assumption that, on the large scale, the universe at any given time is the same everywhere.

Cosmology Study of the content and arrangement of the physical universe.

Coudé focus Optical arrangement in a reflecting telescope whereby light is reflected by two or more secondary mirrors down the polar axis of the telescope to a focus at a place separate from the moving parts of the telescope.

Crater (lunar, Martian) More or less circular depression in the surface of the moon or of Mars.

Crater (meteoritic) Crater on the earth caused by the collision of a meteoroid with the earth and the subsequent explosion of the meteoroid.

Crescent moon One of the phases of the moon when its elongation is less than 90° from the sun and it appears less than half full.

Critical temperature Temperature above which a gas cannot be liquified regardless of pressure.

Crust (of earth) Outer layer of the earth.

Curvature (in cosmology and geometry) Measure of the extent to which the theorems of Euclidean geometry are required to be modified in a given region of space.

D

Dark nebula Cloud of interstellar dust that obscures the light of more distant stars and appears as an opaque curtain.

Daylight savings time Time 1 hour more advanced than standard time, usually adopted in spring and summer to take advantage of long evening twilight.

Declination Angular distance north or south of the celestial equator to some object, measured along an hour circle passing through that object.

Deferent Stationary circle in the Ptolemaic system that represents the path around the earth along which the center of another circle (epicycle) moves; the latter represents the circular path of an object or another epicycle.

Degenerate gas Gas in which the allowable states for the electrons have been filled; it behaves according to different laws from those that apply to perfect gases.

Density Ratio of the mass of an object to its volume.

Descending node Point on the celestial sphere at which an orbit crosses the ecliptic from north to south.

Differential galactic rotation State of rotation of the galaxy whereby it does not rotate as a solid wheel, so that parts adjacent to each other do not always stay close together.

Diffraction Spreading out of light in passing the edge of an opaque body.

Diffraction grating System of closely spaced equidistant slits or reflecting strips which, by diffraction and interference, produce a spectrum.

Diffraction pattern Pattern of bright and dark fringes produced by the interference of light rays, diffracted by different amounts, with each other.

Diffuse nebula Cloud of relatively dense interstellar matter, often in the vicinity of an illuminating star.

Diffusion Scattering of light by many small particles.

Direct motion Orbital or rotational motion from west to east on the celestial sphere or counterclockwise as seen from above the north pole.

Disk (of planet or other object) Apparent extended surface that a planet (or the sun, or a moon, or a star) displays when seen in the sky or is viewed telescopically.

Dispersion Separation of light into its constituent colors.

Distance modulus Difference between the apparent and absolute magnitudes of an object.

Diurnal Daily.

Diurnal circle Apparent path of a star or planet in the sky during a complete day due to the earth's rotation.

Diurnal motion Motion during 1 day.

Doppler broadening Increase in the width of a spectrum line due to Doppler effect caused by thermal motion of the radiating atoms.

Doppler shift Apparent change in wavelength of the radiation from a source due to its relative motion in the line of sight.

Dwarf (star) Main-sequence star (as opposed to a giant or super-giant).

Dynamical parallax Parallax of a binary system, computed with the aid of the mass-luminosity relation and Kepler's harmonic law.

Dyne (dyn) Metric unit of force; the force required to accelerate a mass of 1 g in the amount 1 cm/sec^2.

E

Earthlight Light reflected from earth to moon and back, visible on the dark portion of the moon.

East point Point on the horizon 90° from the north point (measured clockwise as seen from the zenith).

Eccentricity (of ellipse) Ratio of the distance between the foci to the length of the major axis.

Eclipse Cutting off of all or part of the light of one body by another body passing in front of it.

Eclipse path Narrow region or zone on the earth's surface within which an eclipse of the sun may be witnessed as total.

Eclipsing binary star Binary star in which the plane of revolution of the two stars is nearly edge-on to our line of sight, so that the light of one star is periodically diminished by the other passing in front of it.

Ecliptic Apparent annual path of the sun on the celestial sphere.

Ecliptic coordinates Celestial latitude and longitude.

Ecliptic limit Maximum angular distance from a node where the moon can be for an eclipse to take place.

Effective temperature Temperature which a perfect radiator of the same surface brightness would have.

Electromagnetic radiation Radiation consisting of waves propagated through the building up and breaking down of electric and magnetic fields; these include radio, infrared, light, ultraviolet, x rays, and gamma rays.

Electromagnetic spectrum Whole array or family of electromagnetic waves.

Electron Negatively charged subatomic particle that normally moves about the nucleus of an atom.

Element Substance that cannot be decomposed, by *chemical* means, into simpler substances.

Elements (of orbit) Any of several quantities that describe the size, shape, and orientation of the orbit of a body.

Ellipse Closed-plane curve whose every point is the same total distance from two fixed foci within.

Elliptical galaxy Assemblage of stars comparable in size to the Milky Way system but lacking arms and showing an elliptical outline.

Ellipticity Ratio (in an ellipse) of the difference between the major and minor axes to the major axis.

Elongation Amount by which a body's right ascension or celestial longitude differs from that of the sun.

Emission line Discrete bright spectral line.

Emission nebula Diffuse nebula which is excited to radiate by a hot nearby star.

Emission spectrum Spectrum consisting of emission lines.

Encounter (gravitational) Near passing, on hyperbolic orbits, of two objects that influence each other gravitationally.

Energy Ability to do work.

Ephemeris Table that gives the positions of one or a number of celestial bodies at various times during a specified period.

Ephemeris time Kind of time that passes at a strictly uniform rate; used to compute the instants of various astronomical events.

Epicycle Circular orbit of a body in the Ptolemaic system, the center of which revolves about another circle (the deferent).

Equant Stationary point in the Ptolemaic system not at the center of a circular orbit about which a body (or the center of an epicycle) revolves with uniform angular velocity.

Equation of time Difference between apparent and mean solar time.

Equator Great circle on the earth, 90° from its poles.

Equatorial mount Mounting for a telescope, one axis of which is parallel to the earth's axis, so that a motion of the telescope about that axis can compensate for the earth's rotation.

Equinox Point where the celestial equator crosses the ecliptic; also the time of the sun's arrival at one of these points.

Eruptive variable Variable star whose changes in light are erratic or explosive.

Evolutionary cosmology Theory of cosmology that assumes that all parts of the universe have a common age and evolve together.

Excitation Process of imparting to an atom or an ion an amount of energy greater than that it has in its normal or least-energy state.

Excited state Condition of an atom which has more than the minimum amount of energy.

External galaxy Any stellar system comparable to the Milky Way Galaxy.

Extinction Attenuation of light from a celestial body produced by the earth's atmosphere or by interstellar absorption.

Extragalactic Beyond the galaxy.

Extragalactic nebula External nebula to our own Milky Way Galaxy.

Eyepiece Magnifying lens used to view the image produced by the objective of a telescope.

Faculus (pl. faculae) Bright region near the limb of the sun.

Filament Solar prominence seen in projection against the sun's surface.

Fission Breakup of a heavy atomic nucleus into two or more lighter ones.

Flare Sudden outburst of monochromatic emission over a large area of solar surface.

Flare star Member of a class of stars that show occasional, sudden, unpredicted increases in light.

Fluorescence Absorption of light of one wavelength and reemission of it at another wavelength; especially the conversion of ultraviolet into visible light.

Focal length Distance from a lens or mirror to the point where light converged by it comes to a focus.

Focal plane Plane in which the image formed by a lens lies.

Focal ratio (speed) Ratio of the focal length of a lens or mirror to its aperture.

Focus Point behind a lens or before a mirror at which parallel incident rays converge to a common point.

Forbidden lines Spectral lines that are not usually observed under laboratory conditions because they result from atomic transitions that are highly improbable.

Force That which can change the momentum of a body; numerically, the rate at which the body's momentum changes.

Foucault pendulum Experiment first conducted by Jean Foucault in 1851 to demonstrate the rotation of the earth.

*f***/ratio** Ratio of focal length to diameter of a lens or a mirror.

Fraunhofer line Absorption line in the spectrum of the sun or a star.

Frequency Number of waves per second which pass a fixed point.

Full moon That phase of the moon when it is at opposition (180° from the sun) and its full daylight hemisphere is visible from the earth.

G

Galactic center Point in space around which the galaxy is symmetric and around which it rotates.

Galactic cluster Group of several dozen to several thousand stars having common origin and space motion and which are located near the galactic plane.

Galactic disk Components of a galaxy which comprise the flattened, rapidly rotating subsystem.

Galactic equator Imaginary circle on the celestial sphere which bisects the Milky Way.

Galactic latitude Angular distance from the galactic equator.

Galactic longitude Angular distance from the direction of the galactic center, measured eastward along the galactic equator.

Galactic plane Plane of the galactic equator.

Galactic poles Intersections with the celestial sphere of a line through the observer that is perpendicular to the plane of the galactic equator and that contains the galactic center.

Galactic rotation Motion of the galaxy's parts about the galactic center.

Galaxy Large assemblage of gravitationally associated stars; a typical galaxy contains millions to hundreds of billions of stars.

Gamma ray High-energy electromagnetic radiation.

Geocentric latitude Angle between the plane of the earth's equator and the radius to a point on the earth's surface.

Geocentric theory Doctrine that the earth is at the center of the universe.

Giant (star) Star of large luminosity and radius.

Gibbous moon One of the phases of the moon in which more than half, but not all, of the moon's daylight hemisphere is visible from the earth.

Globular cluster One of about 120 large star clusters that form a system of clusters centered on the center of the galaxy.

Globule Small, dense, dark nebula; believed to be a possible protostar.

Granulation "Rice-grain" like structure of the solar photosphere.

Gravitation Attraction of mass for mass.

Gravitational constant (G) Constant of proportionality in Newton's law of gravitation; in metric units G has the value 6.668×10^{-8} dyn cm²/g².

Gravitational energy Energy that can be released by the gravitational collapse, or partial collapse, of a system.

Great circle Circle that is the intersection of the surface of a sphere with a plane passing through its center.

Greatest elongation (east or west) Largest separation in celestial longitude (to the east or west) that an inferior planet can have from the sun.

Greenwich meridian Meridian of longitude passing through the site of the old Royal Greenwich Observatory, near London; origin of longitude on the earth.

Gregorian calendar Calendar (now in common use) introduced by Pope Gregory XIII in 1582.

H

H-I region Region of neutral hydrogen in interstellar space.

H-II region Region of ionized hydrogen in interstellar space.

Harmonic law Kepler's third law of planetary motion: the cubes of the semimajor axes of the planetary orbits are in proportion to the squares of the sideral periods of the planets.

Harvest moon Full moon nearest the time of the autumnal equinox.

Head (of comet) Main part of a comet, consisting of its nucleus and coma.

Helio Prefix referring to the sun.

Heliocentric Centered on the sun.

Heliocentric longitude of the ascending node Angle at the sun between the direction of the vernal equinox and the ascending node of an orbit.

Heliocentric theory Doctrine that the sun is at the center of the universe.

Hertzsprung-Russell (H-R) diagram Plot of absolute magnitude against temperature (or spectral class or color index) for a group of stars.

High-velocity star (or object) Star (or object) with high space motion; generally an object that does not share the high orbital velocity of the sun about the galactic nucleus.

Horizon (astronomical) Great circle on the celestial sphere 90° from the zenith.

Horizon system System of celestial coordinates (altitude and azimuth) based on the astronomical horizon and the direction of the north pole.

Horizontal branch Sequence of stars on the Hertzsprung-Russell diagram of a typical globular cluster of approximately constant absolute magnitude.

Horizontal parallax Angle by which an object appears displaced (after correction for atmospheric refraction) when viewed on the horizon from a place on the earth's equator, as compared to its direction if it were viewed from the center of the earth.

Hour angle Angle measured westward along the celestial equator from the local meridian to the hour circle passing through an object.

Hour circle Great circle on the celestial sphere passing through the celestial poles.

Hyperbola Conic section of eccentricity greater than 1.0.

Hypothesis Tentative theory or supposition, advanced to explain certain facts or phenomena, and which is subject to further tests and verification.

I

Image Optical representation of an object produced by light rays from the object being refracted or reflected by a lens or mirror.

Inclination (of an orbit) Angle between the orbital plane of a revolving body and some fundamental plane—usually the plane of the celestial equator or of the ecliptic.

Inertia Property of resisting acceleration.

Inferior conjunction Configuration of an inferior planet when it has the same longitude as the sun and when it is between the sun and earth.

Inferior planet Planet whose distance from the sun is less than that of the earth.

Infrared radiation Electromagnetic radiation of wavelength longer than the longest (red) wavelengths that can be perceived by the eye, but shorter than radio wavelengths.

Insolation Rate per unit area measured at the ground at which all radiation from the sun is received.

Interferometer (stellar) Optical device making use of the principle of interference of light waves with which small angles can be measured.

International Date Line Arbitrary line on the surface of the earth near longitude 180° across which the date changes by 1 day.

Interstellar dust Small solid particles in the space between the stars.

Interstellar gas Sparse gas in interstellar space.

Interstellar line Line in the spectrum of a star due to the presence of interstellar gas.

Interstellar matter Interstellar gas and dust.

Ion Atom which has become electrically unneutral by the addition or removal of one or more electrons.

Ionization Process by which an atom gains or loses electrons.

Irregular galaxy Galaxy which lacks the symmetry of a spiral or elliptical galaxy.

Irregular variable Variable star whose light variations do not repeat with a regular period.

Island universe Historical synonym for galaxy.

Isotropic Property of being the same in all directions.

J

Jovian planet Any of the planets Jupiter, Saturn, Uranus, and Neptune.

Jupiter Fifth planet from the sun in the solar system.

K

Kepler's laws Three laws, discovered by J. Kepler, that describe the motions of the planets.

Kinetic energy Energy of motion.

Kinetic theory (of gases) Science that treats the motions of the molecules that compose gases.

L

Latitude North-south coordinate on the surface of the earth; the angular distance north or south of the equator measured along a meridian passing through a place.

Law of areas Kepler's second law; the radius vector from the sun to any planet sweeps out equal areas in the planet's orbital plane in equal intervals of time.

Law of the red shifts Relation between the radial velocity and distance of a remote galaxy; the radial velocity is proportional to the distance of the galaxy.

Libration Any of several phenomena by which an observer on earth, over a period of time, can see more than one hemisphere of the moon.

Libration in latitude Libration caused by the fact that the moon's axis of rotation is not perpendicular to its plane of revolution.

Libration in longitude Libration caused by the regularity in the moon's rotation but irregularity in its orbital speed.

Light Electromagnetic radiation that is visible to the eye.

Light curve Graph that displays the variation in light or magnitude of a variable or eclipsing binary star.

Light-year Distance which light will travel in 1 year.

Limb Apparent edge.

Limb darkening Phenomenon where by the sun is less bright near its limb than near the center of its disk.

Limiting magnitude Faintest magnitude that can be observed with a given instrument or under given conditions.

Line broadening Phenomenon by which spectral lines are not precisely sharp but have finite widths.

Line of apsides Line connecting the apses of an orbit; the line along the major axis of the orbit.

Line of nodes Line of intersection of an orbital plane and reference plane, such as the ecliptic.

Linear diameter Actual diameter in units of length.

Local apparent solar time Time as reckoned in any locality by the position of the true sun; equal to the hour angle of the apparent sun plus 12 hours.

Local Group Cluster of galaxies to which our galaxy belongs.

Local mean solar time Time as reckoned in any locality by the position of the mean sun; equal to the hour angle of the mean sun plus 12 hours.

Longitude East-west coordinate on the earth's surface; the angular distance, measured east or west along the equator from the Greenwich meridian, to the meridian passing through a place.

Longitude of the ascending node Angle measured eastward from a reference direction (usually the vernal equinox) in a fundamental plane (usually the plane of the celestial equator or of the ecliptic) to the ascending node of the orbit of a body.

Longitude of the perihelion point Angle at the sun between the line of nodes of an orbit and the direction of its perihelion.

Lorentz contraction Apparent shortening of moving bodies in the direction of their motion.

Luminosity Rate of radiation of electromagnetic energy into space by a star or other object.

Lunar Referring to the moon.

Lunar eclipse Eclipse of the moon.

Lunar ray Bright linear surface marking on the moon.

M

Magellanic Clouds Two neighboring irregular galaxies visible to the naked eye from southern latitudes.

Magnetic field Region of space near a magnetized body within which magnetic forces can be detected.

Magnetic pole One of two points on a magnet (or the earth) at which the greatest density of lines of force emerge. A compass needle aligns itself along the local lines of force on the earth and points more or less toward the magnetic poles of the earth.

Magnifying power Number of times larger (in angular diameter) an object appears through a telescope than with the naked eye.

Magnitude Measure of the amount of light flux received from a star or other luminous object.

Main sequence Sequence of stars on the Hertzsprung-Russell diagram, containing the majority of stars, that run diagonally from the upper left to the lower right.

Major axis (of ellipse) Maximum diameter of an ellipse.

Major planet Jovian planet.

Major semiaxis (or semimajor axis) Half the longest diameter of an ellipse.

Mantle (of earth) Greatest part of the earth's interior, lying between the crust and the core.

Mare Large, dark, comparatively smooth appearing area of the moon.

Mars Fourth planet from the sun in the solar system.

Mass Amount of matter in a body.

Mass-luminosity relation Empirical relation between the masses and luminosities of many (principally main-sequence) stars.

Mass-radius relation (for white dwarfs) Theoretical relation between the masses and radii of white-dwarf stars.

Maximum elongation Configuration of an inner planet at which its apparent distance from the sun is greatest.

Mean solar day Interval between successive meridian passages of the mean sun; average length of the apparent solar day.

Mean solar time Local hour angle of the mean sun plus 12 hours.

Mean sun Fictitious body that moves eastward with uniform angular velocity along the celestial equator, completing one circuit of the sky with respect to the vernal equinox in a tropical year.

Mechanics That branch of physics which deals with the behavior of material bodies under the influence of, or in the absence of, forces.

Mercury Planet nearest to the sun in the solar system.

Meridian (celestial) Great circle on the celestial sphere that passes through an observer's zenith and the north (or south) celestial pole.

Meridian of longitude Imaginary circle on the surface of the earth passing through the poles.

Meridian (terrestrial) Great circle on the surface of the earth that passes through a particular place and the north and south poles of the earth.

Messier Catalogue Catalog of diffuse nebulae, clusters, and galaxies compiled by Charles Messier in 1787.

Meteor Luminous phenomenon observed when a meteoroid enters the earth's atmosphere and burns up; popularly called a shooting star.

Meteorite Portion of a meteoroid that survives passage through the atmosphere and strikes the earth.

Meteoroid Meteoritic particle in space before any encounter with the earth.

Meteor shower Fall of many meteors within a few hours or days apparently originating from the same region of the sky.

Milky Way Galaxy Great system of more than 100 billion stars to which the sun and planets belong.

Minor axis (of ellipse) Smallest or least diameter of an ellipse.

Minor planet (planetoid, asteroid) One of several tens of thousands of small planets, ranging in size from a few hundred miles to less than 1 mile in diameter.

Molecular band Set of regular, closely spaced spectrum lines produced by some species of molecule.

Molecule Combination of two or more atoms bound together; the smallest particle of a chemical compound or substance that exhibits the chemical properties of that substance.

Monochromatic Of one wavelength or color.

N

Nadir Point on the celestial sphere directly below the observer.

Nautical mile Mean length of 1' of arc on the earth's surface along a meridian.

N-body problem Problem of determining the positions and motions of more than two bodies in a system in which the bodies interact under the influence of their mutual gravitation.

Neap tides Tides which occur near the quarter phase of the moon.

Nebula Cloud of interstellar gas or dust.

Neptune Eighth planet from the sun in the solar system.

Neutron Subatomic particle with no charge and with mass approximately equal to that of the proton.

Neutron star Hypothetical star of extremely high density composed entirely of neutrons.

New General Catalogue (NGC) Catalog of star clusters, nebulae, and galaxies compiled by J. L. E. Dreyer in 1888.

New moon Phase of the moon when its longitude is the same as that of the sun.

Newtonian focus Optical arrangement in a reflecting telescope where a flat mirror brings the converging rays from the objective mirror to a focus at the side of the telescope tube.

Newtonian reflector Reflecting telescope in which a plane secondary mirror is used to divert the focus to the side of the tube.

Newton's laws Laws of mechanics and gravitation formulated by Isaac Newton.

Node Intersection of the orbit of a body with a fundamental plane, usually the plane of the celestial equator or of the ecliptic.

Normal spiral Spiral galaxy whose arms attach directly to the nucleus.

North point That intersection of the celestial meridian and astronomical horizon lying nearest the north celestial sphere.

Nova Star that experiences a sudden outburst of radiant energy, temporarily increasing its luminosity by hundreds to thousands of times.

Nuclear transformation Transformation of one atomic nucleus into another.

Nucleus (atomic) Heart of an atom wherein the positive electric charges reside.

Nucleus (of comet) Swarm of solid particles in the head of a comet.

Nucleus (of galaxy) Central concentration of stars, and possibly gas, at the center of a galaxy.

Objective Principal image-forming component of a telescope or other optical instrument.

Objective prism Prismatic lens that can be placed in front of a telescope objective to transform each star image into an image of its spectrum.

Oblateness Measure of the flattening of an oblate spheroid; numerically, the ratio of the difference between the major and minor diameters (or axes) to the major diameter (or axis).

Oblate spheroid Solid formed by rotating an ellipse about its minor axis.

Obliquity of the ecliptic Angle between the plane of the earth's orbit and the celestial equator.

Obscuration (interstellar) Absorption of starlight by interstellar dust.

Occultation Eclipse of a star or planet by the moon or another planet.

Opacity Absorbing power; capacity to impede the passage of light.

Open cluster Comparatively loose or open cluster of stars, containing from a few dozen to a few thousand members, located in the spiral arms or disk of the galaxy: a galactic cluster.

Opposition Configuration in which a planet is most nearly opposite the sun in the sky.

Optical binary Two stars at different distances nearly lined up in projection so that they appear close together, but which are not really dynamically associated.

Optics Branch of physics that deals with light and its properties.

Orbit Path of a body that is in revolution about another body or point.

Orbital elements Set of quantities by which a planet's orbit may be located and described.

Outburst Sudden and brief large increase of intensity in a selected wavelength of solar radio radiation.

P

Parabola Conic section of eccentricity 1.0; the curve of intersection between a circular cone and a plane parallel to a straight line in the surface of the cone.

Paraboloid Parabola of revolution; a curved surface of parabolic cross section. Especially applied to the surface of the primary mirror in a standard reflecting telescope.

Parallactic ellipse Small ellipse that a comparatively nearby star appears to trace out in the sky, which results from the orbital motion of the earth about the sun.

Parallax Apparent difference of direction of an object when seen from the two ends of a base line.

Parallax (stellar) Apparent displacement of a nearby star that results from the motion of the earth around the sun.

Parallel of declination Circle on the celestial sphere everywhere the same angular distance from the celestial equator.

Parallel of latitude Imaginary circle on the surface of the earth everywhere the same angular distance from the equator.

Parsec Distance at which a star's parallax would be 1″ of arc.

Partial eclipse Eclipse of the sun or moon in which the eclipsed body does not appear completely obscured.

Peculiar velocity Velocity of a star with respect to the local standard of rest; that is, its space motion, corrected for the motion of the sun with respect to our neighboring stars.

Penumbra Portion of a shadow from which only part of the light source is occulted by an opaque body.

Penumbra (of sunspot) Less dark outer region of a sunspot.

Perfect cosmological principle Assumption that, on the large scale, the universe appears the same from every place at all times.

Perfect radiator Incandescent body which radiates with the maximum possible efficiency in every wavelength; a blackbody radiator.

Periastron Place in the orbit of a star in a binary star system where it is closest to its companion star.

Perigee Place in the orbit of an earth satellite where it is closest to the center of the earth.

Perihelion Point of an orbit at which a body is nearest the sun.

Period Time interval; for example, the time required for one complete revolution.

Periodic comet Comet whose orbit has been determined to have an eccentricity of less than 1.0, and which is therefore expected to make regularly spaced returns to a given reference point.

Period-luminosity relation Empirical relation between the periods and luminosities of cepheid-variable stars.

Perturbation Deviation, usually small, of a planet from a designated orbit produced by some external agency.

Phase Measure of the extent to which a cyclic process has been completed, as a lunar phase.

Phases of the moon Progression of changes in the moon's appearance during the month that results from the moon's turning different portions of its illuminated hemisphere to our view.

Photographic magnitude Magnitude of an object as measured on the traditional, blue- and violet-sensitive photographic emulsions.

Photometry Measurement of light intensities.

Photon Discrete unit of electromagnetic energy.

Photosphere Sun's radiating surface.

Photovisual magnitude Magnitude corresponding to the spectral region to which the human eye is most sensitive, but measured by photographic methods with suitable green- and yellow-sensitive emulsions and filters.

Planck's radiation law Formula from which can be calculated the intensity of radiation at various wavelengths emitted by a blackbody.

Planet Any of nine solid, nonluminous bodies revolving about the sun.

Planetarium Optical device for projecting on a screen or domed ceiling the stars and planets and their apparent motions as they appear in the sky.

Planetary nebula Shell of gas ejected from, and enlarging about, a certain kind of extremely hot star.

Planetoid Synonym for minor planet and asteroid.

Plasma Highly-ionized gas.

Pluto Ninth planet from the sun in the solar system.

Polar axis Axis of rotation of the earth; also, an axis in the mounting of a telescope that is parallel to the earth's axis.

Population I and II Two classes of stars (and systems of stars), classified according to their spectral characteristics, chemical compositions, radial velocities, ages, and locations in the galaxy.

Population (stellar) Class to which stars may be assigned according to age, motion, and chemical composition.

Postulate Essential prerequisite to a hypothesis or theory.

Potential energy Stored energy that can be converted into other forms; especially gravitational energy.

Precession (of earth) Slow, conical motion of the earth's axis of rotation, caused principally by the gravitational torque of the moon and sun on the earth's equatorial bulge.

Precession of the equinoxes Slow westward motion of the equinoxes along the ecliptic that results from precession.

Pressure broadening Widening of spectrum lines traceable to the high pressure of the gases of the source.

Primary component Larger member of a binary system.

Primary minimum Greater of two minima in an eclipsing binary light curve, produced when the hotter star of the pair is eclipsed by the cooler star (see Secondary minimum).

Prime focus Point in a telescope where the objective focuses the light.

Prime meridian Terrestrial meridian passing through the site of the old Royal Greenwich Observatory; longitude 0°.

Primeval atom Single mass whose explosion (in some cosmological theories) has been postulated to have resulted in all the matter now present in the universe.

Primeval fireball Extremely hot opaque gas that is presumed to comprise the entire mass of the universe at the time of or immediately following the big bang; the exploding primeval atom.

Prism Wedge-shaped piece of glass that is used to disperse white light into a spectrum.

Prolate spheroid Solid produced by the rotation of an ellipse about its major axis.

Prominence Phenomenon in the solar corona that commonly appears like a flame above the limb of the sun.

Proper motion Star's apparent annual motion across the sky.

Proton Heavy subatomic particle that carries a positive charge, and one of the two principal constituents of the atomic nucleus.

Protoplanet (or star or galaxy) Original material from which a planet (or a star or galaxy) condensed.

Ptolemaic system Geocentric scheme devised in the second century to account for the celestial bodies' apparent motions.

Pulsar One of several pulsating radio sources of small angular size that emit radio pulses in very regular short periods.

Pulsating variable Variable star that physically pulsates in size and luminosity.

Q

Quadrature Configuration of a planet in which its elongation is 90°.

Quarter moon Either of the two phases of the moon when its longitude differs by 90° from that of the sun; the moon appears half full at these phases.

Quasar Popular term for quasi-stellar source.

Quasi-stellar galaxy (QSG) Stellar-appearing object of very large red shift presumed to be extragalactic and highly luminous.

Quasi-stellar source (QSS) Stellar-appearing object of very large red shift that is a strong source of radio waves; presumed to be extra-galactic and highly luminous.

R

R Coronae Borealis variables Eruptive variable stars that show sudden and irregular drops in brightness; the class is named for the prototype, R Coronae Borealis.

RR Lyrae variable One of a class of giant pulsating stars with periods less than 1 day; a cluster variable.

RW Aurigae stars Variable stars, generally associated with interstellar matter, that show rapid and irregular light variations.

Radar Technique for observing the reflection of radio waves from a distant object and timing the interval between transmission and reception of the reflected signal.

Radial velocity Component of relative velocity that lies in the line of sight.

Radial velocity curve Plot of the variation of radial velocity with time for a binary or variable star.

Radiant (of meteor shower) Point in the sky from which the meteors belonging to a shower seem to radiate.

Radiation Mode of energy transport whereby energy is transmitted through a vacuum; also the transmitted energy itself, either electromagnetic or corpuscular.

Radiation pressure Transfer of momentum carried by electromagnetic radiation to a body that the radiation impinges upon.

Radio astronomy Technique of making astronomical observations in radio wavelengths.

Radio telescope Telescope designed to make observations in radio wavelengths.

Ray (lunar) Any of a system of bright elongated streaks, sometimes associated with a crater on the moon.

Rayleigh scattering Scattering of light (photons) by molecules of a gas.

Recurrent nova Nova that has been known to erupt more than once.

Reddening (interstellar) Reddening of starlight passing through interstellar dust caused by the dust scattering blue light more effectively than red light.

Red giant Large, cool star of high luminosity; a star occupying the upper right portion of the Hertzsprung- Russell diagram.

Red shift Shift to longer wavelengths of the light from remote galaxies; presumed to be produced by a Doppler shift.

Reflecting telescope Telescope in which the principle optical component (objective) is a concave mirror.

Reflection nebula Diffuse nebula which shines by reflected light.

Reflex motion Apparent contrary motion due to the sun's motion toward the solar apex.

Refracting telescope Telescope in which the principal optical component (objective) is a lens or system of lenses.

Refraction Bending of a beam of light as it passes from one medium to another medium of different density or composition.

Refractor See refracting telescope.

Regression of nodes Consequence of certain perturbations on the orbit of a revolving body whereby the nodes of the orbit slide westward to the fundamental plane (usually the plane of the ecliptic or of the celestial equator).

Relative orbit Orbit of one of two mutually revolving bodies referred to the other body as though the latter were stationary.

Relative parallax Parallax of a nearby object, using for a reference direction a body not infinitely remote.

Relativity Theory formulated by Einstein that describes the relations between measurements of physical phenomena by two different observers who are in relative motion at constant velocity (the special theory of relativity), or at accelerated motion (the general theory of relativity).

Resolution Degree to which fine details in an image are separated or resolved.

Resolving power Measure of the ability of an optical system to resolve or separate fine details in the image it produces; in astronomy, the angle in the sky that can be resolved by a telescope.

Resonance Exaggeration of a small effect by periodic repetition.

Retrograde motion Apparent westward motion of a planet on the celestial sphere or with respect to the stars.

Revolution Motion about a point outside a body.

Right ascension Angular distance east from the vernal equinox expressed in units of time.

Rille (or rill) Crevasse or trenchlike depression in the moon's (or Mars') surface.

Roche limit Least distance at which purely gravitational cohesion can prevent the tidal disruption of a secondary by a primary.

Rotation Motion about a line through a body.

Satellite Lesser body attendant upon a major one.

Saturn Sixth planet from the sun in the solar system.

Scale (of telescope) Linear distance in the image corresponding to a particular angular distance in the sky; say, so many centimeters per degree.

Scarp Lunar cliff.

Schmidt telescope Telescope having a spherical mirror and correcting lens as its principal optical elements.

Science Attempt to find order in nature or to find laws that describe natural phenomena.

Secondary component Smaller of a pair of revolving bodies.

Secondary minimum Lesser of two minima in an eclipsing binary light curve, produced when the cooler star of the pair is eclipsed by the hotter star (compare with Primary minimum).

Secondary mirror (in a reflecting telescope) Auxiliary mirror introduced to locate the focus elsewhere than at the prime focus.

Secular Not periodic.

Secular parallax Mean parallax for a selection of stars derived from the components of their proper motions that reflect the motion of the sun.

Seeing State of atmospheric tranquility.

Seismic waves Vibrations traveling through the earth's interior or along the earth's surface that result from earthquakes.

Seismology Study of earthquakes and the conditions that produce them, and of the internal structure of the earth as deduced from analyses of seismic waves.

Seleno- Prefix referring to the moon.

Semimajor axis Half the major axis of an ellipse.

Semiregular variable Variable star, usually a red giant or supergiant, whose period of pulsation is only approximately constant.

Separation (in a visual binary) Angular separation of the two components of a visual binary star.

Shell star Type of star, usually of spectral class B to F, surrounded by a gaseous ring or shell.

Short-period comets Comets with periods less than about a century.

Shower (meteor) Many meteors, all seeming to radiate from a common point in the sky, caused by an encounter of the earth and a swarm of meteoroids moving together through space.

Sidereal day Interval between two successive meridian passages of the vernal equinox.

Sidereal month Period of the moon's revolution about the earth with respect to the stars.

Sidereal period Period of revolution of one body about another with respect to the stars.

Sidereal time Local hour angle of the vernal equinox.

Sidereal year Period of the earth's revolution about the sun with respect to the stars.

Sign (of zodiac)(astrological term) Any of twelve equal sections along the ecliptic, each of length 30°. Starting at the vernal equinox, and commencing eastward, the signs are Aries, Taurus, Gemini, Cancer, Leo, Virgo, Libra, Scorpio, Sagittarius, Capicornus, Aquarius, and Pisces.

Solar activity Phenomena of the solar atmosphere associated with sunspots, plages, and related phenomena.

Solar antapex Direction away from which the sun is moving with respect to the local standard of rest.

Solar apex Direction toward which the sun is moving with respect to the local standard of rest.

Solar constant Measure of the intensity of solar radiation at the earth's distance from the sun.

Solar motion Motion of the sun, or the velocity of the sun, with respect to the local standard of rest.

Solar parallax Angle subtended by the equatorial radius of the earth at a distance of 1 AU.

Solar system System of the sun and the planets, their satellites, the minor planets, comets, meteoroids, and other objects revolving around the sun.

Solar time Time based on the sun; usually the hour angle of the sun plus 12 hours.

Solar wind Radial flow of corpuscular radiation leaving the sun.

Solstice One of the two points on the ecliptic at which the sun is farthest from the celestial equator; also the moment of the sun's arrival at this point.

South point Intersection of the celestial meridian and astronomical horizon 180° from the north point.

Space motion Velocity of a star with respect to the sun.

Space probe Unmanned interplanetary rocket carrying scientific instruments to obtain data on other planets, Satellites, or on the interplanetary environment.

Specific gravity Ratio of the density of a body or substance to that of water.

Spectral class (or type) Classification of a star according to the characteristics of its spectrum.

Spectral sequence Sequence of spectral classes of stars arranged in order of decreasing temperatures of stars of those classes.

Spectrogram Photograph of a spectrum.

Spectrograph Instrument for photographing a spectrum; usually attached to a telescope to photograph the spectrum of a star.

Spectroheliogram Photograph of the sun obtained with a spectroheliograph.

Spectroheliograph Device used to photograph the sun in light of a single wavelength.

Spectroscopic binary Binary system identified as

such by the cyclic variation of radial velocity of its components.

Spectroscopic parallax Parallax (or distance) of a star that is derived by comparing the apparent magnitude of the star with its absolute magnitude as deduced from its spectral characteristics.

Spectrum Array of colors or wavelengths obtained when light from a source is dispersed, as in passing it through a prism or grating.

Spectrum analysis Study and analysis of spectra, especially stellar spectra.

Spectrum line Monochromatic image of the spectroscope slit.

Speed Rate at which an object moves without regard to its direction of motion; the numerical or absolute value of velocity.

Spherical aberration Defect of optical systems whereby on-axis rays of light striking different parts of the objective do not focus at the same place.

Spicule Narrow jet of rising material in the solar chromosphere.

Spiral arms Arms of interstellar material and young stars that wind out in a plane from the central nucleus of a spiral galaxy.

Spiral galaxy Flattened, rotating galaxy with pinwheel-like arms of interstellar material and young stars winding out from its nucleus.

Sporadic meteor Meteor that does not belong to a shower.

Spring tide Tide at full or new moon.

Stadium Greek unit of length based on the Olympic Stadium; roughly, 1/10 mi.

Standard time Local mean solar time of a standard meridian, adopted over a large region to avoid the inconvenience of continuous time changes around the earth.

Standard meridian Meridians of longitude which are multiples of 15°.

Star Self-luminous sphere of gas.

Star cluster Assemblage of stars held together by their mutual gravitation.

Steady state (theory of cosmology) Theory of cosmology embracing the perfect cosmological principle and involving the continuous creation of matter.

Stellar association Loose galactic cluster.

Stellar evolution Changes that take place in the sizes, luminosities, structures, and so on, of stars as they age.

Stellar magnitude System of indicating the brightnesses of stars on a negative logarithmic scale.

Stellar parallax Angle subtended by 1 AU at the distance of a star; usually measured in seconds of arc.

Subdwarf Star slightly below the main sequence in the Hertzsprung-Russell diagram.

Subgiant Star of luminosity intermediate between those of mainsequence stars and normal giants of the same spectral type.

Summer solstice In the Northern Hemisphere the sun's northernmost point on the ecliptic; also the moment of the sun's arrival at this point.

Sun Star about which the earth and other planets revolve.

Sundial Device for keeping time by the shadow a marker (gnomon) casts in sunlight.

Sunspot Temporary cool region in the solar photosphere that appears dark by contrast against the surrounding hotter photosphere.

Sunspot cycle Semiregular 11-year period with which the frequency of sunspots fluctuates.

Supergiant Star of very high luminosity.

Superior conjunction Configuration in which a planet and the sun have the same longitude with the planet being more distant than the sun.

Superior planet Planet more distant from the sun than the earth.

Supernova Stellar outburst or explosion in which a star suddenly increases its luminosity by from hundreds of thousands to hundreds of millions of times.

Surface gravity Magnitude of the pull of gravity at the surface of a body.

Surveying Technique of measuring distances and relative positions of places over the surface of the earth (or elsewhere); generally accomplished by triangulation.

Synchrotron radiation Radiation emitted by charged particles being accelerated in magnetic fields and moving at speeds near that of light.

Synodic period Interval between successive occurrences of the same configuration of a planet; for example, between successive oppositions or successive superior conjunctions.

T

T Tauri stars Variable stars associated with interstellar matter that show rapid and erratic changes in light.

Tail (of comet) Gases and solid particles ejected from the head of a comet and forced away from the sun by radiation pressure or corpuscular radiation.

Tangential (transverse) velocity Component of a star's space velocity at right angles to the line of sight.

Telescope Optical instrument used to aid the eye in viewing or measuring, or to photograph, distant objects.

Temperature (absolute) Temperature measured in degrees centigrade from absolute zero.

Temperature (centigrade) Temperature measured on a scale where water freezes at 0° and boils at 100°.

Temperature (color) Temperature of a star as estimated from the intensity of the stellar radiation at two or more colors or wavelengths.

Temperature (effective) Temperature of a blackbody that would radiate the same total amount of energy that a particular body does.

Temperature (Fahrenheit) Temperature measured on a scale where water freezes at 30° and boils at 212°.

Temperature gradient Increase or decrease of temperature with depth.

Temperature (Kelvin) Absolute temperature measured in centigrade degrees.

Temperature (kinetic) Measure of the speeds or mean energy of the molecules in a substance.

Temperature (radiation) Temperature of a blackbody that radiates the same amount of energy in a given spectral region as does a particular body.

Terminator Line of sunrise or sunset on the moon.

Terrestial planet Any of the planets Mercury, Venus, Earth, Mars, and sometimes Pluto.

Theory Set of hypotheses and laws that have been well demonstrated as applying to a wide range of phenomena associated with a particular subject.

Thermal energy Energy associated with the motions of the molecules in a substance.

Thermal equilibrium Balance between the input and outflow of heat in a system.

Thermodynamics Branch of physics that deals with heat and heat transfer among bodies.

Thermonuclear energy Energy associated with thermonuclear reactions or that can be released through thermonuclear reactions.

Tides Deformation of the surface of the ocean caused by the gravitational attraction on it of the sun and moon.

Total eclipse Eclipse in which the eclipsed body is wholly obscured.

Train (of meteor) Temporarily luminous trail left in the wake of a meteor.

Transit Instrument for timing the exact instant a star or other object crosses the local meridian. Also, the passage of a celestial body across the meridian; or the passage of one body (say a planet) across the disk of a larger one (say, the sun).

Transverse (tangential) velocity Velocity at right angles to the line of sight.

Triangulation Determination of the distance of a body by observing it from two widely separated points.

Trigonometry Branch of mathematics that deals with the analytical solutions of triangles.

Tropical year Length of time from one vernal equinox to the next.

Turbulence Disorderly, agitated motion.

Turbulent broadening Widening of spectrum lines because of turbulent motion within the source.

Twilight Period of partial light following sunset and preceding sunrise.

Twinkling Dancing of stellar images, due to unsteadiness of the earth's atmosphere.

Two-body problem Problem of predicting the motion of two mutually gravitating bodies.

Tychonic system Hypothetical celestial mechanism proposed by Tycho Brahe to explain the apparent motions of celestial bodies.

U

Ultraviolet radiation Electromagnetic radiation of wavelengths shorter than the shortest (violet) wavelengths to which the eye is sensitive; radiation of wavelengths in the approximate range 100 to 4000 Å.

Umbra (of a shadow cone) Region of total shadow.

Umbra (of a sunspot) Dark, central portion of a sunspot.

Universal time Local mean solar time at the prime meridian.

Universe Totality of all matter and radiation, and the space occupied by same.

Upper transit Arrival of an object at the portion of the celestial meridian which contains the zenith.

Uranus Seventh planet from the sun in the solar system.

V

Van Allen layer Doughnut-shaped region surrounding the earth where many rapidly moving charged particles are trapped in its magnetic field.

Variable star Star that varies in luminosity.

Variation of latitude Slight semiperiodic change in the latitudes of places on the earth that results from a slight shifting of the body of the earth with respect to its axis of rotation.

Vector Quantity that has both magnitude and direction.

Velocity Vector that denotes both the speed and direction a body is moving.

Velocity of escape Speed with which an object must move in order to enter a parabolic orbit about another body (such as the earth), and hence move permanently away from the vicinity of that body.

Venus Second planet from the sun in the solar system.

Vernal equinox Point of intersection of the celestial equator and the ecliptic at which the sun crosses northward; also the time of the sun's crossing.

Vertical circle Any great circle passing through the zenith.

Visual binary star Binary star in which the two components are telescopically resolved.

Volume Measure of the total space occupied by a body.

W

W Ursae Majoris star Any of a class of eclipsing binaries whose components are nearly in contact and hence suffer tidal distortion and loss or transfer of matter.

W Virginis star (type II cepheid) Variable star belonging to the relatively rare class of population II cepheids.

Walled plain Large lunar crater.

Wandering of the poles Semiperiodic shift of the body of the earth relative to its axis of rotation; responsible for variation of latitude.

Wavelength Distance between corresponding points on successive waves of a train.

Weight Measure of the force due to gravitational attraction.

West point Point on the horizon 270° around the horizon from the north point, measured in a clockwise direction as seen from the zenith.

White dwarf Star that has exhausted most or all of its nuclear fuel and has collapsed to a very small size; believed to be a star near its final stage of evolution.

Winter solstice Sun's southernmost point on the ecliptic; also the time of the sun's arrival there.

Wolf-Rayet star One of a class of very hot stars that eject shells of gas at very high velocity.

X-rays Photons of wavelengths intermediate between those of ultraviolet radiation and gamma rays.

X-ray stars Stars (other than the sun) that emit observable amounts of radiation at X-ray frequencies.

Year Period of revolution of the earth around the sun.

Z

Zeeman effect Splitting or broadening of spectral lines due to magnetic fields.

Zenith Point on the celestial sphere directly overhead.

Zenith distance Arc distance of a point on the celestial sphere from the zenith; 90° minus the altitude of the object.

Zodiac Belt of sky centered on the ecliptic and containing the 12 constellations of the zodiac.

Zone of avoidance Region near the Milky Way where obscuration by interstellar dust is so heavy that few or no exterior galaxies can be seen.

Bibliography

Popular

Abetti, G., *The History of Astronomy,* Abelard-Schuman, New York, 1952.

Hoyle, F., *Frontiers of Astronomy,* Harper and Brothers, New York, 1955.

Ley, W., *Watchers of the Skies,* Viking Press, New York, 1963.

Menzel, D. H., *Field Guide to the Stars and Planets,* Houghton Mifflin, Boston, 1964.

Moore, P., *Men and the Stars,* W. W. Norton, New York, 1955.

Page, T., and Page, L. W., *Telescopes: How to Make and Use Them,* Macmillan, New York, 1966.

de Santillana, G., *The Crime of Galileo,* University of Chicago Press, Chicago, 1962.

Struve, O., and Zebergs, V., *Astronomy of the Twentieth Century,* Macmillan, New York, 1962.

Atlases, Handbooks, and Workbooks

How to Use Your Telescope (pamphlet), Edmund Scientific, Barrington, New Jersey, 1959.

Time in Astronomy (pamphlet), Edmund Scientific, Barrington, New Jersey., 1959.

Norton, A. P. and Inglis, J. G., *Norton's Star Atlas,* Sky Publishing, Cambridge, Massachusetts, 1964.

Rey, H. A., *The Stars,* Houghton Mifflin, Boston, 1970.

Elementary Texts

Dixon, R. T., *Dynamic Astronomy,* Prentice-Hall, Englewood Cliffs, New Jersey, 1971.

Inglis, S. J., *Planets, Stars and Galaxies,* 3rd ed., John Wiley and Sons, New York, 1972.

General Texts

Abell, G., *Exploration of the Universe,* Holt, Rinehart and Winston, New York, 1969.

Krogdahl, W. S., *The Astronomical Universe,* Macmillan, New York, 1962.

Menzel, D. H., Whipple, F. L., and de Vaucoleurs, G., *Survey of the Universe,* Prentice-Hall, Englewood Cliffs, New Jersey, 1970.

Jastrow, R., and Thompson, M. H., *Astronomy: Fundamentals and Frontiers,* John Wiley and Sons, New York, 1972.

Brandt, J. C., and Maran, S. P., *New Horizons in Astronomy,* W. H. Freeman, San Francisco, 1972.

Motz, L., and Duveen, A., *Essentials of Astronomy,* Wadsworth, Belmont, California, 1966.

Payne-Gaposchkin, C., and Haramundanis, K., *Introduction to Astronomy,* Prentice-Hall, Englewood Cliffs, New Jersey, 1970.

Index